Advances in Intelligent System

Volume 703

Series editor

Janusz Kacprzyk, Polish Academy of Sciences, Warsaw, Poland
e-mail: kacprzyk@ibspan.waw.pl

The series "Advances in Intelligent Systems and Computing" contains publications on theory, applications, and design methods of Intelligent Systems and Intelligent Computing. Virtually all disciplines such as engineering, natural sciences, computer and information science, ICT, economics, business, e-commerce, environment, healthcare, life science are covered. The list of topics spans all the areas of modern intelligent systems and computing such as: computational intelligence, soft computing including neural networks, fuzzy systems, evolutionary computing and the fusion of these paradigms, social intelligence, ambient intelligence, computational neuroscience, artificial life, virtual worlds and society, cognitive science and systems, Perception and Vision, DNA and immune based systems, self-organizing and adaptive systems, e-Learning and teaching, human-centered and human-centric computing, recommender systems, intelligent control, robotics and mechatronics including human-machine teaming, knowledge-based paradigms, learning paradigms, machine ethics, intelligent data analysis, knowledge management, intelligent agents, intelligent decision making and support, intelligent network security, trust management, interactive entertainment, Web intelligence and multimedia.

The publications within "Advances in Intelligent Systems and Computing" are primarily proceedings of important conferences, symposia and congresses. They cover significant recent developments in the field, both of a foundational and applicable character. An important characteristic feature of the series is the short publication time and world-wide distribution. This permits a rapid and broad dissemination of research results.

More information about this series at http://www.springer.com/series/11156

Bidyut B. Chaudhuri · Mohan S. Kankanhalli
Balasubramanian Raman
Editors

Proceedings of 2nd International Conference on Computer Vision & Image Processing

CVIP 2017, Volume 1

 Springer

Editors
Bidyut B. Chaudhuri
Computer Vision and Pattern
  Recognition Unit
Indian Statistical Institute
Kolkata
India

Balasubramanian Raman
Department of Computer Science
  and Engineering
Indian Institute of Technology Roorkee
Roorkee, Uttarakhand
India

Mohan S. Kankanhalli
School of Computing
National University of Singapore
Singapore
Singapore

ISSN 2194-5357   ISSN 2194-5365   (electronic)
Advances in Intelligent Systems and Computing
ISBN 978-981-10-7894-1   ISBN 978-981-10-7895-8   (eBook)
https://doi.org/10.1007/978-981-10-7895-8

Library of Congress Control Number: 2017963008

This Springer imprint is published by Springer Nature
The registered company is Springer Nature Singapore Pte Ltd.
The registered company address is: 152 Beach Road, #21-01/04 Gateway East, Singapore 189721, Singapore

Preface

The Second International Conference on Computer Vision and Image Processing (CVIP 2017) was organized at Indian Institute of Technology Roorkee (IITR), Greater Noida Extension Center, during September 09–12, 2017. The conference was endorsed by International Association of Pattern Recognition (IAPR) and was primarily sponsored by MathWorks. CVIP 2017 brought together delegates from around the globe in the focused area of computer vision and image processing, facilitating exchange of ideas and initiation of collaborations. Among 175 paper submissions, 64 (37%) were accepted based on multiple high-quality reviews provided by the members of our technical program committee from ten different countries. We, the organizers of the conference, were ably guided by its advisory committee comprising distinguished researchers in the field of computer vision and image processing from seven different countries. A rich and diverse technical program was designed for CVIP 2017 comprising five plenary talks and paper presentations in seven oral and two poster sessions. Emphasis was given to the latest advances in Cybernetic Health, Perception of Visual Sentiment, Reshaping of Human Figures in Images and Videos Using 3D Morphable Models, Vision and Language, and Challenges in Biometric System Development. The papers for the technical sessions were divided based on their theme relating to Computer Vision Applications, Document Image Analysis, Machine Learning and Uncertainty Handling, Surveillance and Security, Summarization, Retrieval and Recognition, and Low-level Computer Vision. This edited volume contains the papers presented in the technical sessions of the conference, organized session-wise. Organizing CVIP 2017, which culminates with the compilation of the volume of proceedings, has been a gratifying and enjoyable experience for us. The success of the conference was due to synergistic contributions of various individuals and groups including the international advisory committee members with their invaluable suggestions, the technical program committee members with their timely high-quality reviews, the keynote speakers with informative lectures, the local organizing committee members with their unconditional help, and our sponsors and endorsers with their timely support. Finally, we would like to thank Springer for agreeing to publish the proceedings in their prestigious Advances in Intelligent Systems and Computing (AISC) series. We hope the technical contributions made by the authors in these volumes presenting the proceedings of CVIP 2017 will be appreciated by one and all.

Kolkata, India   Bidyut B. Chaudhuri
Singapore, Singapore   Mohan S. Kankanhalli
Roorkee, India   Balasubramanian Raman

Committees

General Chairs

Bidyut Baran Chaudhuri, ISI Kolkata, India
Mohan Kankanhalli, NUS, Singapore

Organizing Chairs

Balasubramanian Raman, IIT Roorkee, India
Sanjeev Kumar, IIT Roorkee, India
Partha Pratim Roy, IIT Roorkee, India
Vinod Pankajakshan, IIT Roorkee, India

Program Chairs

Debashis Sen, IIT Kharagpur, India
Sudipta Mukhopadhyay, IIT Kharagpur, India
Dilip Prasad, NTU, Singapore
Mukesh Saini, IIT Ropar, India

Workshop Chairs

Brajesh Kaushik, IIT Roorkee, India
N. Sukavanam, IIT Roorkee, India

Plenary Chairs

Dharmendra Singh, IIT Roorkee, India
R. D. Garg, IIT Roorkee, India

International Advisory Committee

A. G. Ramakrishnan, Indian Institute of Science, Bangalore, India
Arpan Pal, Tata Consultancy Services, Kolkata, India
B. Krishna Mohan, IIT Bombay, India

Gian Luca Foresti, University of Udine, Italy
Jonathan Wu, University of Windsor, Canada
Josep Lladós, Universitat Autònoma de Barcelona, Catalonia, Spain
Michael Blumenstein, Griffith University, Australia
Phalguni Gupta, IIT Kanpur, India
Pradeep Atrey, State University of New York, Albany, USA
Prem K. Kalra, IIT Delhi, India
Santanu Choudhury, IIT Delhi, India
Subhasis Chaudhuri, IIT Bombay, India
Umapada Pal, Indian Statistical Institute, Kolkata, India

Publication Chairs

Debi Prosad Dogra, IIT Bhubaneshwar, India
Rajarshi Pal, IDBRT Hyderabad, India
Biplab Banerjee, IIT Roorkee, India

Area Chairs

Ananda S. Chowdhury, Jadavpur University, India
Arnav Bhavsar, IIT Mandi, India
Christian Micheloni, University of Udine, Italy
Gaurav Bhatnagar, IIT Jodhpur, India
Ibrahim Venkat, Universiti Sains Malaysia, Malaysia
Kidiyo Kpalma, Institut National des Sciences Appliquées de Rennes, France
Maheshkumar H. Kolekar, IIT Patna, India
Pritee Khanna, IIIT Jabalpur, India
Rajiv Ratn Shah, IIIT Delhi, India
Shanmuganathan Raman, IIT Gandhinagar, India
Subrahmanyam Murala, IIT Ropar, India
Vijayan K. Asari, University of Dayton, USA

Technical Program Committee

A. V. Subramanyam, IIIT Delhi, India
Abhishek Midya, NIT Silchar, India
Ajoy Mondal, Indian Statistical Institute, India
Alireza Alaei, Université de Tours, France
Amanjot Kaur, IIT Ropar, India
Amit Kumar Verma, NIT Meghalaya, India
Ananda Chowdhury, Jadavpur University, India
Anil Gonde, Shri Guru Gobind Singhji Institute of Engineering and Technology,
    India
Anindya Halder, North-Eastern Hill University, India
Ankush Mittal, Graphic Era University, India
Ashis Dhara, IIT Kharagpur, India
Aveek Shankar Brahmachari, Stryker Global Technology Center, India

Badrinarayan Subudhi, NIT Goa, India
Bijaylaxmi Das, IIT Kharagpur, India
Dinabandhu Bhandari, Heritage Institute of Technology, India
Dwarikanath Mohapatra, IBM Research, Australia
Enmei, Nanyang Technological University, Singapore
Gan Tian, School of Computer Science and Technology, Shandong University, China
Gao Tao, North China Electric Power University, China
Gaurav Gupta, The NorthCap University Gurgaon, India
Grace Y. Wang, Auckland University of Technology, New Zealand
Guoqiang, Ocean University of China, China
Harish Katti, Indian Institute of Science, Bangalore, India
Hemanth Korrapati, National Robotics Engineering Center, Carnegie Mellon University, USA
Himanshu Agarwal, Maharaja Agrasen Institute of Technology, New Delhi, India
Jatindra Dash, IIT Kharagpur, India
Jayasree Chakraborty, Research Fellow, Memorial Sloan Kettering Cancer Center, USA
K. C. Santosh, Department of Computer Science, the University of South Dakota, USA
Kaushik Roy, West Bengal State University, India
Krishna Agarwal, University of Tromso, Norway
Ma He, Northeastern University, Shenyang, China
Mandar Kale, IIT Kharagpur, India
Manish, University of Nantes, France
Manish Chowdhury, KTH, Sweden
Manish Narwaria, Dhirubhai Ambani Institute of Information and Communication Technology, India
Manoj Kumar, Babasaheb Bhimrao Ambedkar Central University, India
Manoj Thakur, IIT Mandi, India
Meghshyam G. Prasad, Kolhapur Institute of Technology, India
Minakshi Banerjee, RCC Institute of Information Technology, India
Naveen Kumar, NIT Kurukshetra, India
Nidhi Taneja, Indira Gandhi Delhi Technical University for Women, India
P. Shivakumara, University of Malaya, Malaysia
Padmanabha Venkatagiri, National University of Singapore, Singapore
Partha Pratim Kundu, Indian Statistical Institute, India
Prabhu Natarajan, National University of Singapore, Singapore
Puneet Goyal, IIT Ropar, India
S. K. Gupta, IIT Roorkee, India
Sankaraiah Sreeramula, Fusionex International, Malaysia
Santosh Vipparthi, MNIT Jaipur, India
Sarif Kumar Naik, Philips, India
Shrikant Mehre, IIT Kharagpur, India
Sobhan Dhara, NIT Rourkela, India

Subramanyam, IIIT Delhi, India
Suchi Jain, IIT Ropar, India
Sudhish N. George, NIT Calicut, India
Suman Mitra, Dhirubhai Ambani Institute of Information and Communication
    Technology, India
Tanmay Basu, Ramakrishna Mission Vivekananda University, India
Tu Enmei, Rolls-Royce, NTU, Singapore
Vijay Kumar B. G., Australian Centre for Robotic Vision, Australia
Vikrant Karale, IIT Kharagpur, India
Xiangyu Wang, Agency for Science, Technology and Research, Singapore
Ying Zhang, Institute for Infocomm Research, Singapore
Zhong Guoqiang, Ocean University of China, China

Publicity Chairs

Navneet Kumar Gupta, IIT Roorkee, India
Asha Rani, IIT Roorkee, India
Priyanka Singh, State University of New York, Albany, USA
Suresh Merugu, IIT Roorkee, India

Web site

Himanshu Buckchash, Webmaster, IEEE UP Section

Contents

About the Editors

Prof. Bidyut B. Chaudhuri is INAE Distinguished Professor and J. C. Bose Fellow of Computer Vision and Pattern Recognition Unit at Indian Statistical Institute, Kolkata. He received his B.Sc (Hons.), B.Tech, and M.Tech from Calcutta University, India, in 1969, 1972, and 1974, respectively, and Ph.D. from IIT Kanpur in 1980. He did his Postdoc work during 1981–82 from Queen's University, UK. He also worked as a Visiting Faculty at Tech University, Hannover, during 1986–87 as well as at GSF institute of Radiation Protection (now Leibnitz Institute), Munich, in 1990 and 1992. His research interests include digital document processing, optical character recognition; natural language processing including lexicon generation, ambiguity analysis, syntactic and semantic analysis in Bangla and other Indian languages; statistical and fuzzy pattern recognition including data clustering and density estimation; computer vision and image processing; application-oriented research and externally funded project execution; and cognitive science. He is a Life Fellow of IEEE, International Association for Pattern Recognition (IAPR), Third World Academy of Science (TWAS), Indian National Sciences Academy (INSA), National Academy of Sciences (NASc), Indian National Academy of Engineering (INAE), Institute of Electronics and Telecommunication Engineering (IETE), West Bengal Academy of Science and Technology, Optical Society of India, and Society of Machine Aids for Translation. He has published over 400 papers in journals and conference proceedings of national and international repute.

Dr. Mohan S. Kankanhalli is a Dean of School of Computing and Provost's Chair Professor of Computer Science at National University of Singapore. Before that, he was the Vice-Provost (Graduate Education) for NUS during 2014–2016 and Associate Provost (Graduate Education) during 2011–2013. He was earlier the School of Computing Vice-Dean for Research during 2001–2007 and Vice-Dean for Graduate Studies during 2007–2010. He obtained his B.Tech (Electrical Engineering) from IIT Kharagpur in 1986 and M.S. and Ph.D. (Computer and Systems Engineering) from the Rensselaer Polytechnic Institute in 1988 and 1990, respectively. He subsequently joined the Institute of Systems Science. His research interests include multimedia computing, information security, image/video processing, and social media analysis. He is a Fellow of IEEE. He has published over 250 papers in journals and conference proceedings of international repute.

Dr. Balasubramanian Raman is currently an Associate Professor in the Department of Computer Science and Engineering at IIT Roorkee, India. He completed his Ph.D. (Mathematics and Computer Science) from IIT Madras, Chennai, in 2001. His areas of interest include computer vision—optical flow problems, fractional transform theory, wavelet analysis, image and video processing, multimedia security: digital image watermarking and encryption, biometrics, content-based image and video retrieval, hyperspectral and microwave imaging and visualization, and volume graphics. He has published over 100 papers in refereed journals and contributed seven chapters in books.

Moving Target Detection Under Turbulence Degraded Visible and Infrared Image Sequences

Chaudhary Veenu, Kumar Ajay and Sharma Anurekha

Abstract The presence of atmospheric turbulence over horizontal imaging paths introduces time-varying perturbations and blur in the scene that severely degrade the performance of moving object detection and tracking systems of vision applications. This paper proposed a simple and efficient algorithm for moving target detection under turbulent media, based on adaptive background subtraction approach with different types of background models followed by adaptive global thresholding to detect foreground. This proposed method is implemented in MATLAB and tested on turbulence degraded video sequences. Further, this proposed method is also compared with state-of-the-art method published in the literature. The result shows that the detection performance by proposed algorithm is better. Further, the proposed method can be easily implemented in FPGA-based hardware.

Keywords Moving object detection · Imaging under turbulent media
Performance metrics · Background subtraction · Computer vision and target
detection algorithm

C. Veenu (✉) · S. Anurekha
Department of Electronic Science, Kurukshetra University, Kurukshetra 136119
Haryana, India
e-mail: veenuchaudhary790@kuk.ac.in

S. Anurekha
e-mail: anurekhasharma@kuk.ac.in

K. Ajay
Instrumentation Research and Development Establishment, Dehradun 248008
Uttarakhand, India
e-mail: ajay_irde@yahoo.com

© Springer Nature Singapore Pte Ltd. 2018
B. B. Chaudhuri et al. (eds.), *Proceedings of 2nd International Conference on Computer Vision & Image Processing*, Advances in Intelligent Systems and Computing 703, https://doi.org/10.1007/978-981-10-7895-8_1

1 Introduction

Robust and automatic target detection is an important application in computer vision systems and its performance is severely limited, particularly at longer ranges, due to atmospheric degradation such as atmospheric turbulence. Atmospheric turbulence is random, nonlinear, and optical phenomenon that limits the viewing through any atmospheric path, e.g., twinkling of stars, distant lights, shimmering of objects on hot sunny day. These effects are caused due to local reflections and refractions by atmosphere which varies with time, resulting variation in the refractive index [1]. This nonuniform and continuous variation in refractive index results variation of phase of the received optical wave front. These variations in the received distorted wave front cause the image to be focussed at different points in the focal plane of the receiving optics, causing turbulence distortions. This turbulence causes the light from target scene to reach the imaging system with perturbations, i.e., spatiotemporal fluctuations [2].

In general, turbulent medium results target scene in the video sequences to appear blurry and wavering that renders movements in scene even when objects of the scene is stationary. These turbulence-induced movements in the target scene have characteristics that might be similar to those of the real moving objects, resulting significant increase in false detection, thereby degrading the detection performance of the machine vision system. Therefore, it becomes necessary to identify real moving objects in target scene from dynamic background.

Several methods for moving object detection in video sequences have been reported in the literature, but only few dealt with atmospheric turbulence-induced movements [2–8]. Direct background–foreground segmentation methods such as frame differencing [3] and Gaussian mixture model [6] do not perform well in turbulent scenarios, as motions induced by turbulence cause a lot of false detection. Fishbain et al. [2] proposed a method to preserve object motion while eliminating turbulence in real time. In their approach, a reference background image is computed using a temporal median filter. Then, displacement map of individual frames from reference background frame is estimated. Displacement map is then segmented to detect stationary and moving objects. Turbulence compensation is done while preserving moving objects. Oreifej et al. [5] proposed a method to simultaneously recover the stable background and moving objects. The method includes a three-term low rank matrix decomposition approach, with components-background, turbulence-induced motion, and real moving objects. However, this method is computationally very intensive and has high computation and memory requirements. Baldini et al. [7] proposed a method for moving target tracking in dynamic background condition. In their approach, probabilistic method for motion detection is chosen and each detected blob is tracked by matching them with those found by block matching technique. A true object was declared after being tracked for at least three frames. The algorithm has been tested on image sequences of different weather conditions. This method is also highly computationally intensive. Barnich et al. [9] proposed a universal background subtraction algorithm for motion detection that is based on pixel-by-pixel comparison in order to determine whether that pixel belongs to the background and adapts the model. Eli Chen et al. [8] proposed a method based on novel criteria for objects' spatiotemporal properties, to discriminate true objects from false detections, using pixel-based adaptive thresholding technique for foreground detection.

In the present paper, a simple and efficient algorithm is proposed for real moving target detection in turbulence degraded visible and infrared image sequences. The proposed algorithm increases detection performance by increasing true detection and reducing false negative rates significantly. In this paper, the proposed moving target detection method is based on the combination of two types of adaptive background subtraction techniques followed by foreground extraction with adaptive global thresholding. We have tested the proposed method on various video sequences degraded by atmospheric turbulence. The performance analyses of the proposed algorithm were carried out by implementing it in MATLAB (R2012a). Further, this method is also compared with various other reported object detection methods by testing it on various atmospheric degraded image sequences (available at online resources [10]) taken at varying degree of turbulence. Results obtained the performance analysis parameters establishes that the proposed method performs better for moving target detection in turbulence degraded image sequences.

The remainder of this paper is organized as follows: Sect. 2 describes the evolved method and block diagram of the proposed algorithm for moving target detection under turbulence medium. Sections 3 and 4 discuss the proposed performance metrics for results and comparison with state-of-the-art method reported in [8]. Section 5 concludes this paper along with future work.

2   Proposed Algorithm

Background subtraction technique is used to identify moving objects in video frames [11]. A large set of algorithms, reviewed in [12], have been designed to segment the foreground objects from the background of the sequence. The background subtraction methods presented in [12] are based on single method and hierarchy method approaches. However, in the proposed approach for moving target detection under turbulence, we present simultaneously two methods and combine their results to improve the target detection probability and deductions of false alarms.

In the proposed method, each frame of the input video is subtracting from a background image to identify moving pixels. Background image is computed based upon the adaptive background models and is updated in every frame. For turbulence degraded image sequence, background model should be adaptive in nature. This proposed method combines background subtraction techniques based on adaptive background model and temporal mean (moving average) filter background model. The foreground regions are calculated by applying adaptive global thresholding on each absolute background subtraction.

Let $I_n(x, y)$ is the current image frame and $(\mathbf{x}, \mathbf{y})$ is the image coordinates. We define the following two models for estimating background:

The first estimated background model is

$$B_n^1(x, y) = \alpha B_{n-1}(x, y) + (1 - \alpha)I_n(x, y) \tag{1}$$

where $B_n^1(x, y)$ is the first estimated background at nth frame and $\boldsymbol{\alpha}$ is the background learning parameter whose value lies between 0 and 1. Its value is taken to be 0.7 in the proposed method for estimating $B_n^1(x, y)$.

We defined the second estimated background model as:

$$B_n^2(x, y) = mean(I_t(x, y)) \tag{2}$$

where $t = n + 1, \mathbf{n} + 2, \ldots \mathbf{n} + \mathbf{N}$ and $B_n^2(x, y)$ is second estimated background which is based on moving average filter with window size N = 15 at nth frame.

The background subtracted images of background models $B_n^1(x, y)$ and $B_n^2(x, y)$ are defined as

$$BS_n^1(x, y) = \left| I_n(x, y) - B_n^1(x, y) \right| \tag{3}$$

$$BS_n^2(x, y) = \left| I_n(x, y) - B_n^2(x, y) \right| \tag{4}$$

where $BS_n^1(x, y)$ and $BS_n^2(x, y)$ are two background subtracted images based upon estimated background model 1 and background model 2, respectively.

Adaptive threshold is applied on background subtracted images $BS_n^1(x, y)$ and $BS_n^2(x, y)$ in each frame for detecting moving objects. The threshold value for background subtracted image $BS_n^1(x, y)$ is defined as:

$$th\, low = m(BS_n^1(x, y)) - k_1 * std(BS_n^1(x, y)) \tag{5}$$

$$th\, high = m\left(BS_n^1(x, y)\right) + k_1 * std(BS_n^1(x, y)) \tag{6}$$

$$th\, low \leq Th_n^1 \leq th\, high \tag{7}$$

where $\mathbf{m}(BS_n^1(x, y))$ is the global mean of $BS_n^1(x, y)$ image and $\mathbf{std}(BS_n^1(x, y))$ is the standard deviation of $BS_n^1(x, y)$ image. Global mean $(\boldsymbol{m})$ and standard deviation $(\mathbf{std})$ on background subtracted image are computed as follows:

$$m = \frac{1}{PXQ} \sum_{x=0}^{P-1} \sum_{y=0}^{Q-1} BS_n^1(x, y) \tag{8}$$

$$std = \frac{1}{PXQ} \sqrt{\sum_{x=0}^{P-1} \sum_{y=0}^{Q-1} [BS_n^1(x,y) - m]^2} \qquad (9)$$

where P and Q are the number of rows and columns of the image, respectively.

In the calculation of threshold value Th_n^1, k_1 is the bias which can control the adaptation in threshold value and value of k_1 is kept at 2.7 in this proposed method.

Threshold value Th_n^2 for background subtracted image $BS_n^2(x,y)$ is defined by Otsu's threshold value [13] of estimated background model $B_n^2(x,y)$ given as in [14].

Once the background images are computed, we define the foreground images as:

$$F_n^1(x,y) = \begin{cases} 1, & BS_n^1(x,y) = Th_n^1 \\ 0, & \textbf{otherwise} \end{cases} \qquad (10)$$

$$F_n^2(x,y) = \begin{cases} 1, & BS_n^2(x,y) \geq Th_n^2 \\ 0, & \textbf{otherwise} \end{cases} \qquad (11)$$

where $F_n^1(x,y)$ is the foreground image based on first estimated background subtraction technique that uses background modeling defined as per Eq. (1) and $F_n^2(x,y)$ is the foreground image based on the second background subtraction technique that uses background modeling of Eq. (2).

The detected foreground pixels $F_n^1(x,y)$ and $F_n^2(x,y)$ undergo morphological operation of opening and closing with disk-type structure element SE of order 5×5 [14].

The morphological operation of opening [14] is defined as:

$$A_n^1(x,y) = ((F_n^1(x,y)) \, \mathbf{o} \, SE) = (F_n^1(x,y) \ominus SE) \oplus SE \qquad (12)$$

where $A_n^1(x,y)$ are the foreground pixels of opening operation on $F_n^1(x,y)$. The holes will be filled in the pixels of $A_n^1(x,y)$ that are smaller than SE by closing operation, dilation following erosion operation, given as:

$$FG_n^1(x,y) = ((A_n^1(x,y) \bullet SE) = (A_n^1(x,y) \oplus SE) \ominus SE \qquad (13)$$

where $FG_n^1(x,y)$ is the foreground image produced after morphological operations on first foreground image $F_n^1(x,y)$.

Similarly, morphological operation of opening and closing with disk-type structure element SE of order 5×5 is done on second foreground image $F_n^2(x,y)$ defined by:

$$A_n^2(x,y) = ((F_n^2(x,y)) \mathbf{o} \, SE) = (F_n^2(x,y) \ominus SE) \oplus SE \qquad (14)$$

where $A_n^2(x,y)$ are the foreground pixels of opening operation on $F_n^2(x,y)$. The holes will be filled in the foreground image $A_n^2(x,y)$ that are smaller than SE by closing operation, dilation following erosion operation, given as:

$$FG_n^2(x,y) = ((A_n^2(x,y) \cdot SE) = (A_n^2(x,y) \oplus SE) \ominus SE \qquad (15)$$

where $FG_n^2(x,y)$ is the foreground image produced after morphological operations on second foreground image $F_n^2(x,y)$.

Next, we perform logical AND operation between foreground images $FG_n^1(x,y)$ and $FG_n^2(x,y)$ to extract moving pixels from the background. Lastly, the connected component algorithm [14] is used for detecting motion regions in blobs and each region is enclosed in bounding box to present moving object in target scene.

Block diagram of the proposed method of moving target detection in turbulence degraded image sequences is given in Fig. 1. In order to implement the proposed algorithm in MATLAB for detection of moving objects in turbulence-induced movements, we have recorded several video sequences under various atmospheric turbulence conditions. Out of these videos, two video sequences were taken. Each video sequence comprises of 110 frames having a resolution of 760 × 480 pixels. The ground truth of various real moving targets present in these videos is represented as green color bounding boxes where as detected targets after applying the

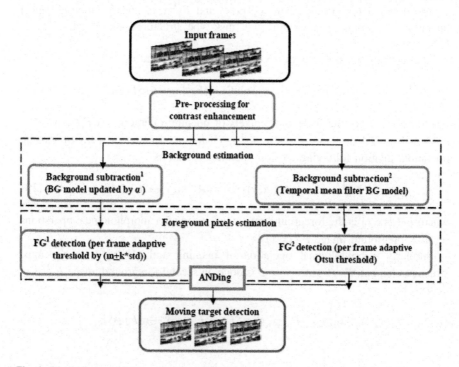

Fig. 1 Block diagram of the proposed method for moving target detection under turbulent media

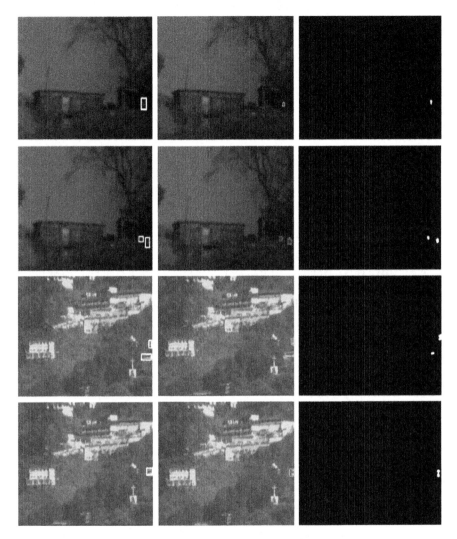

Fig. 2 Sample frames showing ground truth and implementation results of the proposed method on turbulence degraded two video sequences. The first column indicates ground truth information of input frames (green bounding box). Implementation results of the proposed method in column 2 and 3 show detected targets by red bounding boxes and background map, respectively

proposed method is represented as red color bounding boxes. Figure 2 represents sample frames of two video sequences with ground truth information of each frame marked manually as shown in column 1. Column 2 and column 3 present the implementation results of proposed method.

3 Performance Evaluation Metrics

The performance analysis of proposed algorithm has been based upon the performance evaluation metrics to quantitatively analyze the target detection performance in turbulent degraded image sequences. Generally, frame-based metrics [15] are used to measure the detection performance. These metrics are computed by the quantity of parameters such as true positive (TP), false positive (FP) and false negative (FN), defined as follows:

True Positive (TP): Numbers of frames where both ground truth and object detection results agree on the presence of one or more object, and the centroid of at least one or more detected object lies inside the bounding box of ground truth object.

False Positive (FP): Numbers of frames where object detection results contain at least one object, while ground truth either does not contain any object or none of the detected object's centroid falls within the bounding box of ground truth object.

False Negative (FN): Numbers of frames where ground truth contains at least one object, while object detection results either do not contain any object or none of the detected object's centroid falls within the bounding box of ground truth object.

In addition, we calculated statistical parameters—false alarm rate (FAR), precision (P), detection rate (DR), false negative rate (FNR), and F1 presented as follows:

$$
\text{FAR} = \frac{\text{FP}}{\text{TP}+\text{FP}}, \text{P} = \frac{\text{TP}}{\text{TP}+\text{FP}}, \text{DR} = \frac{\text{TP}}{\text{TP}+\text{FN}}, \text{FNR} = \frac{\text{FN}}{\text{TP}+\text{FN}}, \text{F1} = \frac{2*\text{P}*\text{DR}}{\text{P}+\text{DR}}
$$

$$(16)$$

4 Results and Comparison

The moving target detection algorithm proposed in this paper method was tested on recorded, atmospherically degraded video sequences and compared with the state-of-the-art method published in the literature [8] for moving target detection under similar imaging conditions. The video sequences used for evaluating the detection performance of the proposed method with published work are taken from resources available online at [10].

We have taken four video sequences, two with moderate strength of turbulence and other two with higher strength of turbulence. Sample frames from four video sequences are presented in Fig. 3. Each row in Fig. 3 shows frames from original recorded video sequences, while column 1 shows input frame with ground truth information. In Fig. 3, column 2 shows the detected targets using the proposed method and column 3 shows the results obtained using Eli Chen et al.'s method [8]. The ground truth moving targets were marked by green bounding boxes, and the

Fig. 3 Sample frames from video showing ground truth information and moving target detection results obtained using the proposed method and published method of [8]. The first column indicates ground truth of four video sequences; 1, 2, 3, and 4 with green bounding box. The second column shows the target detection using proposed method, and last column shows the results obtained using the method proposed by Chen et al. [8]. Detected moving objects are marked by red bounding boxes

detected moving objects using the proposed method and those obtained using Eli Chen et al. method are marked by red bounding boxes.

From the results shown in Fig. 3, it can be observed that the proposed algorithm performs better than the published method of [8] under both moderate and high turbulence conditions. It can be seen from the results given in Fig. 3 that the method used by Eli Chen et al. [8] results in some miss detection, which are detected using proposed algorithm.

We have carried out the performance analysis of the proposed algorithm with published method of [8] by computing numbers of true detection, false alarms, and miss detections over all the four videos taken from online resources of [10]. For quantitative comparisons, TP, FP, and FN are evaluated by two approaches: number of frame-based approach [15] and approach mentioned in Eli Chen et al. [8] work. The results are summarized in Tables 1 and 2.

We have also calculated the statistical parameters DR, P, FNR, and F1 from the results of Tables 1 and 2 for the proposed method and published method of [8]. Every parameter was averaged for two video sequences at both moderate turbulence strength and strong turbulence strength. The results are summarized in Table 3. It can be concluded from these parameters that the proposed algorithm produced better detection performance irrespective of the evaluation approach.

The performance metric F1 describes the harmonic mean of false alarm (FP) and miss detection (FN) with regard to TP (as shown in formula of F1 by Eq. 16), thus describing detection performance in the range of 0–1. A higher value of F1 means better detection performance of moving object detection algorithm. It can be seen from the graph given in Fig. 4 that the proposed method has higher F1, under both moderate and strong turbulence effects as compared to the state-of-the-art method used in [8], enabling better detection performance with proposed method.

Table 1 Dataset of the proposed method and published method of [8] on number of frame-based approach

Video sequence	Strength of turbulence	Proposed method			Published method of [8]		
		TP	FP	FN	TP	FP	FN
1	Moderate	185	226	45	49	0	72
2	Moderate	357	145	205	287	59	343
3	Strong	373	358	22	204	42	301
4	Strong	98	359	33	12	9	141

Table 2 Dataset of the proposed method and published method of [8] on approach mentioned in [8]

Video sequence	Strength of turbulence	Proposed method			Published method of [8]		
		TP	FP	FN	TP	FP	FN
1	Moderate	6	1	3	1	0	6
2	Moderate	15	0	6	7	0	13
3	Strong	20	2	0	9	0	11
4	Strong	0	0	6	2	1	0

Table 3 Performance metrics of the proposed method and published method of [8]. Each value was averaged for two video sequences

Strength of turbulence	Evaluation approach	Proposed method				Published method of [8]			
		P	DR	FNR	F1	P	DR	FNR	F1
Moderate	No. of frame-based approach	0.58	0.72	0.28	**0.63**	0.92	0.34	0.57	**0.47**
	Approach mentioned in [8]	0.93	0.69	0.31	**0.79**	1	0.25	0.76	**0.39**
Strong	No. of frame-based approach	0.36	0.85	0.16	**0.5**	0.72	0.27	0.76	**0.39**
	Approach mentioned in [8]	0.79	1	0	**0.88**	0.5	0.23	0.78	**0.31**

5 Conclusion and Future Work

This paper proposes simple and efficient method for detecting moving objects in long-range, turbulence degraded videos. We have tested this method on both visual and infrared video sequences degraded by turbulence. We have also compared the detection performance of our approach with the state-of-the-art method in varying degree of turbulence. Results obtained establish that the proposed algorithm performance is better under both moderate strength and higher strength of turbulence degraded video sequences. The method presented is generic in nature used for aerial-, marine-, and ground-based scenarios. We plan to further use this simple and effective target detection algorithm in video stabilization for removing atmospheric turbulence of long-range imaging systems.

Fig. 4 F1 of the proposed method compared with state-of-the-art method of [8]. A higher value of F1 means better detection performance

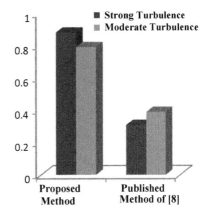

References

1. Roggermann, M.C., Welsh, B, " Imaging through turbulence", Cap.3, CRC Press, USA, pp 57–115 (1996).
2. B. Fishbain, L. P. Yaroslavsky and I.A. Ideses, "Real time stabilization of long range observation system turbulent video", J. Real Time Image Proc. 2, 11–22, 2007.
3. Y. Benezeth, P.M. Jodoin, B. Emile, H. Laurent, and C. Rosenberger, "Comparative study of background subtraction algorithms", J. Electron Imaging 19, 033003 (2010).
4. O. Haik and Y. Yitzhaky, "Effects of image registration on automatic acquisition of moving objects in thermal; video sequences cdegraded by atmosphere", Appl Opt. 46, 8562–8572 (2007).
5. O. Oreifej, L. Xin and M. Shah, "Simultaneous video stabilization and moving object detection in turbulence", IEEE Trans. Pattern Anal. Mach. Intell. 35, 450–462 (2013).
6. C. Stauffer and W. Grimson, "Adaptive background mixture models for real time tracking", in Proceedings of the IEEE Conference on Computer Vision and Pattern Recoginition, pp 246–252, 1999.
7. G. Baldini, P. Campadelh, D. Cozzi, and R. Lanzarotti, "A simple and Robust method for moving target tracking", in Proceedings of International Conference of Signal Processing, Pattern Recognition and Applications, (ACTA, 2012), 108–112.
8. E. Chen, O. Haik and Y. Yitzhaky, "Detecting and tracking moving objects in long distance imaging through turbulent medium", Appl Opt. 53, 1181–1190 (2014).
9. O. Barnich and M. Van Droogenbroeck, "ViBe a universal background subtraction algorithm for video sequences", IEEE Trans. Image Process. 20, 1709–1724 (2011).
10. OnlineResource1: http://www.ee.bgu.ac.il/~itzik/DetectTrackTurb/.
11. S. Cheung and C. Kamath, "Robust techniques for background subtraction in urban traffic video", Proc. SPIE 5308, 881–892. (2004).
12. Andrew Sobral, "A comprehensive review of background subtraction algorithms evaluated with synthetic and real videos", Computer Vision and Image Understanding, 4–21 (2014).
13. Otsu N., "A threshold selection method from gray-level histograms", IEEE Transactions on Systems, Man, and Cybemetics, Vol. 9, No. 1, 1979, pp. 62–66.
14. R.C. Gonzalez and R.E. Woods, Digital Image Processing, 3rd Ed, (Prentice-Hall, 2008).
15. Faisal Bashir and Fatih Porikli, "Performance evaluation of object detection and tracking systems", CVPR (2006).
16. MathWorks Inc: http://www.mathworks.com

Effective Denoising with Non-local Means Filter for Reliable Unwrapping of Digital Holographic Interferometric Fringes

P. L. Aparna, Rahul G. Waghmare, Deepak Mishra
and R. K. Sai Subrahmanyam Gorthi

Abstract Estimation of phase from the complex interference field has become an emerging area of research for last few decades. The phase values obtained by using `arctan` function are limited to the interval $(-\pi, \pi]$. Such phase map is known as wrapped phase. The unwrapping process, which produces continuous phase map from the wrapped phase, becomes tedious in presence of noise. In this paper, we propose a preprocessing technique that removes the noise from the interference field, thereby improving the performance of naive unwrapping algorithms. For de-noising of the complex field, real part and imaginary parts of the field are processed separately. Real-valued images (real and imaginary parts) are processed using non-local means filter with non-Euclidian distance measure. The de-noised real and imaginary parts are then combined to form a clean interference field. MATLAB's `unwrap` function is used as unwrapping algorithm to get the continuous phase from the cleaned interference field. Comparison with the Frost's filter validates the applicability of proposed approach for processing the noisy interference field.

Keywords Holographic interferometry · Non-local means · Non-Euclidian distance · Phase unwrapping · Image de-noising

1 Introduction

In interferometric techniques, physical quantities are encoded into phase, and hence reliable phase estimation becomes the major task. These techniques include synthetic aperture radar for surface topography, magnetic resonance imaging for mapping of

P. L. Aparna
National Institute of Technology, Surathkal 575025, India

R. G. Waghmare · D. Mishra
Indian Institute of Space Science and Technology, Trivandrum 695547, India

R. K. Sai Subrahmanyam Gorthi (✉)
Indian Institute of Technology, Tirupati 517506, India
e-mail: rkg@iittp.ac.in

© Springer Nature Singapore Pte Ltd. 2018
B. B. Chaudhuri et al. (eds.), *Proceedings of 2nd International Conference on Computer Vision & Image Processing*, Advances in Intelligent Systems and Computing 703, https://doi.org/10.1007/978-981-10-7895-8_2

internal structure of the body, digital holographic interferometry and moiré for in-plane and out-off plane deformation assessment, fringe projection profilometry for 3D reconstruction of the object, digital holographic microscopy for study of micro-scopic biological objects, and many others. The term *phase* changes from method to method, e.g., in fringe projection profilometry, the phase means the phase of the sinusoids of the fringe pattern, whereas in holography, phase means the phase of the actual light wave. However, following equation represents most general form for the complex interference field [1].

$$\Gamma(m, n) = a(m, n)exp\,(j\,\phi(m, n)) + \eta(m, n) \tag{1}$$

where $\Gamma(m, n)$ represents the interference field, $a(m, n)$ is the real amplitude, and $\phi(m, n)$ is the phase of the interference field. $\eta(m, n)$ is the noise and is assumed to be white Gaussian with zero mean and variance σ_η^2.

Several noise filtering techniques for the reduction of speckle noise in digital holo-graphic interferometry for phase unwrapping have been developed in the past. For example, Sukumar et al. [2] proposed Kalman filter for denoising in post-processing step to restore the unwrapped phase without any noise. Several other techniques such as Fourier transform profilometry [3], windowed Fourier transform profilometry [4], wavelet transform profilometry by [5] have been proposed over last three decades for the analysis of these fringe patterns. The phase map generated by most of these methods is noisy and wrapped. This requires careful selection of the combination of proper noise filtering [6–8] and phase unwrapping algorithms [9–12].

The noise filtering algorithms developed so far assumes the noise to be addi-tive Gaussian. In digital holographic interferometry (DHI) and synthetic aperture radar (SAR), the interference fields are generated using coherent sources that pro-duce speckle pattern on the surface of the object under test. Speckle noise is a type of noise where an undesirable signal gets multiplied with the original image, as opposed to additive Gaussian noise where noise gets added to the intended signal giving rise to a degraded image. Due to this reason, speckle noise is known to be multiplicative noise. If the original image is represented by $g(m, n)$ and speckle noise by $\eta(m, n)$, then the degraded observation $h(m, n)$ is given by:

$$h(m, n) = g(m, n) \,.\, \eta(m, n) \tag{2}$$

where (m, n) indicate the pixel location.

Speckle noise suppression has been a challenging task for long time. De-noising methods began with techniques relying on local statistics. Lee et al. [6] and Kuan et al. [13] proposed adaptive speckle suppression filters by making use of the local statistics of the degraded image and the multiplicative model of the noise. Exponentially weighted kernel for speckle suppression was proposed by Frost et al. [7] which uses the property of coefficient of variation. The enhanced versions proposed by Lopes et al. [14] filter images independently in homogeneous, heterogeneous, and isolated point regions.

The contribution of this paper is to propose NLM with non-Euclidian distance measure [15] for speckle noise suppression in interferometric fringes and to demonstrate that this denoising helps in effective reconstruction of continuous phase through simple MATLAB's unwrapping function (unwrap). The results obtained by proposed approach have been compared to the performance of (i) denosing followed by unwrapping with NLM with Euclidian distance (which is equivalent to modeling the noise as additive) and (ii) application of simple adaptive filter (Frost filter) and unwrapping. The results, depicted in Sect. 3, demonstrate the effectiveness of the proposed approach.

In Sect. 2, we propose an image denoising algorithm using NLM with non-Euclidian distance measure on interferometric fringes along with the guidelines for pre- and post-processing. We model the noise in interference fields as (i) additive, and apply NLM filter with Euclidian distance (henceforth, we refer to this approach as NLM) for de-noising which is well adapted to additive Gaussian noise and (ii) multiplicative, and apply NLM filter with non-Euclidian distance (proposed approach) which is well adapted to multiplicative noise and show that modeling the noise as multiplicative performs better. Section 3 presents experimental results and observations. Finally, we conclude our paper in Sect. 4.

2 Methodology

2.1 Preprocessing

In DHI, multiplication of object wave field before deformation with the complex conjugate of that after deformation generates the reconstructed interference field [16]. The phase of this field possesses the information about the object deformation. For de-noising the interference field, we consider real and imaginary parts of the field separately. Each part is represented by the real-valued image taking values from the interval $[-1, 1]$. These images are then normalized to the interval $[0, 1]$ by simple mathematical manipulation. Both the images are then processed using proposed approach and then converted back to original range $[-1, 1]$. Finally, these images are combined to get the clean reconstructed interference field. The digital holographic interferometric image can be represented using,

$$\Gamma(m, n) = R(m, n) + iI(m, n) \tag{3}$$

where, $R(m,n)$ and $I(m,n)$ are the real and imaginary parts of the complex pixel at location (m,n). Phase is then calculated using `arctan` function and unwrapped using MATLAB's `unwrap` function.

2.2 Non-local Means Filter

Mathematically, for a given image Y, the filtered value at a point is calculated as a weighted average of all the pixels in the image, following the equation,

$$NLM[Y(p)] = \sum_{\forall q \in Y} w(p,q)Y(q) \tag{4}$$

For $0 \le w(p,q) \le 1$,

$$\sum_{\forall q \in Y} w(p,q) = 1 \tag{5}$$

where p is the point to be filtered and q represents each of the pixels in the image. The weights are based on the similarity between the neighborhood N_p and N_q of pixels p and q. N_i is defined as a square neighborhood centered at pixel i with a user-defined radius R_{sim}. The weight $w(p,q)$ is calculated as

$$w(p,q) = \frac{1}{Z(p)} e^{-\frac{d(p,q)}{\sigma^2}} \qquad Z(p) = \sum_{\forall q} e^{-\frac{d(p,q)}{\sigma^2}} \tag{6}$$

$$d(p,q) = \|(Y(N_p) - Y(N_q))\|^2_{R sim} \tag{7}$$

The distance $d(p, q)$ represents Euclidian distance measure and the weight $w(p, q)$ is based on Euclidian distance.

2.3 NLM with Non-Euclidian Distance

In this version of NLM filter, denoising process is expressed as a weighted maximum likelihood estimation problem [15]. The distance measurement in this version is done as given in Eq. 8 which can be well adapted to multiplicative noise,

$$d(p,q) = log\left(\frac{A_p}{A_q} + \frac{A_q}{A_p}\right) \tag{8}$$

where A_p and A_q are pixel amplitudes of each of the pixels in N_p and N_q, respectively.

We apply NLM with both distance measurements (Eqs. 7 and 8) to denoise holographic fringes and plot the unwrapped estimated phase and compare the results with that of Frost filter.

Figure 1 summarizes the algorithm to compute the de-noised pixel value $Y(q)$ of a pixel q with NLM filter.

3 Experimental Results

The reliability of the proposed approach is verified using reconstructed interference field generated from a real-time holographic experiment performed by Waghmare et al. [17]. The complex interference field is separated into two real-valued images corresponding to real and imaginary part of the complex field. The real and imaginary parts of the complex image, being *sine* and *cosine* parts, lie in the interval $[-1, 1]$. Since the proposed algorithms works on images, we need to normalize the range from $[-1, 1]$ to $[0, 1]$ using simple mathematical manipulation.

These images are then processed with proposed method. We consider a search window W_{search} with radius R_{search} of the image in the neighborhood of the given pixel to be denoised and move the similarity window W_2 (Fig. 1) with radius R_{sim} through this search window and take the weighted average of pixels in the search window using Eq. 6, as it would be time consuming for the similarity window to move through the entire image. For our experiments, we have set the dimensions of similarity and search windows to 5×5 and 11×11, respectively. Finally, the reconstructed interference field is formed from the real and imaginary components by converting range of the cleaned images from $[0, 1]$ to $[-1, 1]$.

Figure 2 depicts results on phase pattern of size 200×200 generated using MATLAB's peak function. Speckle noise with gradually increasing variance (*var*) was added to the phase pattern generated, and it was observed that up till a variance *var* $= 0.25$, the proposed approach is able to give satisfactory unwrapped phase results with MATLAB's naive unwrap function.

Figure 3 displays the performance of the proposed method. Figure 3a, b show the real and imaginary parts of holographic images degraded by noise, respectively. Figure 3c, d shows the real and imaginary parts after de-noising using Frost filter. Figure 3e, f shows the denoised images with NLM method. Figure 3g, h shows the denoised images by proposed method(weighted maximum likelihood estimation-based NLM). Figure 4 depicts the wrapped phase maps and 3D plots of the unwrapped phase maps generated using MATLAB's unwrap function from the de-noised fringe patterns by Frost filter, NLM (Euclidian), and proposed approach, respectively. Figure 5 displays the results for another DHI dataset, where the phase

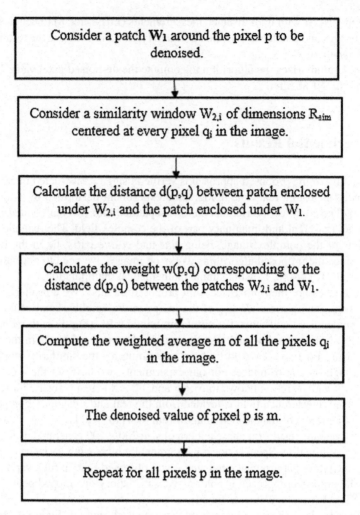

Fig. 1 Non-local means filter algorithm

map underlying in the fringe pattern is rapidly varying. The experimental results depicted in this section validate the reliability of the proposed method in the real-time scenarios as a preprocessing step.

It is observed that there are some discontinuities in phase estimation plots for Frost filter because of the inefficient noise removal of the Frost filter. Modeling noise as

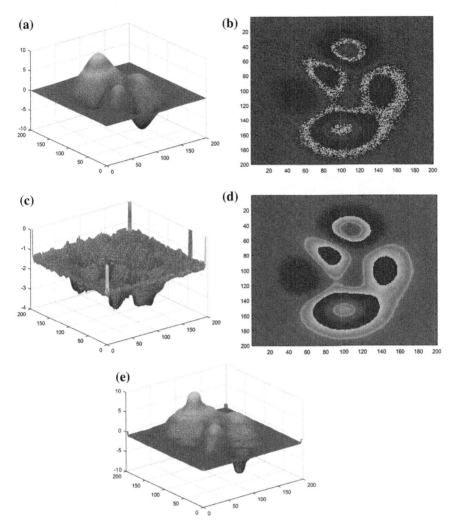

Fig. 2 **a** 3D meshplot of original phase generated using peaks **b** corresponding fringe pattern with noise var = 0.25 **c** Unwrapped phase from noisy fringe pattern **d** wrapped phase with proposed approach and **e** unwrapped phase with proposed approach

additive and applying NLM filter with Euclidian distance measure also has some discontinuities, whereas modeling noise as multiplicative and processing with the proposed method produces continuous and unwrapped phase map from the noisy interference fringes. Thus modeling the noise as multiplicative yields better results.

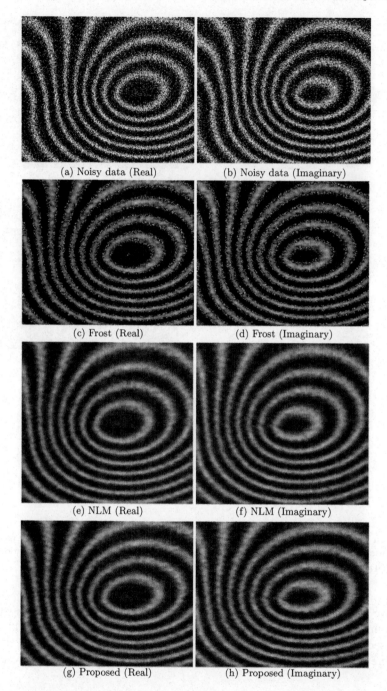

Fig. 3 Performance of Frost Filter, NLM, and proposed approach. First row shows the noisy images corresponding to real and imaginary parts of the complex interference field. Second row shows the denoised images by Frost filter, third row shows denoised images by NLM method, whereas fourth row represents the denoised images by proposed method

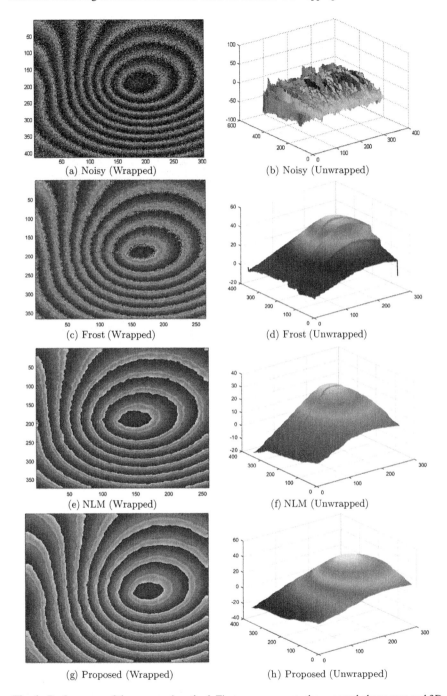

(a) Noisy (Wrapped)

(b) Noisy (Unwrapped)

(c) Frost (Wrapped)

(d) Frost (Unwrapped)

(e) NLM (Wrapped)

(f) NLM (Unwrapped)

(g) Proposed (Wrapped)

(h) Proposed (Unwrapped)

Fig. 4 Performance of the proposed method. First row represents the wrapped phase map and 3D plot of the unwrapped phase map generated using `arctan` and `unwrap` functions of MATLAB from the noisy fringes. Second row shows that of Frost filter, whereas third row shows that of proposed method

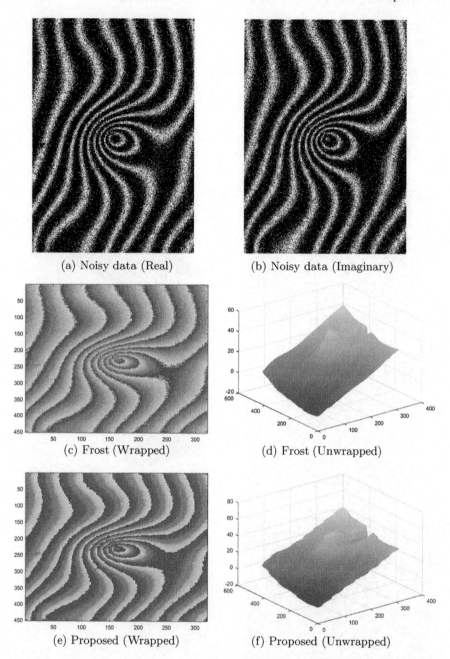

(a) Noisy data (Real)   (b) Noisy data (Imaginary)

(c) Frost (Wrapped)   (d) Frost (Unwrapped)

(e) Proposed (Wrapped)   (f) Proposed (Unwrapped)

Fig. 5 Performance of Frost filter and proposed approach. First row shows the noisy images corresponding to real and imaginary parts of the complex interference field. Second row shows the wrapped and unwrapped phases after denoising with Frost filter, third row represents the wrapped and unwrapped phase after denoising with proposed method

4 Conclusion

We propose the weighted maximum likelihood denoising for holographic fringes which uses non-Euclidian distance measure for the NLM filter as a pre-processing step. Through experiments, it was found that proposed approach can tolerate speckle noise of variance up to 0.25. The performance of the proposed method is compared with Frost filter for noise removal. It was observed through experimental results that the modified-NLM filter outperforms the Frost filter when the images are corrupted by real speckle noise. We have shown that modeling the noise as multiplicative rather than additive gives better results. We have also shown that the continuous and unwrapped phase map can be generated by using `arctan` and `unwrap` function of the MATLAB on the cleaned interference field, which is more accurate than the one generated from noisy interference field.

References

1. Gorthi, Sai Siva and Rastogi, Pramod, "Fringe projection techniques: Whither we are?," *Opt. Laser. Eng.*, **48**, 133–140, (2010).
2. P. R. Sukumar, R. G. Waghmare, R. K. Singh, G. R. K. S. Subrahmanyam and D. Mishra, "Phase unwrapping with Kalman filter based denoising in digital holographic interferometry," *Adv. in Computing, Communications and Informatics*, 2256–2260, (2015).
3. M. Takeda and H. Ina and S. Kobayashi, "Fourier-transform method of fringe-pattern analysis for computer-based topography and inteferometry," *J. Opt. Soc. Am. A*, **43(1)**, 156–160, (1982).
4. Kemao Qian, "Windowed Fourier transform for fringe pattern analysis," *Appl. opt.*, **43(13)**, 2695–2702, (2004).
5. L. R. Watkins and S. M. Tan and T. H. Barnes, "Determination of interferometer phase distributions by use of wavelets," *Opt. Lett.*, **24(13)**, 905–907, (1999).
6. J. S. Lee, "Digital image enhancement and noise filtering by use of local statistics," *IEEE Trans. Pattern Anal. Mach. Intell.*, **2(2)**, 165–168, (1980).
7. V. S. Frost, J. A. Stiles, K. S. Shanmugan, and J. C. Holtzman, "A model for radar images and its application to adaptive digital filtering of multiplicative noise," *IEEE Trans. Pattern Anal. Mach. Intell.*, **4(2)**, 157–166, (1982).
8. Francisco Palacios, Edison Gonalves, Jorge Ricardo, Jose L Valin, "Adaptive filter to improve the performance of phase-unwrapping in digital holography", *Opt. Commun.*, **238(46)**, 245–251, (2004).
9. Goldstein, Richard M and Zebker, Howard A and Werner, Charles L, "Satellite radar interferometry: Two-dimensional phase unwrapping," *Radio Science*, **23(4)**, 713–720, (1988).
10. Huang MJ, Sheu W, "Histogram-data-orientated filter for inconsistency removal of interferometric phase maps," *Opt. Eng.*, **44(4)**, 45602(1–11), (2005).
11. Yongguo Li, Jianqiang Zhu, Weixing Shen, "Phase unwrapping algorithms, respectively, based on path-following and discrete cosine transform," *Optik*, **119(11)**, 545–547, (2008).
12. E. Zappa, G. Busca, "Comparison of eight unwrapping algorithms applied to Fourier-transform profilometry," *Opt. Laser. Eng.*, **46(2)**, 106–116, (2008).
13. D. T. Kuan, A. A. Sawchuk, T. C. Strand, and P. Chavel, "Adaptive noise smoothing filter for images with signal dependent noise," *IEEE Trans. Pattern Anal. Mach. Intell.*, **7(2)**, 165–177, (1985).
14. A. Lopes, R. Touzi, and E. Nezzy, "Adaptive speckle filters and scene heterogeneity," *IEEE Trans. Geosci. Remote Sens.*, **28(6)**, 992–1000, (1990).

15. Deledalle, Charles-Alban, Loc Denis, and Florence Tupin, "Iterative weighted maximum like-lihood denoising with probabilistic patch-based weights" *IEEE Transactions on Image Processing*, **18(12)**, 2661–2672, (2009).
16. Sai Siva Gorthi and Pramod Rastogi, "piece-wise polynomial phase approximation approach for the analysis of reconstructed interference fields in digital holographic interferometry," *J. Opt. A: Pure Appl. Opt.*, **11**, 1–6 (2009).
17. R. G. Waghmare, D. Mishra, G. Sai Subrahmanyam, E. Banoth, and S. S. Gorthi, "Signal tracking approach for phase estimation in digital holographic interferometry," *Applied optics*, **53(19)**, 4150–4157, (2014).

Iris Recognition Through Score-Level Fusion

Ritesh Vyas, Tirupathiraju Kanumuri, Gyanendra Sheoran and Pawan Dubey

Abstract Although there are many iris recognition approaches available in the literature, there is a trade-off as which approach is giving the most reliable authentication. In this paper, score-level fusion of two different approaches, XOR-SUM Code and BLPOC, is used to achieve better performance than either approach individually. Different fusion strategies are employed to investigate the effect of fusion on genuine acceptance rate (GAR). It is observed that fusion through sum and product schemes provides better result than that through minimum and maximum schemes. For further improvement, sum and product schemes are more explored through weighted sum with different weights. The best GAR and equal error rate (EER) values are 98.83% and 0.95%, respectively. Performance of proposed score-level fusion is also compared with existing approaches.

Keywords Iris recognition · Score-level fusion · Genuine acceptance rate (GAR)

1 Introduction

Biometrics is providing large solutions to the security issues related to knowledge (PIN or password) or token (smart card, ATM card etc.)-based approaches. Biometrics refer to the analysis of a person's behavioral and physiological traits. Behavioral characteristics deal with the behavior of an individual and may include signature, voice, or gait while palmprint, fingerprint, face, and iris, which are associated with

R. Vyas (✉) · T. Kanumuri · G. Sheoran · P. Dubey
National Institute of Technology Delhi, Delhi, India
e-mail: ritesh.vyas@nitdelhi.ac.in

T. Kanumuri
e-mail: ktraju@nitdelhi.ac.in

G. Sheoran
e-mail: gsheoran@nitdelhi.ac.in

P. Dubey
e-mail: pawandubey@nitdelhi.ac.in

© Springer Nature Singapore Pte Ltd. 2018
B. B. Chaudhuri et al. (eds.), *Proceedings of 2nd International Conference on Computer Vision & Image Processing*, Advances in Intelligent Systems and Computing 703, https://doi.org/10.1007/978-981-10-7895-8_3

the body parts, embrace into the physiological properties. Among all physiological traits, iris is best suited for personal authentication because the spatial patterns or features present in human iris are highly idiosyncratic to an individual [5, 20].

Human iris is a ring-shaped strip lying between the pupillary (or inner) and limbic (or outer) boundaries of an eye ball [2]. There is certain textural information present in each iris which makes it distinguishable from the other one. An iris recognition system necessarily has few elementary steps like preprocessing of the eye image, normalization, feature extraction, and matching. Preprocessing includes segmenting the iris boundaries from the whole eye image. Normalization is required in order to form a fixed-size template from the iris, so that different iris images can be compared irrespective of the varying lighting conditions. Feature extraction and matching are two such steps which control the accuracy of an iris recognition system. Bowyer et al. [2] have reviewed many state-of-the-art feature extraction approaches and image matchers for iris. Roots of iris recognition were laid down in 1993 when Daugman [3] presented the first automatic iris recognition system. In this paper, phase of 2D Gabor filter coefficients was encoded to form the iriscode. Many commercial iris recognition systems are based on Daugmans iriscode. Miyazawa et al. [11] in 2008 used 2D discrete Fourier transform of the image and its phase component for image matching. In 2011, Farouk [7] represented the iris as labeled graphs. Then elastic graph matching (EGM) was used for matching two iris images. Rai and Yadav [13] used HAAR wavelet along with 1D log Gabor filter for extracting iris features. Khalighi et al. [8] achieved feature extraction through non-subsampled contourlet transform (NSCT) and gray level cooccurrence matrix (GLCM). Umer et al. [16] proposed feature extraction through multiscale morphological operations. Residue images formed after the application of multiscale morphological features for different orientations are accumulated to form the feature vector.

Recently, Bansal et al. [1], proposed local principal independent components (LPIC)-based feature extraction. They have defined four discriminating features namely energy feature (EF), sigmoid feature (SF), effective information (EI), Hanman transform (HT) which can be used with different classifiers to give good recognition rate. Umer et al. [17] have used texture code cooccurrence matrix (TCCM) and its features for representing the iris.

The key discussion of this paper is about score-level fusion of texture- and transform-based approaches for iris recognition: XOR-SUM Code and BLPOC. XOR-SUM Code [15] is an efficient technique for palmprint recognition. Idea behind its use in iris recognition is that Gabor filter at different orientations is able to capture textural information of iris lying in respective directions. But, this technique is highly affected by the Gabor parameters. While, band limited phase-only correlation (BLPOC) [11] uses the phase information of 2D discrete Fourier transform (DFT) of iris images for matching different images. It is observed that Gabor filter is not able to capture the textural features of some of the iris images. For matching of such images, BLPOC can be a good option. Therefore, this paper investigates the fusion of above-mentioned schemes. Experiments of the proposed technique on IITD database show that it is highly effective for iris recognition.

Organization of the rest of this paper is as follows. Section 2 explains the pre-processing and normalization of iris images. Implementation of XOR-SUM Code and BLPOC is discussed in Sect. 3. Section 4 describes the fusion techniques employed. Experimental setup and results are elucidated in Sect. 5. Finally, conclusion is presented in Sect. 6.

2 Preprocessing and Normalization

Preprocessing of an eye image is necessary to isolate the region of interest (ROI) (i.e., the rectangular iris template) from the given eye image. For ROI extraction, technique of [18] is employed here. The process starts with contrast enhancement so that low-resolution images can be improved before processing (Fig. 1b). After binarization, specular reflections are removed from the pupil (Fig. 1d) because these specular reflections may cause spurious circles during the circle detection step. So, their removal is necessary before starting the segmentation process. The pupil center and pupil radius are detected by fitting a matrix over the connected circular pupil area (Fig. 1e). Distance of edge points from the pupil center which is repeated maximum number of times is selected as the iris radius. After getting all these parameters, iris part is segmented from rest of the image (Fig. 1f). Daugman's rubber sheet model [3–5] is used for transforming the circular iris into rectangular template of size 64 × 512 (Fig. 1g).

3 Implementation of Used Approaches

3.1 XOR-SUM Code

In XOR-SUM Code [15], Gabor filter at different orientations is employed to extract line features of palmprint. Similar idea is also applied to iris images [19] because iris template is also having certain structures which are lying in different directions. Process of feature extraction using XOR-SUM Code is depicted in Fig. 2.

Feature extraction through Gabor filter is highly sensitive to its parameter selection. Expression for a 2D Gabor filter is written in (1).

$$\psi(x, y, \sigma, f, \theta) = \frac{1}{\sqrt{2\pi\sigma^2}} \exp\left\{-\frac{x^2+y^2}{\sigma^2}\right\} \times \exp(2\pi i f(x\cos\theta + y\sin\theta)) \tag{1}$$

where f denotes the frequency of sinusoidal gratings. Optimal values of rest of the parameters are given in Table 1, which are achieved empirically. This work uses four orientations of Gabor filter, i.e., $N = 4$ (Fig. 3) because higher number of orientations will result in more redundancies.

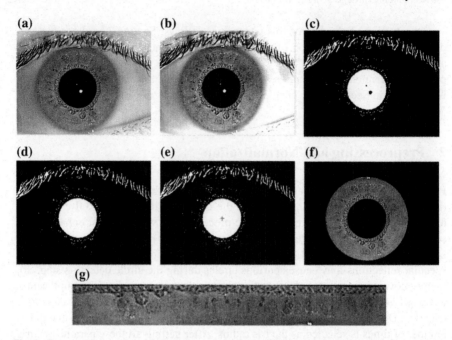

Fig. 1 Preprocessing and normalization process **a** original image, **b** contrast enhancement, **c** binarization, **d** removal of specular reflections, **e** detection of pupil center and radius, **f** segmented iris, **g** normalized iris

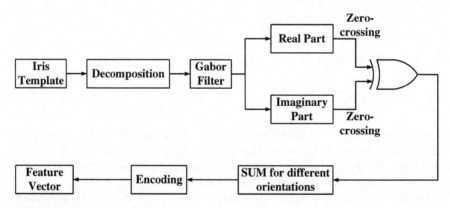

Fig. 2 XOR-SUM Code process

Table 1 Optimal 2D Gabor filter parameters

Size	f	σ	θ	p
15×15	0.1833	2.809	$p*/N$	0, 1, …, N −1

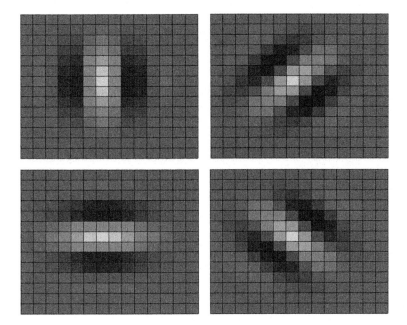

Fig. 3 Spatial representation of Gabor filters in four different orientation

The sum output S of XOR operation of real and imaginary Gabor filtered image is encoded into bits, to generate XOR-SUM Code (*XSC*), using (2).

$$XSC(n) = \left\{ \begin{array}{ll} 1 & if \quad n \leq S < n + \frac{N+1}{2} \\ 0 & otherwise \end{array} \right\} \quad (2)$$

where $n = 1, 2, \ldots, \lceil (N+1)/2 \rceil$ is the number of bits used for encoding the XOR-SUM S. Each *XSC* will result in a 3-bit code for $N = 4$ orientations. Finally, scores are calculated using Hamming distance classifier.

3.2 BLPOC

Band-limited phase-only correlation (BLPOC) [11] is an image matching technique which is independent of intensive parametric optimization as in the case of Gabor-based approaches. It simply uses phase information of 2D discrete Fourier transforms (DFTs) of different iris images.

If there are two iris images p and q of size $M \times N$ and $P(k_1, k_2)$ and $Q(k_1, k_2)$ represent their 2D-DFTs, then their cross-phase spectrum is defined as in (3).

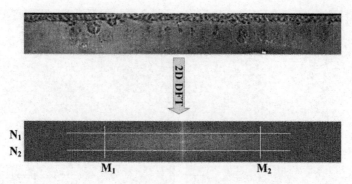

Fig. 4 Band limiting in frequency spectrum of an iris image

$$R_{P,Q}(k_1, k_2) = \frac{P(k_1, k_2)\overline{Q(k_1, k_2)}}{\left|P(k_1, k_2)\overline{Q(k_1, k_2)}\right|} \tag{3}$$

where $\overline{Q(k_1, k_2)}$ is the complex conjugate of $Q(k_1, k_2)$. Phase-only correlation (POC) is defined as the 2D inverse DFT of cross-phase spectrum, i.e., $r_{p,q}$. This POC function gives a peak if two similar images are being matched, and for non-similar images, the peak is not significant.

Further, as it is evident from Fig. 4, that the complete phase information of the spectrum is not meaningful. Therefore, frequency band of the iris spectrum is limited to only significant information range, i.e., from N_1 to N_2 and from M_1 to M_2. In this paper, these four parameters are chosen experimentally and the optimal parameter set $(M_1, M_2, N_1, N_2) = (110, 350, 13, 52)$ for an iris template size of 64×512.

This BLPOC concept helps in more distinctive peak when matching two similar images, while it does not affect the peak much for non-similar images (Figs. 5 and 6).

4   Score-Level Fusion

Aim of score-level fusion is to improve the classification performance than that can be achieved through single classifier. Although feature-level fusion is also an option, sometimes it is not practical because of the varying dimensions of the feature vector extracted through different techniques.

Before fusing the scores, it is required to normalize them so that they can lie within same range. In this paper, both the scores, i.e., Hamming distance scores of XOR-SUM Code and BLPOC scores are already in the range of 0–1. But there is one difference in the scores, i.e., Hamming distance scores are small for similar images and large for non-similar images while BLPOC scores are large for similar images

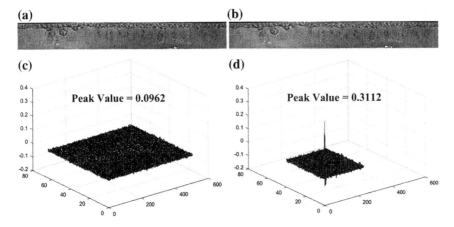

Fig. 5 BLPOC example for similar images (**a**), **b** two iris samples from same subject, **c** simple POC function **d** BLPOC function

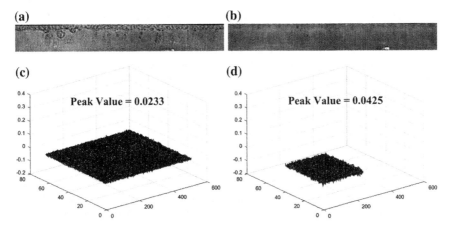

Fig. 6 BLPOC example for non-similar images (**a**), **b** two iris samples from different subjects, **c** simple POC function **d** BLPOC function

and small for non-similar images. Therefore, before combining them, BLPOC scores are normalized by subtracting from 1.

Experiments performed with different fusion strategies [12] show that there is significant improvement in performance. The fusion strategies employed here are minimum (MIN), maximum (MAX), sum (SUM), product (PROD), and weighted sum (WS). Different parameters showing improvement in combined performance are discussed under next section.

Table 2 Performance comparison of XOR-SUM Code and BLPOC with different fusion schemes

Matching scores	GAR (%)	EER (%)	DI
XOR-SUM Code	97.64	1.68	3.74
BLPOC	96.86	2.19	2.29
Fusion_MIN	98.1	1.75	4.29
Fusion_MAX	97.86	3.39	2.32
Fusion_SUM	98.4	0.97	2.99
Fusion_PROD	98.62	1.57	4.23

5 Experimental Setup and Results

The proposed combination of matching scores is investigated with the benchmark iris dataset, IITD iris V1.0 [9]. This dataset contains 2240 grayscale iris images captured from 224 subjects at the rate of 10 images per subject. First five images of a person belong to left eye and next five belong to right eye. So in total, there are 448 classes. Dimensions of each image is 320×240, and all images are stores in bitmap format. This dataset is considered to be a challenging one because of large number of occlusion affected image, i.e., a large number of images are having noise factors like eyelashes, eyelids, reflections.

Because of the large amount of noise, some images were not segmented properly. Classes with unsegmented iris are very few in numbers, so they can be discarded from inclusion in further processing without affecting the recognition performance significantly. The detail of discarded classes is: right class of subject 17, left class of subject 20, left class of subject 74, and right class of subject 156.

The performance evaluation consisted of 4440 genuine and 2458650 imposter scores. The quantitative analysis of performance contains calculation of three parameters, i.e., genuine acceptance rate (GAR), equal error rate (EER), and decidability index (DI). Parameters for the four basic fusion strategies are shown in Table 2. It is apparent from the table that recognition rate of every fusion scheme is higher than that of any of the individual techniques, i.e., XOR-SUM Code and BLPOC. Receiver operator characteristics (ROC) curves for all the four basic fusion schemes are shown in Fig. 7 (left).

Table 2 clearly demonstrates that fusion strategy plays a critical role in the improvement of the classifier performance. Fusion schemes SUM and PROD are showing the maximum improvement in the recognition rate. Therefore, these two strategies are further combined through weighted sum scheme for further enhancing the performance. Expression for finding weighted sum scores is given in (4).

$$score_{ws} = w_1 * score_{PROD} + w_2 * score_{SUM} \tag{4}$$

Table 3 Evaluation parameters for different weights

S. No.	w_1	w_2	GAR (%)	EER (%)	DI
1	0.1	0.9	98.62	1.06	3.04
2	0.15	0.85	98.69	1.23	3.07
3	0.2	0.8	98.78	1.49	3.1
4	**0.25**	**0.75**	**98.47**	**0.95**	**3.13**
5	0.3	0.7	98.57	1.03	3.17
6	0.35	0.65	98.66	1.21	3.21
7	0.4	0.6	98.78	1.55	3.25
8	0.45	0.55	98.52	1.04	3.29
9	0.5	0.5	98.71	1.22	3.34
10	0.55	0.45	98.83	1.71	3.4
11	0.6	0.4	98.69	1.16	3.45
12	**0.65**	**0.35**	**98.83**	**1.69**	**3.52**
13	0.7	0.3	98.69	1.32	3.59
14	0.75	0.25	98.54	1.21	3.67
15	0.8	0.2	98.47	1.18	3.76
16	0.85	0.15	98.47	1.21	3.86
17	0.9	0.1	98.59	1.3	3.97

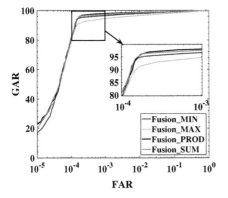 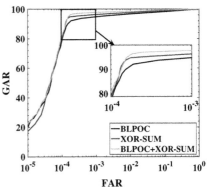

Fig. 7 ROC curves, left: different fusion schemes, right: comparative curves

Table 4 Comparison of performance

Approaches	GAR	EER	DI
Umer et al. [16]	98.37	0.7	–
Barpanda et al. [14]	91.9	8.34	2.18
Dhage et al. [6]	97.81	–	–
Masek [10]	94.06	–	–
Proposed score-level fusion	**98.78**	**1.49**	**3.1**

The weights w_1 and w_2 can be selected empirically. Table 3 shows the evaluation parameters for different weights. Figure 7 (right) shows the performance of weighted sum approach with respect to BLPOC and XOR-SUM approach through ROC curve. The proposed iris recognition using score-level fusion is compared with some previously reported approaches in Table 4.

6 Conclusions

This paper presents a score-level fusion of two approaches, XOR-SUM Code and BLPOC, for iris recognition. XOR-SUM Code is using Gabor filter for extraction of textural information, and BLPOC uses phase information of 2D-DFT which is illumination invariant. Their scores are fused for getting the improved recognition performance. Weighted sum of two fusion strategies, Fusion_SUM and Fusion_PROD gives better recognition rate and equal error rate at optimum values of weights $(w_1, w_2) = (0.55, 0.45)$ and $(w_1, w_2) = (0.25, 0.75)$, respectively. It is evident through the performed experiments that the proposed score-level fusion is outperforming the state-of-the-art approaches of iris recognition.

Acknowledgements The authors would like to thank Indian Institute of Technology Delhi for providing free access to their iris database.

References

1. Bansal, M., Hanmandlu, M., Kumar, P.: IRIS based authentication using local principal independent components. Optik (Stuttg). 127, 4808–4814 (2016)
2. Bowyer, K.W., Hollingsworth, K., Flynn, P.J.: Image understanding for iris biometrics: A survey. Comput. Vis. Image Underst. 110(2), 281–307 (2008)
3. Daugman, J.: High Conf Visual Recog of Persons by a test of statistical significance. IEEE Trans. Pattern Anal. Mach. Intell. 15(11), 1148–1161 (1993)
4. Daugman, J.: Demodulation By Complex-Valued Wavelets for Stochastic Pattern Recognition. Int. J. Wavelets, Multiresolution Inf. Process. 01(1), 1–17 (2003)
5. Daugman, J.: How iris recognition works. Circuits Syst. Video Technol. IEEE Trans. 14(1), 21–30 (2004)
6. Dhage, S.S., Hegde, S.S., Manikantan, K., Ramachandran, S.: DWT-based Feature Extraction and Radon Transform Based Contrast Enhancement for Improved Iris Recognition. In: Int. Conf. Adv. Comput. Technol. Appl. vol. 45, pp. 256–265 (2015)
7. Farouk, R.M.: Iris recognition based on elastic graph matching and Gabor wavelets. Comput. Vis. Image Underst. 115(8), 1239–1244 (2011)
8. Khalighi, S., Pak, F., Tirdad, P., Nunes, U.: Iris Recognition using Robust Localization and Nonsubsampled Contourlet Based Features. J. Signal Process. Syst. 81(1), 111–128 (2014)
9. Kumar, A., Passi, A.: Comparison and combination of iris matchers for reliable personal authentication. Pattern Recognit. 43, 1016–1026 (2010), http://www4.comp.polyu.edu.hk/~csajaykr/IITD/Database_Iris.htm
10. Masek, L., Kovesi, P.: MATLAB Source Code for a Biometric Identification System Based on Iris Patterns (2003), http://www.peterkovesi.com/studentprojects/libor/sourcecode.html

11. Miyazawa, K., Ito, K., Aoki, T., Kobayashi, K., Nakajima, H.: An effective approach for Iris recognition using phase-based image matching. IEEE Trans. Pattern Anal. Mach. Intell. 30(10), 1741–1756 (2008)
12. Park, H.A., Park, K.R.: Iris recognition based on score level fusion by using SVM. Pattern Recognit. Lett. 28(15), 2019–2028 (2007)
13. Rai, H., Yadav, A.: Iris recognition using combined support vector machine and Hamming distance approach. Expert Syst. Appl. 41(2), 588–593 (2014)
14. Sankar, S., Majhi, B., Sa, P.K.: Region based feature extraction from non-cooperative iris images using triplet half-band fi lter bank. Opt. Laser Technol. 72, 6–14 (2015)
15. Tamrakar, D., Khanna, P.: Palmprint verification with XOR-SUM Code. Signal, Image Video Process. 9(3), 535–542 (2013)
16. Umer, S., Dhara, B.C., Chanda, B.: Iris recognition using multiscale morphologic features. Pattern Recognit. Lett. 65, 67–74 (2015)
17. Umer, S., Dhara, B.C., Chanda, B.: Texture code matrix-based multi-instance iris recognition. Pattern Anal. Appl. 19(1), 283–295 (2016)
18. Vyas, R., Kanumuri, T., Sheoran, G.: An Approach for iris segmentation in constrained environments. In: Panigrahi, B.K., Hoda, M.N., Sharma, V., and Goel, S. (eds.) Nature Inspired Computing (Proceedings of CSI 2015), Advances in Intelligent Systems and Computing, vol 652. pp. 99–107. Springer Singapore (2018)
19. Vyas, R., Kanumuri, T., Sheoran, G.: Iris Recognition using 2-D Gabor Filter and XOR-SUM Code. In: 2016 1st India International Conference on Information Processing (IICIP), pp. 1–5 (2016). https://doi.org/10.1109/IICIP.2016.7975369
20. Wildes, R.: Iris recognition: an emerging biometric technology. Proc. IEEE 85(9), 1348–1363 (1997)

A Novel Pattern Matching Approach on the Use of Multi-variant Local Descriptor

Deep Suman Dev and Dakshina Ranjan Kisku

Abstract The objective of pattern matching problem is to find the most similar image pattern in a scene image by matching to an instance of the given pattern. For pattern matching, most distinctive features are computed from a pattern that is to be searched in the scene image. Scene image is logically divided into sliding windows of pattern size, and all the sliding windows are to be checked with the pattern for matching. Due to constant matching between the pattern and the sliding window, the matching process should be very efficient in terms of space, time and impacts due to orientation, illumination and occlusion must be minimized to obtain better matching accuracy. This paper presents a novel local feature descriptor called Multi-variant Local Binary Pattern (MVLBP) for pattern matching process while LBP is considered as base-line technique. The efficacy of the proposed pattern matching algorithm is tested on two databases and proved to be a computationally efficient one.

1 Introduction

Pattern matching [1–3] is one of the fundamental research segments in the field of computer vision and pattern recognition and seeks most similar image portions (local window [4]) by matching to an instance of the pattern in a comparatively large image where the pattern may be or may not be present in the same or different orientation. Pattern matching has fascinating applications [1–3] which include robot

D. S. Dev (✉)
Department of Information Technology, Neotia Institute of Technology,
Management and Science, Jhinga, Diamond Harbour Road, Amira, South 24 Parganas,
Kolkatta 743368, West Bengal, India
e-mail: deepsumandev@yahoo.co.in

D. R. Kisku
Department of Computer Science and Engineering, National Institute of Technology
Durgapur, Durgapur 713209, West Bengal, India
e-mail: drkisku@cse.nitdgp.ac.in

© Springer Nature Singapore Pte Ltd. 2018
B. B. Chaudhuri et al. (eds.), *Proceedings of 2nd International Conference on Computer Vision & Image Processing*, Advances in Intelligent Systems and Computing 703, https://doi.org/10.1007/978-981-10-7895-8_4

37

Fig. 1 Pattern matching process

vision, object tracking, image localization, image classification, shape matching, boundary information extraction, video indexing, traffic monitoring, face tracking, face detection. In pattern matching application, priori information about the target pattern must be available from which most distinctive features can be computed. Pattern matching process is shown in Fig. 1.

To achieve high degree of matching accuracy as well as to reduce computational complexity is still a major challenge in pattern matching problem. A good number of fast pattern matching algorithms are proposed as alternative to full-search algorithms exploited and discussed in [5, 6]. In [5, 6], Haar-like features are used for pattern representation and matching is performed using strip sum and image square sum techniques. However, this pattern matching technique is not much proved to be a computationally efficient one in terms of memory space and time requirement, and therefore, overall performance gets slow down for large dataset. In [7], branch-and-bound approach has been used to maximize a large class of classifiers functions efficiently on sub-images having good speed. In [8], a fast full-search pattern matching method has been proposed. In [9], a fast pattern matching algorithm which can handle arbitrary 2D affine transformations to minimize sum of absolute differences (SAD) error has been proposed. It uses sub-linear algorithm that randomly examines a small number of pixels, and further it is accelerated by branch-and-bound scheme. In [10], rotation-invariant texture classification using feature distributions is being proposed and found very efficient for pattern recognition using features with rotation-invariant property. In [11], both shape and texture data are considered to represent a face image and the basic LBP-based face recognition process has been proposed. Various other local texture features such as modified census transform, MBLBP, LBP histogram and locally assembled binary feature have been introduced in [12–15], respectively.

This paper proposes a novel pattern matching technique based on local features called Multi-variant Local Binary Pattern (MVLBP) which is constructed from basic LBP. It computes decimal equivalent of binary number for all source pixels by thresholding neighbouring pixels of source pixel to implement graphical inter-relationship between source and neighbouring pixels of it. For each feature point, feature vector is calculated to obtain histograms for all training images and followed by feature matching process for similarity measurement. The proposed work addresses the problems related to changes in orientation, illumination and occlusions in the image as object may or may not be found rigid and may not be invasive in nature.

The rest of the paper is organized as follows: Sect. 2 discusses motivation of the work. Section 3 presents MVLBP-based pattern matching process. Experimental results are given in Sect. 4. Concluding remarks are made in the last section.

2 Motivation

In local feature-based pattern matching process, local descriptor needs to be robust across substantial variations due to illumination, orientation, low resolution, occlusions and other factors. Image pattern should retain significant amount of spatial information, shape and texture information for pattern matching. Local Binary Pattern (LBP) [16] with rotation-invariant property is one of the local descriptors that collects information of how neighbourhood pixels around a source pixel are correlated with the source pixel and extracts features to obtain textural and shape information from image in pattern matching. For each source pixel, LBP operates on a 3 × 3 adjacent neighbour area with 8 neighbourhood pixels to form an 8-bit binary sequence. If intensity value of source pixel is found to be less than or equal to the intensity value of a neighbourhood pixel, then 1 else 0 is placed in the binary sequence. Whereas the MB-LBP [13] approach encodes rectangular regions using LBP operator. The MBLBP features can capture large-scale structure by comparing central rectangle's average intensity value with all neighbourhood intensity values. In order to use the structural properties of local descriptors and apply them in pattern matching, a novel variant called Multi-variant LBP (MVLBP) is proposed exploited from LBP.

3 Pattern Matching Methodology—The MVLBP Approach

As the pattern matching approach needs matching across substantial variations of image properties such as illumination, orientation, occlusion, therefore, the present pattern matching techniques must be adaptable ones while local structural information is used for matching and enhanced performance. Local descriptors like LBP and its rotation-invariant operators are found robust to various object recognition problems, and they are able to represent the pattern in quite efficient way. MVLBP is one such LBP-based local descriptor which can be used for feature extraction to obtain local texture and shape information from an image pattern for pattern matching problem. The process is simple and has higher discriminative power, which is robust against monotonic grey value or scale changes caused due to illumination variations.

MVLBP is a holistic approach which takes 3×3 block size of neighbourhood area of pixels as a circle with radius 1. Each source pixel's initial intensity value would be updated with decimal equivalent of binary value being calculated by thresholding its neighbour pixel values with it. The updated pixel value basically depends on neighbours' original intensity values along with the sequence of neighbour pixels being taken for thresholding with the pixel whose value is going to be updated. The algorithm gives a wide range of options to choose variants based on neighbour pixel movements. Out of 32 variants depending upon 8 neighbour pixels, MVLBP chooses variants of those neighbours whose intensity value is found greater than the source pixel. So, to obtain maximum spatial and textural information by tracking minor to minor variations in the image and to show graphical interrelationship between source and its neighbourhood pixels, a particular sequence cannot be an optimal solution for all source pixels in an image. In order to obtain optimal weighted value, the selection of neighbour along with its variant may not be same for all source pixel as intensity value of all pixels along with their neighbour is found different. For that, all possible sequences (named as variants) formed in zigzag (i.e. horizontal or vertical or circular) fashion should be checked to get optimal solution for each source pixel. Maximization of weighted value not only helps to get optimal solution but also helps to identify the most distinctive fiducial points for pattern matching which in turn increase the matching accuracy of the process.

In this process, initially the images are resized to 128×128 size and it needs 3×3 matrix to calculate the weighted value of each source pixel. Using MVLBP, $[(128 - 3) + 1]^2 = 15{,}876$ different features for each image can be formed to train if we use image of dimension 128×128 pixels.

In a 3×3 neighbour area of pixels for each source pixel, 8 neighbours' are being considered to form 8-bit binary sequence (for each neighbour, 1-bit binary data is associated in the sequence). For each pair of pixel comparison, binary '1' or '0' is placed in the process of weighted value calculation in binary. For 8-bit binary weighted value, there must be 8 comparisons. As, binary bit position count starts from '0', for that comparison count 'c' is started from '0' and ends at '7'.

$$X = \text{Binary bit for each pair of pixel comparison} = \begin{cases} 1, & if \quad I_{pixel's\_neighbour} \geq I_{pixel} \\ 0, & otherwise \end{cases}$$

$$(1)$$

$$Weighted\ Value(w) = \sum_{c=7}^{0} (X)2^c \tag{2}$$

where c = comparison count

Initially, w_s (optimal weighted value for any source pixel) = 0;

for each neighbour 'n', of a source pixel 's' as starting pixel of variants; where $n = 1,...,8$

$$Number\ of\ Variants\ of\ n(n_v) = \begin{cases} 6, & if\ n\ is\ diagonal\ neighbour \\ 2, & otherwise \end{cases}$$

$$for\ all\ n_v\ of\ n,\ w_s = \max(w_{v_i}^n, w_s);\ if\ I_n \geq I_s \tag{3}$$

where $w_{v_i}^n$ = *weighted value of source with respect to nth neighbour's variant v_i and i varies from 1 to 6 for diagonal neighbour and 1 to 2 for other neighbours.*

Table 1 shows an example in which a pixel with intensity value '55' has been updated with 32 rotation sequences by various pixel movement adopted to generate multiple variants of LBP, where we try to select an optimal value and it's corresponding structure for matching. Figure 2 shows an example of how a source pixel's value is updated by MVLBP for a source image of size 4 × 4 and pattern matching process with a 3 × 3 pattern. The MVLBP algorithm is shown in Algorithm 1.

Lemma 1 *Optimal weighted value calculation for source pixel is entirely dependent on surrounding neighbour pixel's intensity value of the source pixel.*

Table 2 shows that rotation starting with the neighbour pixel whose value is greater than that of the source pixel value is giving maximum weighted decimal value for the source pixel compared to all possible neighbour pixels with their variants. For the source pixel '55', neighbour 82 with variants V17 and V18 and neighbour 73 with variants V19 and V20 returns better weighted value for '55' where from maximum value will be chosen. Here, weighted value returned by either 82-V17 or 73-V20 can be selected for weighted value of '55'.

Lemma 2 *Selection of an appropriate variant gives better weighted value for source pixel which helps to find discriminative fiducial points on the image.*

In MVLBP, at most 32 variants can be constructed. These give wide range of options to choose accurate variant for getting optimal weighted value for all source pixels. Out of 32 variants, V17 and V20 variants are giving best choice for '55'. No

Table 1 Different variants of MVLBP

Variant number	Variant type	Rotation paths	Weighted binary value	Weighted decimal equivalent value of binary	Pixel movement		
V1	Horizontal	41-73-42-82-15-51-46-13	$(01010000)_2$	$(80)_{10}$	41	73	42
					82	**55**	15
					51	46	13
V2	Horizontal	42-73-41-15-82-13-46-51	$(01001000)_2$	$(72)_{10}$	41	73	42
					82	**55**	15
					51	46	13
V3	Horizontal	51-46-13-82-15-41-73-42	$(00010010)_2$	$(18)_{10}$	41	73	42
					82	**55**	15
					51	46	13
V4	Horizontal	13-46-51-15-82-42-73-41	$(00001010)_2$	$(10)_{10}$	41	73	42
					82	**55**	15
					51	46	13
V5	Vertical	41-82-51-73-46-42-15-13	$(01010000)_2$	$(80)_{10}$	41	73	42
					82	**55**	15
					51	46	13
V6	Vertical	42-15-13-73-46-41-82-51	$(00010010)_2$	$(18)_{10}$	41	73	42
					82	**55**	15
					51	46	13
V7	Vertical	51-82-41-46-73-13-15-42	$(01001000)_2$	$(72)_{10}$	41	73	42
					82	**55**	15
					51	46	13
V8	Vertical	13-15-42-46-73-51-82-41	$(00001010)_2$	$(10)_{10}$	41	73	42
					82	**55**	15
					51	46	13
V9	Horizontal and vertical	41-73-42-15-82-51-46-13	$(01001000)_2$	$(72)_{10}$	41	73	42
					82	**55**	15
					51	46	13
V10	Horizontal and vertical	41-82-51-46-73-42-15-13	$(01001000)_2$	$(72)_{10}$	41	73	42
					82	**55**	15
					51	46	13
V11	Horizontal and vertical	42-73-41-82-15-13-46-51	$(01010000)_2$	$(80)_{10}$	41	73	42
					82	**55**	15
					51	46	13

(continued)

Table 1 (continued)

Variant number	Variant type	Rotation paths	Weighted binary value	Weighted decimal equivalent value of binary	Pixel movement		
V12	Horizontal and vertical	42-15-13-46-73-41-82-51	$(00001010)_2$	$(10)_{10}$	41	73	42
					82	55	15
					51	46	13
V13	Horizontal and vertical	51-46-13-15-82-41-73-42	$(00001010)_2$	$(10)_{10}$	41	73	42
					82	55	15
					51	46	13
V14	Horizontal and vertical	51-82-41-73-46-13-15-42	$(01010000)_2$	$(80)_{10}$	41	73	42
					82	55	15
					51	46	13
V15	Horizontal and vertical	13-15-42-73-46-51-82-41	$(00010010)_2$	$(18)_{10}$	41	73	42
					82	55	15
					51	46	13
V16	Horizontal and vertical	13-46-51-82-15-42-73-41	$(00010010)_2$	$(18)_{10}$	41	73	42
					82	55	15
					51	46	13
V17	Circular	82-41-73-42-15-13-46-51	$(10100000)_2$	$(160)_{10}$	41	73	42
					82	55	15
					51	46	13
V18	Circular	82-51-46-13-15-42-73-41	$(10000010)_2$	$(130)_{10}$	41	73	42
					82	55	15
					51	46	13
V19	Circular	73-42-15-13-46-51-82-41	$(10000010)_2$	$(130)_{10}$	41	73	42
					82	55	15
					51	46	13
V20	Circular	73-41-82-51-46-13-15-42	$(10100000)_2$	$(160)_{10}$	41	73	42
					82	55	15
					51	46	13
V21	Circular	15-13-46-51-82-41-73-42	$(00001010)_2$	$(10)_{10}$	41	73	42
					82	55	15
					51	46	13
V22	Circular	15-42-73-41-82-51-46-13	$(00101000)_2$	$(40)_{10}$	41	73	42
					82	55	15
					51	46	13

(continued)

Table 1 (continued)

Variant number	Variant type	Rotation paths	Weighted binary value	Weighted decimal equivalent value of binary	Pixel movement		
V23	Circular	46-51-82-41-73-42-15-13	$(00101000)_2$	$(40)_{10}$	41	73	42
					82	**55**	15
					51	46	13
V24	Circular	46-13-15-42-73-41-82-51	$(00001010)_2$	$(10)_{10}$	41	73	42
					82	**55**	15
					51	46	13
V25	Circular	41-73-42-15-13-46-51-82	$(01000001)_2$	$(65)_{10}$	41	73	42
					82	**55**	15
					51	46	13
V26	Circular	41-82-51-46-13-15-42-73	$(01000001)_2$	$(65)_{10}$	41	73	42
					82	**55**	15
					51	46	13
V27	Circular	51-46-13-15-42-73-41-82	$(00000101)_2$	$(5)_{10}$	41	73	42
					82	**55**	15
					51	46	13
V28	Circular	51-82-41-73-42-15-13-46	$(01010000)_2$	$(80)_{10}$	41	73	42
					82	**55**	15
					51	46	13
V29	Circular	13-15-42-73-41-82-51-46	$(00010100)_2$	$(20)_{10}$	41	73	42
					82	**55**	15
					51	46	13
V30	Circular	13-46-51-82-41-73-42-15	$(00010100)_2$	$(20)_{10}$	41	73	42
					82	**55**	15
					51	46	13
V31	Circular	42-15-13-46-51-82-41-73	$(00000101)_2$	$(5)_{10}$	41	73	42
					82	**55**	15
					51	46	13
V32	Circular	42-73-41-82-51-46-13-15	$(01010000)_2$	$(80)_{10}$	41	73	42
					82	**55**	15
					51	46	13

other variant's starting pixel is giving best value to select. So by maximizing weighted pixel value for all pixels in the image it is possible to identify most discriminative fiducial points for pattern matching.

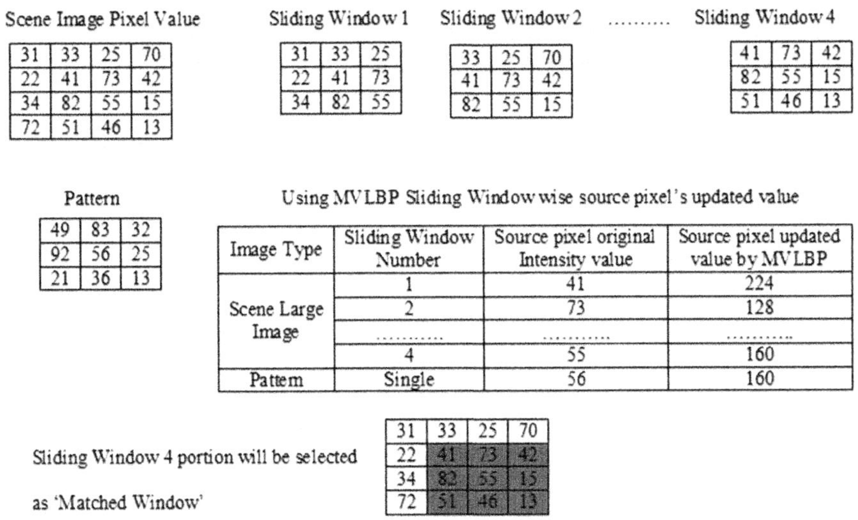

Scene Image Pixel Value

31	33	25	70
22	41	73	42
34	82	55	15
72	51	46	13

Sliding Window 1

31	33	25
22	41	73
34	82	55

Sliding Window 2 ....

33	25	70
41	73	42
82	55	15

Sliding Window 4

41	73	42
82	55	15
51	46	13

Pattern

49	83	32
92	56	25
21	36	13

Using MVLBP Sliding Window wise source pixel's updated value

Image Type	Sliding Window Number	Source pixel original Intensity value	Source pixel updated value by MVLBP
Scene Large Image	1	41	224
	2	73	128
	....	....	....
	4	55	160
Pattern	Single	56	160

Sliding Window 4 portion will be selected

as 'Matched Window'

31	33	25	70
22	41	73	42
34	82	55	15
72	51	46	13

Fig. 2 Pattern matching process using MVLBP

Table 2 Starting pixel-wise updated decimal value for source pixel '55'

Sl no	Intensity value of starting pixel	Variant no	Weighted decimal value	Sl no	Intensity value of starting pixel	Variant no	Weighted decimal value
1	82	V17	160	17	15	V22	40
2	73	V20	160	18	46	V23	40
3	82	V18	130	19	13	V29	20
4	73	V19	130	20	13	V30	20
5	41	V1	80	21	51	V3	18
6	41	V5	80	22	42	V6	18
7	42	V11	80	23	13	V15	18
8	51	V14	80	24	13	V16	18
9	51	V28	80	25	13	V4	10
10	42	V32	80	26	13	V8	10
11	42	V2	72	27	42	V12	10
12	51	V7	72	28	51	V13	10
13	41	V9	72	29	15	V21	10
14	41	V10	72	30	46	V24	10
15	41	V25	65	31	51	V27	5
16	41	V26	65	32	42	V31	5

Algorithm 1: MVLBP

Let I be the set of images and $I = [I_1, I_2, \ldots, I_n] \in P$, where P is the total set of images.

For each Image in I
{

    Read the image into W;
    Resize the image into 128×128size;
    For each source pixel in W
    {

        Consider matrix size of 3×3;
        For all 8 neighbour pixels of the source pixel
        {

           If $I_{neighbour} \geq I_{source}$
           {
           For all variants
           {
           Concatenate all calculated value either '1' or '0' by comparing source pixel with neighbours present within the area of 3×3 where source pixel is at center location to form the weighted value in binary for the source pixel.
           Convert the binary value to its equivalent decimal value and store within an array 'mvlbp_new_pixel'.
           }
           }
        }
        Find the maximum from array 'mvlbp_new_pixel' and replace the source pixel value with the calculated maximum value in new_image$_{mvlbp}$ of image W;
        return(15,876 features);
    }
}

---

4 Evaluation

The proposed pattern matching algorithm MVLBP is tested on two publicly available databases, namely the COIL-100 [17] and the Caltech 101 [18] databases. In the pattern matching process, scanning of input image is performed first, followed by resizing of image to 128 × 128 pixels. Then after dividing the image into a group of sub-regions of size 3 × 3 each, weighted decimal value is calculated with respect to neighbourhood pixel for each matrix. Histogram of each block with the weighted value is generated followed by concatenation of all histograms into a concatenated histogram which represents a feature vector. The measurement of similarity is done by calculating similarity proximity with Euclidean distance between pattern histogram and all possible candidate histograms. A candidate window is selected based on the similarity score while it is compared with a threshold determined heuristically.

4.1 Databases

The Columbia Object Image Library (COIL-100) database [17] is a publicly available database which contains colour images of 100 objects. The colour images are collected by placing the objects on a motorized turntable against a black background. The turntable is rotated through 360° to vary the object pose with respect to a fixed colour camera. Images of the objects are taken at pose intervals of 5°. This corresponds to 72 poses per object.

On the other hand, the Caltech 101 database [18] comprises of open-source images of 101 different categories with most categories having minimum of 40 images. All images are with dimension of 300 × 200 pixels. Most images have little or no clutter. The objects tend to be centred in each image.

4.2 Experimental Results

The MVLBP algorithm is tested on two databases described in Sect. 4.1. Table 3 shows the results determined on the COIL-100 and the Caltech 101 databases. Accuracy curves are given in Fig. 3 which exhibits the trade-off between training samples and accuracy. The algorithm gives a wide range of options to choose variants based on neighbour pixel movements.

From Table 2 it is seen that when intensity value of neighbouring pixel is found greater than intensity value of source pixel, these variants can produce better weighted value for source pixel. So by maximizing weighted pixel value for all pixels in the image, it is possible to identify the most discriminative fiducial points for pattern matching which in turn increase the matching accuracy of the process.

With having less training compared to MBLBP and rotation-invariant LBP, MVLBP has better matching accuracy. When training samples get increase gradually, the matching accuracy also increases. However, compared to MBLBP and LBP, the proposed MVLBP algorithm shows a consistent rise in matching accuracy and has better matching accuracy compared to other two existing local descriptors, viz., LBP and MBLBP with respect to training samples/percentage.

Table 3 Table shows pattern matching accuracies of MBLBP, LBP and MVLBP on COIL-100 and Caltech 101 databases

COIL-100 database				Caltech 101 database			
Training percentage	MBLBP	LBP	MVLBP	Training percentage	MBLBP	LBP	MVLBP
10	50.14	55.61	59.13	10	52.14	54.98	57.53
20	71.23	67.68	69.45	20	60.53	66.68	70.24
30	77.42	78.25	79.46	30	68.07	72.55	77.47
40	80.69	81.23	87.79	40	77.58	80.33	86.41
50	87.34	84.54	92.99	50	84.69	81.94	92.38
60	90.56	88.02	94.98	60	88.63	86.54	95.67
70	94.54	93.64	97.12	70	94.35	94.51	98.14
80	96.25	97.21	99.74	80	97.58	96.71	99.71
90	100	100	100	90	100	100	100

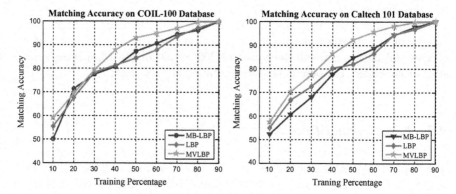

Fig. 3 Matching accuracies of MBLBP, LBP and MVLBP on the COIL-100 (left) and the Caltech 101 (right) databases are shown

Training percentage represents number of images tested or trained compared to total number of images present in the database, and matching accuracy would be $\frac{(T-U)}{T} \times 100\%$; where T = total number of database images and U = number of unrecognized images.

5 Conclusion

In this paper, a novel and efficient pattern matching algorithm MVLBP has been presented. As the proposed MVLBP-based pattern matching method takes all 8 neighbour pixels close to a source pixel as reference, not the entire image is taken as reference; therefore it can handle minor changes of object's pose variations in the image, and it does not require very rich text information on the pattern which makes MVLBP features very effective in handling blurred or low-resolution images and also robust to occlusion. MVLBP features are also found to be scale and illumination-invariant. MVLBP gives an option to choose the optimal value for pixels by varying all possible 32 variants. MVLBP-based pattern matching shows gradual and monotonic increasing accuracy with increasing training percentage of samples. The proposed MVLBP-based pattern matching approach can be used for other tasks like object attribute classification, video-based pattern recognition and unconstrained pattern detection with the challenges of arbitrary pose variations, occlusions, illumination.

References

1. Lewis, J.P.: Fast Template Matching. In: Vision Interface 95, Canadian Image Processing and Pattern Recognition Society Conference, pp. 120–123. Quebec City, Canada (1995).
2. Mahalakshmi, T., Muthaiah, R., Swaminathan, P.: An Overview of Template Matching Technique in Image Processing. Research Journal of Applied Sciences, Engineering and Technology. 4(24), 5469–5473 (2012).
3. Zhao, W., Chellappa, R.: Face Recognition: A Literature Survey. ACM Computing Surveys. 35(4), 399–458 (2003).
4. Ouyang, W., Zhang, R., Cham, W.K.: Fast Pattern Matching Using Orthogonal Haar Transform. In: IEEE Conference on Computer Vision and Pattern Recognition (CVPR), pp. 3050–3057. IEEE Press (2010).
5. Li, Y., Li, H., Cai, Z.: Fast Orthogonal Haar Transform Pattern Matching via Image Square Sum. IEEE Transactions on Pattern Analysis and Machine Intelligence. 36(9), 1748–1760 (2014).
6. Ouyang, W., Cham, W.K.: Fast Algorithm for Walsh Hadamard Transform on Sliding Windows. IEEE Transactions on Pattern Analysis and Machine Intelligence. 32(1), 165–171 (2010).
7. Lampert, C., Blaschko, M., Hofmann, T.: Beyond Sliding Windows: Object Localization by Efficient Subwindow. In: Computer Vision and Pattern Recognition, IEEE Press (2008).
8. Tombari, F., Mattoccia, S., Stefano, L.D.: Full Search-equivalent Pattern Matching with Incremental Dissimilarity Approximations. IEEE Transactions on Pattern Analysis and Machine Intelligence. 31(1), 129–141 (2009).
9. Korman, S., Reichman, D., Tsur, G., Avidan, S.: FAsT Match: Fast Affine Template Matching. In: Computer Vision and Pattern Recognition, IEEE Press (2013).
10. Pietikäinen, M., Ojala, T., Xu, Z.: Rotation-invariant Texture Classification using Feature Distributions. Pattern Recognition. 33(1), 43–52 (2000).
11. Rahim, M.A., Hossain, M.N., Wahid, T., Azam, M.S.: Face Recognition using Local Binary Patterns (LBP). Global Journal of Computer Science and Technology Graphics and Vision. 13 (4), 1–9 (2013).
12. Froba, B., Ernst, A.: Face Detection with the Modified Census Transform. In: 6th IEEE International Conference on Automatic Face and Gesture Recognition, pp. 91–96. IEEE Press (2004).
13. Zhang, L., Chu, R., Xiang, S., Liao, S., Li, S.Z.: Face Detection based on Multi-block LBP Representation. In: IAPR International Conference on Biometrics, pp. 11–18. LNCS 4642, Springer, Berlin, Heidelberg (2007).
14. Zhang, H., Gao, W., Chen, X., Zhao, D.: Object Detection using Spatial Histogram Features. Image and Vision Computing. 24(4), pp. 327–341 (2006).

15. Yan, S., Shan, S., Chen, X., Gao, W.: Locally Assembled Binary (LAB) Feature with Feature-centric Cascade for Fast and Accurate Face Detection. In: IEEE Computer Society Conference on Computer Vision and Pattern Recognition, pp. 1–7 (2008).
16. Ayoob, M.R., Kumar, R.M.S.: Face Recognition using Symmetric Local Graph Structure. Indian Journal of Science and Technology. 8(24), 1–5 (2015).
17. Nene, S.A., Nayar, S.K., Murase, H.: Columbia Object Image Library (COIL-100). Technical Report, CUCS-006-96 (1996).
18. Wang, G., Zhang, Y., Fei-Fei, L.: Using Dependant Regions or Object Categorization in a Generative Framework. In: IEEE Conference on Computer Vision and Pattern Recognition, IEEE Press (2006).
19. Muhammad, S, Khalid, A, Raza, M, Mohsin, S: Face Recognition using Gabor Filters. Journal of Applied Computing Science and Mathematics. 11, pp. 53–57 (2011).

GUESS: Genetic Uses in Video Encryption with Secret Sharing

Shikhar Sharma and Krishan Kumar

Abstract Nowadays, video security systems are essential for supervision everywhere, for example video conference, WhatsApp, ATM, airport, railway station, and other crowded places. In multi-view video systems, various cameras are producing a huge amount of video content which makes it difficult for fast browsing and securing the information. Due to advancement in networking, digital cameras, and media, interactive sites, the importance of privacy and security is rapidly increasing. Hence, nowadays the security of digital videos become an emerging research area in the multimedia domain; especially when the communication happens over the Internet. Cryptography is an essential practice to protect the information in this digital world. Standard encryption techniques like AES/DES are not optimal and efficient in case of videos. Therefore, a technique is immediately required, which can provide the security to video content. In this paper, we address the video security-related issues and their solutions. An optimized version of the genetic algorithm is employed to solve the aforementioned issues through modeling the simplified version of genetic processes. It is used to generate a frame sequence such that the correlation between any two frames is minimized. The frame sequence determines the randomization in order of frames of a video. The proposed method is not only fast but also more accurate to enhance the efficiency of an encryption process.

Keywords Video encryption · Genetic algorithm · Secret sharing

1 Introduction

Conventionally, the video is a collective of the numerous scene while a scene made up through a set of shots. Moreover, these shots were an unbreakable series of the frames which is continuously recorded by the camera in a regular interval of time.

S. Sharma (✉) · K. Kumar
Department of Computer Science & Engineering, NIT, Srinagar, Uttarakhand, India
e-mail: shikhar01.cse14@nituk.ac.in

K. Kumar
e-mail: kkberwal@nituk.ac.in

© Springer Nature Singapore Pte Ltd. 2018
B. B. Chaudhuri et al. (eds.), *Proceedings of 2nd International Conference on Computer Vision & Image Processing*, Advances in Intelligent Systems and Computing 703, https://doi.org/10.1007/978-981-10-7895-8_5

So, such audiovisual information was usually accessed in an identical amount of time. However, during last decade, a very big amount of video content has been recorded by multiple cameras. In addition to this, rapid growth in the network as well as computing infrastructure and frequent utilization of digital video technology, a good number of multimedia real-time applications are forthwith required. Thus, a huge amount of audiovisual data is fast generated for various applications [1–5] such video conferencing, videos surveillance security systems, and WhatsApp [6–9].

A video call can be defined as an experience that two people make the communication with one another using moving video image and audio, and essentially additional features, i.e., Web video conferencing. Web video conferencing may be employed in a Web training seminar or Webinar where the sender's video image live with you in a Web browser. You can send your video image as well. However, it is not most likely, because Webinars tend to communicate one-way video to you [10]. However, such audiovisual content is not safe over the Internet until we send the encrypted message/videos instead. Therefore an essential security framework is urgently required.

Video surveillance security systems are going to be omnipresent in urban life. With the underlying principle to detect crime in urban areas, video camera setup is being set up in these areas that provide the much secure environment to the society. Sometimes, people won't commit a crime or will commit them elsewhere, because they afraid to catch by such active surveillance or being identified later on from video recordings [11]. Therefore, authorized personnel may not only removed from the surveillance video but also edited into the video itself. It can only be retrieved with a secret key.

A perceptual watermarking approach [12] for compressed domain video is suggested to deal with the huge payload problem in the surveillance security systems where a particular signature is embedded into the header of the video for authentication. They counted all the privacy information into the video without affecting its visual quality. This model can monitor the unauthorized persons in a restricted environment and provides the privacy of the authorized persons. Moreover, it allows the privacy information to be revealed in a secure and reliable way. However, this approach failed to make a secure and reliable communicate over the Internet in the real time.

Secret Sharing: The secret schemes are ideal for storing highly sensitive and important information as multimedia data such as missile launch codes, numbered bank accounts, and encryption keys. The information pieces must be held highly confidential. However, it is also critical, so should not be lost. Conventional models for encryption are ill-suited to achieve high levels of confidentiality and reliability at a time due to storage of encryption keys. One must hold a single copy of the key at one location for maximum secrecy or hold multiple copies of the key at different locations for greater reliability. Consequently, increasing the reliability of the key through storing multiple copies leads the lower confidentiality as offering an additional opportunity for attack vectors. There are more chances for a copy to fall into the wrong hands.

This problem can address using secret sharing schemes in order to achieve the highest levels of confidentiality and reliability. This refers to the techniques to distribute single secret among a group of people or participants. Each has an allocated share of the secret. The secret reconstruction is only possible when a sufficient number of possibly different types, of shares, is combined together. Individual shares are of no use on their own. In 1979, the schemes [13, 14] for secret sharing was introduced as a solution for safeguarding the cryptographic keys where a secret S is divided among n shares. The secret S shared among n participants or shareholders where reconstruction of the secret S is possible only with any m or more than m shares. However, the key advantage of the secret sharing schemes is, the reconstruction of the secret S becomes next to impossible, in the case of fewer than m shares.

Security of Key: cryptography comprises the nuts and bolts to protect information from undesirable individuals by conversion into a form non-recognizable or unreadable by the attackers while in transmission and during storage. Data cryptography is the rushing of the data content, such as "image," "text," "audio," "video" and so forth to make these data unintelligible, invisible or unreadable during storage or transmission called encryption. The main principle of cryptography is keeping information secure from unauthorized person or attackers.

A reverse technique of data encryption is data decryption, to recover the original data. With the rapid growth in various multimedia technologies, more and more multimedia data are produced and transmitted in numerous fields like the commercial, medical, and military fields; this sensitive data should not be leaked or accessed by general users. Hence, security of multimedia data has become more demanding technology. Based on the key, two types of cryptographic schemes are suggested [15] as follow:

Symmetric Key Cryptography is one of the most important types of cryptography where a key is shared between both communicating parties (receiver and sender), i.e., the same key is used for encryption and decryption. The main advantage of this scheme is it used for private encryption of data in order to achieve high performance. For example, Advanced Encryption Standard (AES), International Data Encryption Algorithm (IDEA), Data Encryption Standard (DES).

Asymmetric Key Cryptography is one of the most important types of cryptography where the same key is not shared between both communication parties (receiver and sender), i.e., two different keys are used in the process of encryption and decryption. In asymmetric key cryptography, encryption key refers to the public key (everyone can access) and the decryption key refers to the private key (the only receiver can access). For example, Rivest Adi Shamir and (RSA), Diffie–Hellman. From the literature review [15], it was observed that the characteristics feature that determines the strength of the key is not quantifiable. However, matrices should be employed to compare and evaluate the performance of the cryptographic algorithms. Symmetric or asymmetric as type, key size, and number of rounds; the complexity of the algorithm, integrity and authentication of the message as functions are considered for the main characteristics. Nowadays, attackers have the understanding to test the strength of the algorithm or try brute force attack and differential cryptanalysis to break the secured approach. The various parameters are used to validate the effect of the attacks which are based on the key length and complexity of the algorithm from which key is generated. In order to generate a more complex key, key generation processes should be involved with more complexity. It will become very difficult for a cryptanalyst to attack the key.

Genetic Algorithm: (GA) can be understood as a family of computational models. Its working principles are evolution and natural selection. These sets of algorithms convert the particular problem in a specific domain as a model through a chromosome-like data structure. The model evolves the chromosomes using recombination, selection, and mutation operators. There are numerous applications [16] available in the domain of computer security where it is used to find the optimal solutions to a specific problem. A randomly selected population of chromosomes is usually the origin of the process of a genetic algorithm.

From the last decade, numerous cryptographic schemes have been explored; some techniques out of them are based on GA. Kumar et al. [17] suggested GA-based encryption using a secret key for the encryption process, crossover operator and pseudorandom sequence generator (by nonlinear feed forward shift register). Pseudorandom sequence decides the crossover point which helps to obtain the fully encrypted data. Later on, they used mutation after encryption [18] where encrypted data become hidden by masking with the steno-image. Another GA-based model was proposed for generating the pseudorandom numbers [19] where encryption process follows the working of the crossover and mutation operator. It uses the concept of genetic algorithms and pseudorandom binary sequence. In addition to this, nine parameters of linear congruential generators are used in a key generation procedure. However, this model comprises the confidentiality issue.

Later on, Qiao, et al. [20] stated that strong key may be generated by a good randomness quality of the numbers. It can be selected on or above the threshold value. The randomness of the sample was checked by the coefficient of correlation. Moreover, a new symmetrical block ciphering approach named improved cryptography inspired by GA was presented [21] where a random process generated the session key which was an enhancement of GA-inspired cryptography. Its results are more efficient than the other encryption techniques, such as video shuffling and shredding [22]. However, the existing state of the art approaches fails to meet the requirements of the real-time applications, and generating more complex and secure key which creates difficulty for the attackers. Therefore, an integrated model is required to promptly address these issues.

In this paper, we used two abstracts (steganography, random sequence generator) to address the above limitations. *Steganography* is the practice of concealing a message, image, file, or video within another message, image, file, or video [23–25]. This paper uses steganography as its first phase to conceal the actual frames into dummy frames having random values, thus forming glitch frames. *Random sequence generator* is a random sequence for a permutation of numbers between 1 to N. Since it is random, so the permutation cannot be predicted, and which is impossible to reproduce sequentially and reliably in polynomial time. Hence, a random number generator is used to generate keys and a genetic algorithm is used to make the key more complex. The key selection is to entirely depend on the value of the fitness function of the different strings generated by the random number. The permutation act as key GA may be applied. The salient features of our work are delineated as follow:

- We formulated the video encryption and the video decryption problem as a GA-based optimization problem.
- The proposed method manipulates images instead of bit manipulation as used in traditional method providing better efficiency and speed.
- Secret sharing scheme is employed on video frames which provide more security to video over the Internet.

The rest of the paper is structured as follows. Section 1 introduced the secret sharing schemes and importance of GA in encryption. The proposed video encryption system is described in Sect. 2. The experiments and results are discussed in Sect. 3. In the end, the work has been concluded in Sect. 4.

2 Proposed Model

This encryption problem can be solved using GA where the frames of the video represent as the chromosomes. Numerous different positions of each chromosome are encoded as bits, characters, or numbers according to problem attributes. Sometimes, these positions are known as genes. During evolution, these may be changed randomly within a range and then a set of chromosomes are referred as population. An evaluation function is employed to measure the *goodness* of each chromosome. In order to simulate the natural reproduction and mutation of species, two basic operators, i.e., crossover and mutation are used during an evaluation. For survival and combination, the selection of chromosomes is biased toward the fittest chromosomes [16].

Moreover, phenotypes, creatures or individuals, are also known as the population of candidate solutions; these are progressed toward better solutions to an optimization problem. Each such candidate solution comprises a set of properties in form of chromosomes or genotype which can be mutated and altered. In the next iteration of the algorithm, a new generation of candidate solutions may be used. The algorithm usually terminates with the condition of either a satisfactory fitness level has been reached for the population or a maximum number of generations have been produced. A typical genetic algorithm requires *genetic representation* and *fitness function* to evaluate of the solution domain. The various components of our proposed model are shown in Fig. 1.

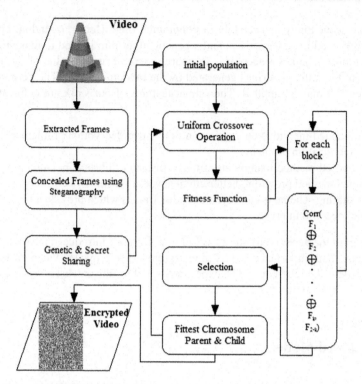

Fig. 1 Various component of our model

At sender's end:, Firstly, a video is broken down into N number of frames. These frames are concealed into random images of size 640 × 480. Secondly, a permutation size of N (number of frames extracted from the video) is used to generate the initial population of two chromosomes. Here, each chromosome represents a sequence in which frames will undergo with XOR operation. Block size is used as another parameter for XOR operation which represents the chunk of frames; on which XOR operation is performed. Thirdly, the result of XOR operation of the frames is then stored in permutation given by child chromosome. Finally, the offspring is reproduced using uniform crossover operation on initial population. These offspring acts as a key for decryption of the video if the parent is used for encryption or vice versa. For example the encryption and decryption process for 4 frames using block size, $k = 2$ as shown in Fig. 2.

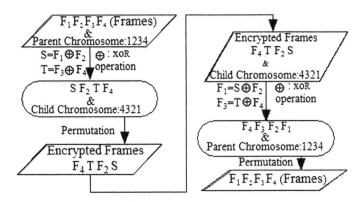

Fig. 2 Illustration for four encrypted and decrypted frames of a Video

```
program Encryption (Output)               program Fitness_Calculation (Output)

var                                       var
Frames: IntegerArray;                     Frames: IntegerArray;
k: Integer;                               k: Integer;
C: IntegerArray;                          val: Integer;
Parent: IntegerArray;                     temp: Integer;
Minfit: Integer;                          begin
Temp: Integer;                            val: 0;
begin                                     For each block of size, k
Minfit = 1;                               temp: 0;
If Number of Frames is not divisible by   For each Frame in block
block size, k  then                       temp: temp XOR Frames_{n};
Return Failure;                           end
end                                       val: Max( val,temp );
For each Chromosome, C_{i}                end
temp: Fitness(Frames,C_{i},k);           Return val;
If Fitness value is less than Minfit then  end
Parent: C_{i};
Minfit: temp;
end
end
Encrypt Frames using Parent as Frame Sequence
key: (k,Parent,Child(parent));
end
```

At receiver's end: the key represents the sequence in which XOR operation should be performed on the frames of transmitted video using the same block size parameter. The pseudo-code for encryption and calculating fitness of a chromosome is discussed as above.

Fitness of chromosomes is defined as the maximum correlation between frame obtained after XOR operation and any frame within the same block. Thus, the problem maps into a minimization problem which focuses on minimizing the fitness of a chromosome. The fittest chromosome represents that frames are minimally correlated with each other. If any one of them is missing, original frames cannot be restored. If child chromosome is fittest, then its parent is used in the decryption process or vice versa.

Encryption: asymmetric key is used in the model which comprises three parameters a block size, parent and child permutation of fittest chromosome. The structure of the key can be represented as (k, P, C), where k is block size, P is permutation used in encryption, and C is a permutation that to be used with P for decryption.

Decryption: the transmitted video is again broken down into frames. Using Child permutation, C from key as the sequence in which XOR operation is performed on frames using block size parameter given in key, the concealed frames are restored in parent permutation. Using parent permutation, P, the concealed frames can be ordered correctly. These frames can be used to extract the original frame information, and hence, the actual video frames are restored.

3   Experiment and Results

The work has been implemented and analyzed. The implementation has been done on the standard dual-core desktop with 2.7GHz processor using MATLAB R2013a. The fitness value of fittest chromosome is obtained. We used 10 video samples. The size of the initial population is 10 and iterated for 10 times.

3.1   Qualitative Analysis

Each frame is resized with dimensions of 200×160 after breaking down a video into N frames, and then operations are performed on them as discussed in Figs. 1 and 2. Here, only a few frames of a video are represented to illustrate the concealing, encryption and decryption operations as shown in Fig. 3. As it can be seen from Fig. 3. that the concealing and encryption are too hard to be decrypted using brute force.

3.2   Quantitative Analysis

The minimum correlation between actual frames and decrypted frames is calculated to demonstrate the similarity between them, and the correlation between secret frames is also calculated as tabulated in Table 1. To enable comparison between previous works and our work, same experimental methodology is used. Table 1. shows the results of different works based on their correlation value. They are the following reasons to attain the correlation between frames, which is used as the parameter:

- Correlation denotes the similarity between two digital signals, that is, if two signals are more similar then correlation is 1, otherwise 0 for two dissimilar signals.
- If there exist two frames such that they are more similar, then the concept of secret sharing becomes obsolete.
- If the correlation is 0, then both frames are important as decryption is not possible if either one is missing as per the concept of secret sharing [14].

From Table 1, it is observed that both encrypted and decrypted frames are similar to each other. While the previous works offer good encryption and decryption

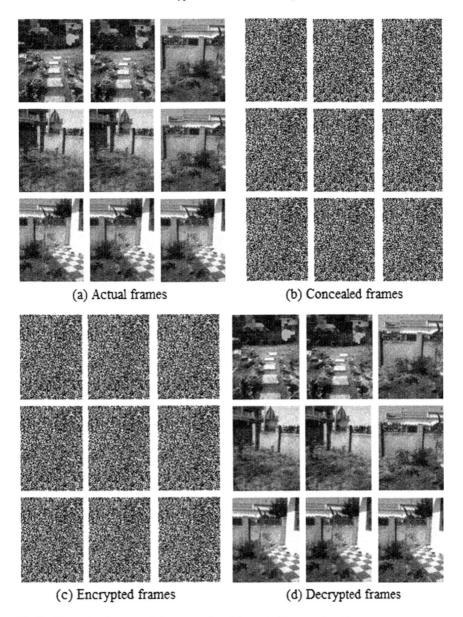

(a) Actual frames   (b) Concealed frames

(c) Encrypted frames   (d) Decrypted frames

Fig. 3   Illustration for concealed, encrypted, and decrypted frames of a video

Table 1 Comparison of correlation between frames

Method	Correlation value between actual and decrypted frames	Correlation value between encrypted frames
Simmons et al. [15]	0.98	0.86
Zhang et al. [12]	0.93	0.42
Kumar et al. [18]	0.85	0.09
Tragha et al. [21]	0.88	0.07
Qiao et al. [20]	0.84	0.08
GUESS($k = 25$)	0.89	0.03
GUESS($k = 35$)	0.91	0.01
GUESS($k = 40$)	0.87	0.04

correlation, but correlation between frames makes them vulnerable to attack. While in our work, it indicates encrypted frames are dissimilar making impossible to break it. Therefore, the model is more secure and reliable.

3.3 Computational Complexity

The proposed method was implemented for video shots of 100–700 frames with the varying duration 3–25 s. In order to reduce the computational cost, each frame is scale-down to 200×160. The time taken by the proposed method in the encryption and decryption of the 10 video samples is shown in Table 2.

The proposed GUESS model is compared with similar work [19] that was carried out for hundreds of samples. Its population varies greatly as compared to

Table 2 Computational time comparison

Number of frames	Block size	Time (s)
325	25	42.575
125	25	16.375
625	25	81.875
225	15	29.250
250	50	32.000
400	40	52.800
410	10	53.710
380	20	49.780
375	25	49.500
300	20	37.800

any other. The key size of the proposed method is a variable which increases greatly with the size of video sample. It helps to create the complex key. The computational time reported by Key Gen. [19] is about 78 s to encrypt an image; in comparison to this, the proposed model takes only about 130 ms to encrypt a frame. Hence, the proposed approach meets the requirement for the real-time applications such as YouTube, Video Broadcasting, WhatsApp.

4 Conclusion

In this work, we proposed a GUESS model which is based on the GA and secret sharing schemes. The model was used to generate frame sequence such that the correlation between any two frames is minimized. The frame sequence determines the randomization as a size of the number of video frames. The proposed method minimizes the time required as well as enhancing the efficiency of encryption process which meets the requirements of the real-time applications. For example YouTube, Video broadcasting, transmission of highly classified data by concealing it in frames. The efficiency of the proposed method is so promising that even ordinary data can be encrypted. Consequently, the proposed GUESS model takes lesser time and is also more accurate and efficient.

References

1. K., Krishan, et al. "Eratosthenes sieve based key-frame extraction technique for event summarization in videos." MTAP (2017), pp. 1–22.
2. Kelahmetoglu, et al., *Efficient utility of WhatsApp: from computer screen to the surgeon's hand to determine maxillofacial traumas,* JCS, 26, 4, (2015), pp. 1437.
3. G. A. Spanos, et al., *Performance Study of a Selective Encryption Scheme for the Security of Networked Real-Time Video,* ICCCN, (1995), pp. 2–10.
4. K., Kumar, et al., *Equal Partition based Clustering approach for Event Summarization in Videos,* SITIS'16, 2016, pp. 119–126.
5. K., Sridhar, et al., *System method and article of manufacture with integrated video conferencing billing in a communication system architecture,* U.S. Patent No. 5,867,494. (1999).
6. M., Sari, *What Up with WhatsApp,* TABJ, (2014), pp. 28.
7. A., Mohamed, et al., *An overview of video encryption techniques,* IJCTE, 2, 1, (2010), pp. 103–110.
8. W., Chung, et al., *Fast encryption methods for audiovisual data confidentiality,* ITISOP, 2001.
9. C., Shi, et al., *A fast MPEG video encryption algorithm,* ACM ICM'98.
10. M., Gough, *Video conferencing over IP: Configure, secure, and troubleshoot,* Elsevier, (2006).
11. A. Senior et al., *Enabling video privacy through computer vision,* IEEE Security & Privacy, 3, 3, (2005), pp. 50-57.
12. W. Zhang, S. S. Cheung and Minghua Chen, *Hiding privacy information in video surveillance system,* IEEE ICIP, (2005), pp. 868–871.
13. G.R. Blakley, *Safeguarding Cryptographic Keys,* Proc. Am. Federation of Information Processing Soc, 48, (1979), pp. 313-317.
14. A. Shamir, *How to Share a Secret,* Comm. ACM, 22, 11, (1979), pp. 612–613.

15. S. J., Gustavus, *Symmetric and asymmetric encryption,* ACM Computing Surveys (CSUR) 11, 4, (1979), pp. 305-330.
16. L., Wei, *Using genetic algorithm for network intrusion detection,* Proceedings of the United States Department of Energy Cyber Security Group, 1, (2004), pp. 1–8.
17. A. Kumar, N. Rajpal, *Application of Genetic Algorithm in the Field of Steganography,* Journal of Information Technology, 2, 1, (2004), pp. 12-15.
18. A. Kumar, et al., *New Signal Security System for Multimedia Data Transmission Using Genetic Algorithms,* NCC'05, IIT Kharagpur (2005) pp. 579–583.
19. A. Soni, et al., *Key Gen. Using Genetic Algorithm for Image Encryption,* (2013).
20. L., Qiao, et al., *A new algorithm for MPEG video encryption,* ICISST, 1997.
21. A. Tragha, et al., *ICIGA: Improved Cryptography Inspired by Genetic Algorithms,* ICHIT'06, (2006).
22. S. Wang, et al., *A secure steganography method based on genetic algorithm,* JIHMSP, 1, 1, (2010), pp. 28–35.
23. Hasso, et al., *Steganography in Video Files,* IJCSI, 13, 1, (2016), pp. 32.
24. K. Krishan, et al., *Event BAGGING: A novel event summarization approach in multi-view surveillance videos,* IEEE IESC'17, 2017.
25. K. Kumar, et al., "SOMES: An efficient SOM technique for Event Summarization in multi-view surveillance videos," Springer ICACNI'17, 2017.

Learning-Based Fuzzy Fusion of Multiple Classifiers for Object-Oriented Classification of High Resolution Images

Rajeswari Balasubramaniam, Gorthi R. K. Sai Subrahmanyam and Rama Rao Nidamanuri

Abstract In remote-sensing, multi-classifier systems (MCS) have found its use for efficient pixel level image classification. Current challenge faced by the RS community is, classification of very high resolution (VHR) satellite/aerial images. Despite the abundance of data, certain inherent difficulties affect the performance of existing pixel-based models. Hence, the trend for classification of VHR imagery has shifted to object-oriented image analysis (OOIA) which work at object level. We propose a shift of paradigm to object-oriented MCS (OOMCS) for efficient classification of VHR imagery. Our system uses the modern computer vision concept of superpixels for the segmentation stage in OOIA. To this end, we construct a learning-based decision fusion method for integrating the decisions from the MCS at superpixel level for the classification task. Upon detailed experimentation, we show that our method exceeds in performance with respect to a variety of traditional OOIA decision systems. Our method has also empirically outperformed under conditions of two typical artefacts, namely unbalanced samples and high intra-class variance.

Keywords Multi-classifier system · Object-oriented image analysis
Segmentation · Superpixels · Classification · Fusion

1 Introduction

Early research pertaining to the classification of remote-sensing images had focused on the principle of similar objects having similar spectral properties. Therefore, the spectral information at pixel level had been the main feature considered during

R. Balasubramaniam · R. R. Nidamanuri
Indian Institute of Space Science and Technology, Thiruvananthapuram, India

G. R. K. Sai Subrahmanyam (✉)
Indian Institute of Technology, Tirupati, India
e-mail: rkg@iittp.ac.in

© Springer Nature Singapore Pte Ltd. 2018
B. B. Chaudhuri et al. (eds.), *Proceedings of 2nd International Conference on Computer Vision & Image Processing*, Advances in Intelligent Systems and Computing 703, https://doi.org/10.1007/978-981-10-7895-8_6

classification. However, limitations with regard to urban land cover classification were found in the usage of generalised training libraries for classifying man-made classes.

To solve the shortcomings of pixel-based classification methods and satisfy the current requirements of classification of very high resolution imagery, the recent trends have shifted towards object-oriented image analysis (OOIA) [2]. When it comes to very high resolution images, there are certain drawbacks of OOIA technique as well. Due to very high resolution, a single object in a scene may be categorised as multiple objects due to high intra-object variance. Also, due to the arbitrary nature of the objects formed, it is not possible to perform neighbourhood operations over them. In this situation, performance capabilities of OOIA drop.

Though OOIA currently provides the most feasible solution for remote-sensing image classification problem, each of its stages is open to discretion. In this paper, we have tried to improve the different stages of OOIA and design a robust standard OOIA model that can be used for efficient image classification. We have proposed an object-oriented multi-classifier system with a fuzzy decision fusion technique (OOMCS), which utilises the advantages of contextual information from OOIA and also the capabilities of accurate classification of multi-classifier techniques to its advantage. We use quick shift-based superpixels for the segmentation stage; this has shown excellent boundary adherence and is amongst the best of superpixelization schemes with lowest under-segmentation error [1]. To this end, we have designed learning-based fuzzy decision fusion system for effectively combining the decisions from multiple classifiers. Our experimentations show that it empirically outperforms the existing state-of-the-art OOIA methods and performs exceptionally well during frequently encountered artefacts of unbalanced situation and high intra-class variance. Thus, the overall contributions of this paper include:

- Segmenting an image into superpixels as the primary stage of remote-sensing image classification, which is a trade-off between the pixels in the pixel-based classification and objects in object-oriented image analysis.
- Introduce a paradigm shift in object-oriented image analysis by using multi-classifier systems to improve its classification performance.
- Design of a novel learning-based fuzzy decision fusion system to effectively combine the capabilities of the multiple classifiers used in the classification stage.
- Extensive evaluation of the proposed methodologies and demonstration of its capabilities in the artefact cases of sparse samples and high intra-class variance.

Section 2 describes the methodology and the algorithms involved, in detail. Section 3 describes the implementation details; In Sect. 4, we describe the results obtained over various data sets and discuss its performance on special cases. Final section concludes with discussions for plausible variations for further development over the algorithm to improve its performance.

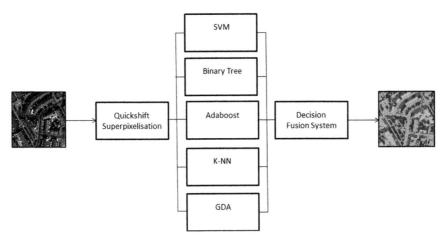

Fig. 1 Proposed scene parsing framework

2 Proposed Methodology

The OOMCS with fuzzy fusion scheme can be broadly split into three stages. A simplified block diagram is represented in Fig. 1. First stage generates superpixels using quick shift algorithm. It is followed by a feature extractor and a set of widely used classifiers which run over the feature vectors to provide their decisions. Finally, a fusion mechanism that assigns priority weights to classifiers is based on their performance over the training samples. This system mainly focuses on improving its performance with minimal training. Each stage is described in detail in the following subsections.

2.1 Superpixel Generation

2.1.1 Quick Shift Algorithm

Quick shift algorithm produces segments without the user specifying the number of segments required and is predominantly mode seeking over intensity and spatial vectors which is ideal for satellite image segmentation. It is a mode-seeking algorithm with a time complexity of $\mathcal{O}(dN^2)$ [6].
Given N data points,

$$\mathbf{x}_1, \mathbf{x}_2, \ldots, \mathbf{x}_N \epsilon \mathbf{X}(R^d)$$

a mode-seeking clustering algorithm begins by computing the parzen density estimate:

$$P(x) = 1/N \sum_i K(\mathbf{x} - \mathbf{x}_i) \tag{1}$$

where $\mathbf{x} \epsilon R^d$

N denotes the number of pixels within the kernel and

$K(\mathbf{x})$ is a Gaussian window or any other window [4]. The mode of the density $P(\mathbf{x})$, need not be measured using the gradient or quadratic lower bound (as in the case of mean shift algorithm). In quick shift, each point $\mathbf{x}_i$ is moved to the nearest neighbour for which there is an increment in the density $P(\mathbf{x})$ [7]. Thus, the trajectory in the case of quick shift is given by:

$$y_i = argmax_{j:\, d(\mathbf{x}_j, \mathbf{x}_i) < \tau} \left(\frac{P_j - P_i}{d_{ij}} \right) \tag{2}$$

Note that, $(P_j - P_i)/d_{ij}$ is a numerical approximation of the gradient of P in the direction of $\mathbf{x}_j - \mathbf{x}_i$ where

y_i—is the trajectory of vector i.

P_i—is the probability density of i.

τ—denotes the threshold that determines the maximum size of the superpixel.

d_{ij}—L_2 distance of pixel $\mathbf{x}_i$ from $\mathbf{x}_j$.

This approximation helps in reducing the time complexity compared to mean shift algorithm [7]. Figure 2 represents a sample aerial imagery which has been classified by quick shift-OOIA [5]. The centre column depicts the superpixels generated. It can be observed that the superpixels formed are more or less uniform unlike typical oversegmentation algorithms. Classification of the above image has been performed over four classes, namely buildings (yellow), roads (cyan), vegetation (blue) and shadows (red).

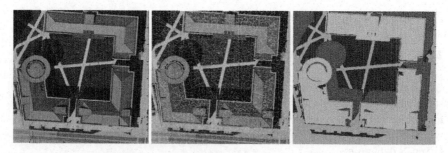

Fig. 2 Aerial image classification by quick shift-OOIA

2.2 Multiple Classifier Decisions

The MCS that we have designed contains five classifier modules, namely one versus all RBF-SVM, binary decision tree, Adaboost classifier, K-nearest neighbours and Gaussian discriminant analyser. The choice of classifiers are based on several factors such as using the kernel trick of SVMs, efficient handling of redundant attributes binary classification tree, resistance to overfitting by Adaboost forest.

2.3 Fuzzy Rank-Based Fusion

The existing fusion systems perform majority voting or simple averaging to combine decisions from multiple classifiers. We introduce a novel technique for combination of decisions called fuzzy ranking. The proposed approach is based on the intuition that the training accuracies can naturally determine weights assigned to each classifier over individual classes. The graphical depiction of the approach is shown in Fig. 3.

Elements of the class-wise histogram matrix (w) are the predicted probabilities which are the weights w_{ij}. w_{ij} represents the weight of classifier i for class j. The flow of these weights for final decision-making have been pictorially is depicted in Fig. 3. Once the rank matrix is generated during training phase, the next step is to assign scores to classify a test sample. Let DC_i represent the decision of classifier i, this decision is transformed into a one-hot vector notation, wherein 1 is placed at the position of the decision class in the vector whose length is equal to the number

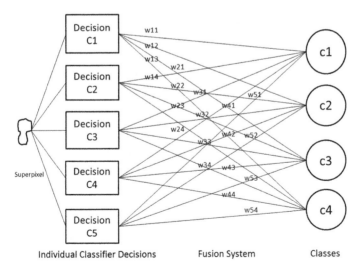

Fig. 3 Fusion model

Algorithm 1 Decision Fusion

1: **procedure** FUZZY RANK GENERATION
2:     ncl ← *no.ofclassifiers*
3:     nc ← *no.ofclasses*
4:     *Initialize* rank matrix ← zeros(size(nclxnc))
5:     *loop*:
6:     **for** $i = 1 : $ ncl **do**
7:         **for** $j = 1 : $ nc **do**
8:             rank matrix(i,j) ← $\frac{TruePositives(i,j)}{TrainingSamples(j)}$
9:             $w(i,j) \leftarrow \frac{rankmatrix(i,j)}{\sum_{j=1}^{nc}\{rankmatrix(i,j)\}}$.
10:        **end**
11:    **end**
12:    $S_c = w \cdot DC$
13:    $O_c \leftarrow \sum_{j=1}^{nc}\{S_c(:,j)\}$
14:    Decision outcome ← max O_c

---

of classes. Mathematically, the fusion can be represented as a weighed decision of all classifiers, where the weights have been learnt from the training samples. This weighed decision is obtained as shown in line 13 and 14 of Algorithm 1. The '·' in line 12 refers to the element-wise dot product of the corresponding matrices. Here, O_{cj} represents the *j*th class score. Therefore, the final decision will be the class which is assigned the highest score (Tables 1 and 2).

3   Implementation

Extensive experimentation has been performed to test the capabilities of this model under varying conditions. The model has been tested over data sets with varying spatial resolution. The data sets are chosen to focus on urban land cover classification. The major classes considered in the study are vegetation/barren land, roads, buildings and shadow. Shadow is an important class, especially for very high resolution images. Special artefact cases (unbalanced situation and high intra-class variance) are tested upon. Also, the model is evaluated with a very small percentage of training samples. Performance of the model is analysed through multiple statistical tests. Finally, a comparison is done with individual classifier performances. Upcoming sections describe in detail the experimentations performed.

3.1   Locations and Data sets

We have chosen three data sets, each containing five tiles (images). The images are three band optical images. The spatial extent per image is 512×512 pixels and

Table 1 Data set 1-Jaipur, India

Classifiers	Vegetation		Roads		Buildings		Shadow		Overall accuracy (%)
	Precision	Recall	Precision	Recall	Precision	Recall	Precision	Recall	
RBF-SVM	0.6018	1.0000	0.7321	0.8666	0.9443	0.6000	1.0000	0.4000	71.66
Binary tree	1.0000	0.7333	0.6777	0.7333	0.3886	0.8000	1.0000	0.9333	80.00
Adaboost	0.6480	0.6660	0.3848	0.7333	0.2333	0.3330	0.3333	0.3330	51.66
KNN	0.9333	0.8000	0.7381	0.7333	0.7936	0.8000	0.9333	0.8000	78.33
GDA	0.7381	0.9333	0.6666	0.6666	0.6667	1.0000	1.0000	0.7333	83.33

Table 2 Data set 2-Madrid, Spain

Classifiers	Vegetation		Roads		Buildings		Shadow		Overall accuracy (%)
	Precision	Recall	Precision	Recall	Precision	Recall	Precision	Recall	
RBF-SVM	0.5541	0.8800	0.8199	0.7600	1.0000	0.6800	0.7428	0.5200	72.57
Binary tree	0.5952	0.6660	0.6039	0.6400	0.6699	0.7600	0.6366	0.5600	65.85
Adaboost	0.5866	0.5666	0.6388	0.5600	0.5333	0.7200	0.6028	0.7600	65.85
KNN	0.6916	0.8600	0.7032	0.8000	0.9100	0.6800	0.8333	0.7600	76.66
GDA	0.5866	0.6060	0.8083	0.8000	1.0000	0.6000	0.7238	0.8800	71.66

Table 3 Data set description

Location	Dimensions	No. of tiles	Spatial resolution (cm/pixel)
Jaipur, India	$512 \times 512 \times 3$	5	50
Madrid, Spain	$512 \times 512 \times 3$	5	30
Vaihingen, Germany	$512 \times 512 \times 3$	5	9

the radiometric resolution is 8 bits. The locations chosen are Jaipur, India; Madrid, Spain; and Vaihingen, Germany. A summary of the data set description is specified in Table 3.

3.2 Features

Feature extraction is done over the superpixels generated by quick shift segmentation. The features extracted can be broadly split into intensity-based features and texture-based features. Intensity-based features include sample mean and standard deviation across the superpixel. Entropy is a statistical measure of randomness that can be used to characterise the texture of the input region (Tables 4 and 5).

$$\mu_j = \frac{\sum_i x_{ij}}{N} \tag{3}$$

$$\sigma_j = \frac{\sum_i (x_{ij} - \mu_j)^2}{N} \tag{4}$$

$$\xi = - \sum (p. * log_2(p)) \tag{5}$$

where

x_i pixels present in the given superpixel.
N total no. of pixels in the given superpixel.
μ_j mean pixel intensity in band j.
σ_j Std. deviation of intensity in band j.
ξ entropy.
p histogram counts of intensity values over all bands.

Hence, in our study, the feature vector over every superpixel is a $\mathfrak{R}^7$ vector containing the intensity (over all 3 bands) and texture information. Superpixels do not hold any geometric information without further processing; hence, shape and geometry-based features have not been considered (Fig. 4).

3.3 Design Specification

The first stage of the model is the segmentation process. In the quick shift algorithm, we first transform the input image from RGB colour space to L-A-B colour space. The L-A-B space is used to emulate human perception of colour. The feature vector over which nonparametric density estimation is done is a combination of spectral and spatial features. The spectral features include 3-dimensional L-A-B vector and the spatial feature is a 2-dimensional (x, y) positional vector. Hence, the size of the total feature vector is five dimensions. Segment localisation happens over the gradient contour in the feature vector space. The kernel is a Gaussian kernel with a window size of 2 pixels wide which is set to obtain fine-scale density estimation over the defined image feature vector space. The threshold distance is a constraint over the size of the segment, which is set as 10 pixel radius (τ) and the ratio factor to trade-off between spectral and spatial features while segmentation is set as 0.5. Once the superpixels are generated, a 7-dimensional feature vector is extracted over individual superpixels, which contain intensity-based and texture-based information. The z-score standardization is performed over the feature vectors as a method of feature normalisation. Table 6 represents the segmentation results for different data sets.

4 Results and Discussions

The performance of OOMCS with fuzzy fusion scheme has been tested through various statistical measures. It has consistently shown elevated performance as compared to its individual classifier counterparts. The statistics accounted for, include, precision, recall and accuracy. Also, F-measure, a test for statistical significance is performed for the special case of unbalanced samples and high intra-class variance. A statistically significant outcome for the F-measure suggests that the results did not happen by chance. Another important objective of the model is to work with a very small percentage of training samples. Hence, accuracy assessment for varying percentage of training samples has been performed. The accuracy assessment plots for different data sets have been shown in Fig. 5.

The results after the above-mentioned analysis on multiple data sets have been summarised in Tables 1, 2, 4 and 5. The performance of the OOMCS-fusion scheme is clearly superior to the other classifiers. The classified results of a sample image from each data set by the best classifier and our method is depicted in Fig. 4. Four classes have been considered for the study. The classes include vegetation (blue), roads (cyan), buildings (yellow) and shadow (red). From the accuracy assessment curves, it can be observed that the performance of the model increases with increase in resolution and attains close to perfection with just 25–30 training samples per class. It can be seen from Fig. 4 that visually, building class which has high correlation with road class has been better identified in all data sets using the multi-classifier

Table 4 Data set 3-Vaihingen, Germany

Classifiers	Vegetation		Roads		Buildings		Shadow		Overall accuracy (%)
	Precision	Recall	Precision	Recall	Precision	Recall	Precision	Recall	
RBF-SVM	0.9582	0.9500	0.8750	0.9000	0.9375	0.7000	0.8154	1	88.75
Binary tree	1.0000	0.8500	0.8250	0.7500	0.7832	0.8500	0.9582	1	86.25
Adaboost	1.0000	0.8500	0.6666	0.9500	0.6250	0.4000	0.9582	1	80.00
KNN	0.9583	0.9500	0.9500	0.9500	0.9500	0.7500	0.8450	1	91.25
GDA	1.0000	0.9500	0.7595	0.7500	0.7178	0.7500	1.0000	1	86.25

Table 5 Our method

Data sets	Vegetation		Roads		Buildings		Shadow		Overall accuracy (%)
	Precision	Recall	Precision	Recall	Precision	Recall	Precision	Recall	
Jaipur, India	0.6710	1.0000	0.5714	0.6660	0.8330	0.8000	1.0000	0.5330	75.00
Madrid, Spain	0.6199	0.8066	0.8344	0.8800	1.0000	0.8000	0.7028	0.6800	79.47
Vaihingen, Germany	1.0000	0.9500	0.9166	0.9000	0.9285	0.9000	0.9166	1.0000	92.50

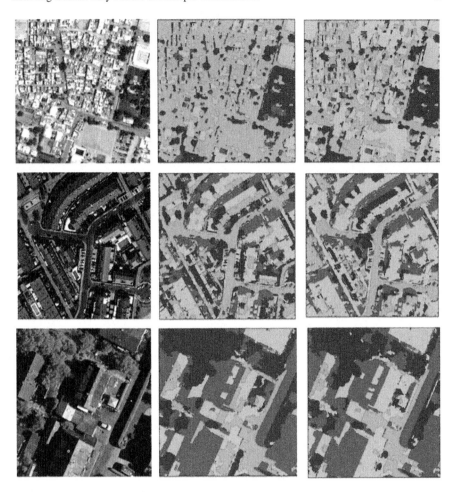

Fig. 4 Left column: Satellite imagery, Centre column: Best classifier, Right column: Fusion method

Table 6 Segmentation details

Data set	Average no. of regions	Time (s)
Jaipur, India	1345	3.863
Madrid, Spain	1087	3.856
Vaihingen, Germany	878	3.798

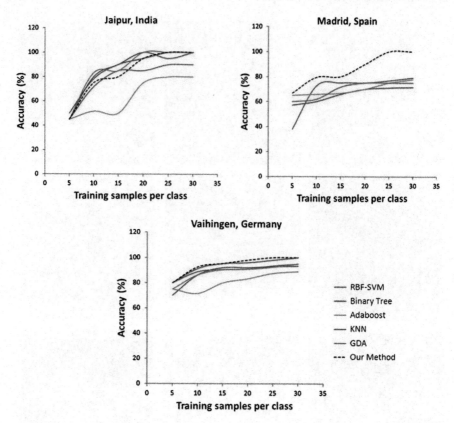

Fig. 5 Accuracy assessment

Table 7 Comparison of classifiers based on F-measure

Decision systems	Sparse samples (Jaipur-shadow)	High within class variation (Jaipur-roads)
	F-measure	F-measure
RBF-SVM	0.65	0.666
Binary tree	0.571	0.1666
Adaboost	0	0.2
KNN	0.8	0.5454
GDA	0.88	0
Our method	0.9088	0.8333

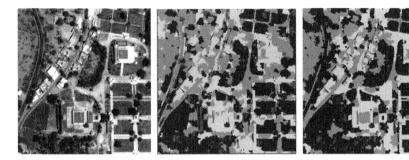

Fig. 6 Left column: Satellite imagery (Bottleneck cases), Centre column: Best classifier result, Right column: Fusion method result

system as compared to the best classifier performance. The statistic precision is a measure of correctness, and the recall is a measure of completeness.

Apart from the general analysis, a special case study of two artefacts has been done. First, the presence of very less percentage of a particular class. The sample image shown in Fig. 6 shows the two artefacts. The shadow class (red) is the example of sparse samples, by sparse we mean, the proportion of the class in the given image is very less. The second case study is the test of the ability of the model to incorporate high intra-class variance for decision-making. The road class (cyan) of Fig. 6 is the example of having high intra-class variance, and this can be clearly seen in the imagery. Our model handles both the issues with high degree of flexibility due to its robust fusion system. Performance of the model is evaluated through the F-measure test. As shown in Table 7, our method scores the highest as compared to any other method. As a part of analysis, we have also identified that KNN and GDA are robust for sparse sample scenario and SVM and KNN are robust for high intra-class variance situations. Finally, to visualise the results of classification during such bottlenecks, a sample result is shown in Fig. 6.

5 Conclusion

In this paper, we have introduced a new paradigm shift to object level multi-classifier decision fusion using a novel fuzzy rank-based fusion system. This system uses the concept of superpixels over which decision fusion is applied. Class score for a test superpixel is generated through the fusion of these weights. This system has maintained its performance capabilities over a wide range of resolutions from 50 to 9 cm. Hence, its scope of application is very broad and will find its use in many real-time applications. This model can be modified as per requirement. For instance, in our study, the main focus of interest was to improve the true positive rate with just a small set of training samples. Hence, the rank matrix was generated using the true positive rate of the training samples. Therefore, depending on applications, the nature of rank matrix can be varied to solve different purposes. If, for example, false positive rate has to be reduced, that factor can be used in generating the rank matrix and fusion can be done. The flexibility and the simplicity of the fusion method is the major advantage of our proposed system.

Acknowledgements The Vaihingen data set was provided by the German Society for Photogrammetry, Remote Sensing and Geoinformation (DGPF) [3].

References

1. Achanta, R., Shaji, A., Smith, K., Lucchi, A., Fua, P., Susstrunk, S.: Slic superpixels compared to state-of-the-art superpixel methods. Pattern Analysis and Machine Intelligence, IEEE Transactions on 34(11), 2274–2282 (2012)
2. Blaschke, T., Lang, S., Lorup, E., Strobl, J., Zeil, P.: Object-oriented image processing in an integrated gis/remote sensing environment and perspectives for environmental applications. Environmental information for planning, politics and the public 2, 555–570 (2000)
3. Cramer, M.: The dgpf-test on digital airborne camera evaluation–overview and test design. Photogrammetrie-Fernerkundung-Geoinformation 2010(2), 73–82 (2010)
4. Fulkerson, B., Vedaldi, A., Soatto, S., et al.: Class segmentation and object localization with superpixel neighborhoods. In: ICCV. vol. 9, pp. 670–677. Citeseer (2009)
5. Nussbaum, S., Menz, G.: eCognition Image Analysis Software, pp. 29–39. Springer Netherlands, Dordrecht (2008)
6. Vedaldi, A., Fulkerson, B.: VLFeat: An open and portable library of computer vision algorithms. http://www.vlfeat.org/ (2008)
7. Vedaldi, A., Soatto, S.: Quick shift and kernel methods for mode seeking. In: Computer vision–ECCV 2008, pp. 705–718. Springer (2008)

Image Retrieval Using Random Forest-Based Semantic Similarity Measures and SURF-Based Visual Words

Anindita Mukherjee, Jaya Sil and Ananda S. Chowdhury

Abstract In this paper, we propose a novel image retrieval scheme using random forest-based semantic similarity measures and SURF-based bag of visual words. A patch-based representation for the images is carried out with SURF-based bag of visual words. A random forest, which is an ensemble of randomized decision trees, is applied next on a set of training images. The training images accumulate into different leaf nodes in each decision tree of the random forest as a result. During retrieval, a query image, represented using SURF-based bag of visual words, is passed through each decision tree. We define a query path and a semantic neighbor set for such query images in all the decision trees. Different measures of semantic image similarity are derived by exploring the characteristics of query paths and semantic neighbor sets. Experimental results on the publicly available COIL-100 image database clearly demonstrate the superior performance of the proposed content-based image retrieval (CBIR) method with these new measures over some of the similar existing approaches.

Keywords Semantic similarity measures · Random forest · Query path
SURF · Visual words

1 Introduction

Content-based image retrieval (CBIR) has emerged over the years as a popular area of interest for researchers in the computer vision and the multimedia communities. The principal aim of CBIR is to organize digital picture archives from a thorough

A. Mukherjee
Dream Institute of Technology, Kolkata, India

J. Sil
IIEST Shibpur, Howrah, India

A. S. Chowdhury (✉)
Jadavpur University, Kolkata, India
e-mail: aschowdhury@etce.jdvu.ac.in

© Springer Nature Singapore Pte Ltd. 2018
B. B. Chaudhuri et al. (eds.), *Proceedings of 2nd International Conference on Computer Vision & Image Processing*, Advances in Intelligent Systems and Computing 703, https://doi.org/10.1007/978-981-10-7895-8_7

analysis of their visual content [1]. Bag of visual words (BoVW) framework has become popular for modeling the image content [2]. In this model, an image is represented as a collection of elementary local features like SURF [3] or SIFT [4]. These local descriptors are then quantized by k-means algorithm to build a bag of visual words. In recent past, there have been efforts to improve the BoVW model. For example in [5], Bouachir et al. have used a fuzzy c-means-based approach to improve the retrieval performance. The authors in [6] have developed an affinity-based visual word assignment model. They have also proposed a new measure of dissimilarity using a penalty function. For probabilistic similarity measures in image retrieval, please see [7]. While these methods have shown promises, still there remains a wide scope to better the retrieval performance. One factor which has highly contributed to this scope is lack of proper semantic similarity measures. Notion of semantic similarity plays a pivotal role in the image content modeling and retrieval [8]. We have come across interesting works, where initial notions of semantic image similarity based on random forest are developed [9–11]. In this paper, we propose an image retrieval scheme using random forest-based semantic similarity measures and SURF-based BoVW model. The rationale behind using SURF features is its much faster execution time as compared to that of SIFT [3]. The main contribution of this paper is the design and detailed analysis of random forest-based new semantic similarity measures. Experimental comparisons on the publicly available COIL-100 [12] image database clearly show the merit of our approach.

2 Proposed Method

In this section, we describe in detail the proposed method. The section contains three parts. In the first part, we discuss the SURF-based BoVW model for image representation. We then describe how random forest is used for training. Finally, we derive novel semantic similarity measures based on random forest.

2.1 *Image Representation Using SURF-Based BoVW*

We first discuss basics of SURF features following [3]. We then mention how patch-based image representation is done using SURF-based BoVW. SURF uses Hessian Matrix to detect interest points. The Hessian Matrix $H(\mathbf{x}, \sigma)$ for any point $\mathbf{x} = (x, y)$ in an image I at a scale σ is mathematically expressed as:

$$H(\mathbf{x}, \sigma) = \begin{bmatrix} L_{xx}(\mathbf{x}, \sigma) & L_{xy}(\mathbf{x}, \sigma) \\ L_{yx}(\mathbf{x}, \sigma) & L_{yy}(\mathbf{x}, \sigma) \end{bmatrix} \tag{1}$$

In Eq. (1), $L_{xx}(\mathbf{x}, \sigma)$ marks the convolution of the Gaussian second-order derivative $\frac{\delta^2}{\delta x^2} g(\sigma)$ with the image I at point $\mathbf{x}$ and so on. Integral images are used to efficiently obtain these computationally intensive convolutions. The above interest points are found across different scales (σ values). For the extraction of interest point descriptors (representation of neighborhood of any interest point, i.e., a patch), SURF uses sum of Haar wavelet responses. In the present problem, 64-dimensional SURF vectors are used to represent several patches in each training image. These local SURF descriptors need to be quantized to build the visual vocabulary. We apply k-means algorithm (with $k = 500$) to achieve this goal. Each cluster is treated as a unique visual word, and the collection of such visual words form the visual vocabulary [2]. Each image is then represented using a histogram of these visual words. Thus, at the end of this step, we have a 500-dimensional BoVW vector representing each training image.

2.2　Random Forest-Based Training

Here, we discuss how random forest can be used for training in this context of image retrieval. The rationale behind the choice of random forest is its very high accuracy and capability to handle large volume of data. Random forest is an ensemble classifier of decision trees with bagging (randomizing the training set) capability [13]. It votes for the most popular class among the individual trees. The information gain I for the jth node in a decision tree is given by:

$$I = H(S_j) - \sum_{i=L,R} \frac{|S_j^i|}{|S_j|} H(S_j^i)$$

(2)

In Eq. (2), $H(S)$ denotes the entropy of a node S, which for a discrete set of C labels is given by: $- \sum_{c \in C} p(c) log_2(p(c))$ and $|S_j|$ denotes the number of training images in the node S_j. So, $|S_j^L|$ and $|S_j^R|$, respectively, represent the number of training images in the left child and the right child of the node S_j. In this problem, we use the 500-dimensional BoVW vector and the class label for each training image as the two inputs to the random forest. At the end of this training phase, the training images are grouped into various leaf nodes in different decision trees.

2.3　Random Forest-Based Semantic Similarity Measures

Though random forest is mostly applied for classification, following [9], we have used it here to derive measures of semantic image similarity. In this section, we discuss three such measures. During the retrieval stage, a query image passes through each decision tree. Let m denote a training image, q denote a query image, and

t denote a decision tree in a random forest. Further, let M and T, respectively, denote the total number of training images and the total number of decision trees in the random forest. We now have the following definitions and expressions.

Definition 1 The **semantic neighbor set** $SNS(q,t)$ is defined following [9] as the set of training images present at the leaf node into which a query image q falls in a decision tree t.

Definition 2 The frequency-based similarity measure **sm1(m, q)** is defined as the number of trees $(t, t \in [1, T])$ in the random forest a training image m appears in $SNS(q, t)$. So, we mathematically express $sm1(m, q)$ as:

$$sm1(m, q) = \sum_{t=1}^{T} \phi_m(t) \tag{3}$$

Here, $\phi_m(t) = 1$ if $m \in SNS(q, t)$ and is 0 otherwise $(1 \le m \le M)$.

Note that since $sm1$ is based on frequency, we do not normalize it.

Definition 3 A **query path** $p_k(q, t)$ of length k for a query image q in a tree $t(1 \le t \le T)$ is denoted by a sequence of nodes $n_0(t), n_1(t), \ldots, n_{(k-1)}(t)$, where $n_0(t)$ is the root, $n_i(t)$ is the ith intermediate node $(1 \le i \le (k-2))$, and $n_{(k-1)}(t)$ is a leaf node in the tree t. Here, q falls into $n_{(k-1)}(t)$ and the training images which are present in $n_{(k-1)}(t)$ form $SNS(q, t)$.

In Fig. 1, we show a typical query path in a decision tree using a sequence of red lines. The red oval marks the leaf node where the query image falls in this tree and xi in the same figure denotes the ith $(1 \le i \le 500)$ element of the 500-dimensional BoVW vector. The label 5 in the leaf node indicates that the probability of class 5 is maximum at this node. However, we have determined the SNS, i.e., training images of different classes (and not just that of the highest class) which have accumulated in such nodes for the evaluation of our proposed semantic measures.

Definition 4 Let the set of $(k - 1)$ features on the query path $p_k(q, t)$ be denoted by $f(t) = \{f_1(t), f_2(t), \ldots, f_{(k-1)}(t)\}$ and the set of weights (relative importance) of these k features be denoted by $\alpha(t) = \{\alpha_1(t), \alpha_2(t), \ldots, \alpha_{(k-1)}(t)\}$. Here, f_i connects $(n_i, n_{(i+1)}), (1 \le i \le (k-1))$, and so on. A query path-based similarity measure **sm2(m, q)** between a training image m and a query image q is defined as the summation over all trees the product of weights of all features appearing in a path in each tree. We mathematically express $sm2(m, q)$ as:

$$sm2(m, q) = \sum_{t=1}^{T} \prod_{i=1}^{(k-1)} \alpha_i(t) \tag{4}$$

We have actually used a normalized version of $sm2(m, q)$, defined as $sm2(m, q) = sm2(m, q)/max_{\forall m, 1 \le m \le M} sm2(m, q)$.

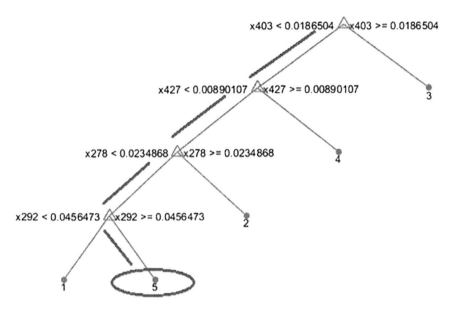

Fig. 1 A typical query path in a decision tree

Definition 5 Further, let the set of k features $f_i(t)$, $(1 \leq i \leq (k-1))$ on the query path $p_k(q, t)$ be at respective levels $l_i(t)$, $(1 \leq i \leq (k-1))$. Since each tree is essentially a binary tree, we define another query path-based similarity measure **sm3(m, q)** between a training image m and a query image q as the summation over all trees the product of level modulated weights of all features appearing in a path in each tree. So, $sm3(m, q)$ can be mathematically expressed as:

$$sm3(m, q) = \sum_{t=1}^{T} \prod_{i=1}^{(k-1)} \alpha_i(t) \times \frac{1}{2^{l_i(t)}} \tag{5}$$

We have actually used a normalized version of $sm3(m, q)$, defined as $sm3(m, q) = sm3(m, q)/max_{\forall m, 1 \leq m \leq M} sm3(m, q)$, where M denotes the total number of training images.

3 Complexity Analysis

In this section, we analyze the complexity of the construction of the random forest classifier and the computation of the three semantic measures. Let M, $|B|$, and $d(t)$ denote the number of training images, number of elements in the BoVW vector representing an image, and depth of a decision tree t. Note that in a random forest, as a result of bagging, each decision tree is constructed with a randomly chosen sub-set of the number of elements in the BoVW vector $(f(t) = \{f_1(t), f_2(t), \ldots, f_{(k-1)}(t)\}$, $(k-1) \leq |B|)$. At each node in a tree t, we have to compute the information gain for different $f_i(t)'s, (1 \leq i \leq (k-1))$. Then, the maximum cost of constructing a deci-sion tree becomes $O(M|B|d(t))$. With a total of T such decision trees in the ran-dom forest and a maximum depth D, where $D = max_{\forall t, 1 \leq t \leq T}(d(t))$, the maximum cost of constructing the random forest is $O(M|B|DT)$. The maximum value of the length k of a query path $p_k(t)$ in a decision tree t is $d(t)$, which in the worst case can be D. So, the worst-case complexity for evaluating $sm1(m, q)$ (please see the defi-nition in Eq. (3)) is $O(TD)$. The cost of using the weights $\alpha_i(t)$ for corresponding features $f_i(t), (1 \leq i \leq (k-1))$ in a decision tree t is $O(1)$. Note that these weights are already computed using Eq. (2) at the time of the construction of the decision trees. In the worst case, we need to use this $(D-1)$ times for all T trees. So, the worst-case complexity for evaluating $sm2(m, q)$ (please see the definition in Eq. (4)) is $O(T(D-1)) \approx O(TD)$. Similarly, the cost of evaluation of the levels $l_i(t)$ for any feature $f_i(t)$ in a decision tree t is also $O(1)$. In the worst case, we need to evaluate this $(D-1)$ times. So, the worst-case complexity for evaluating $sm3(m, q)$ (please see the definition in Eq. (5)) is $O(T(D-1)) + O(T(D-1)) = O(T(D-1)) \approx O(TD)$. The overall worst-case complexity of construction of the random forest and evaluation of any semantic measure is $O(M|B|DT) + O(TD) = O(M|B|DT)$.

4 Experimental Results

We use the publicly available COIL-100 image database [12] for experimentation. The database contains a total of 7200 images with 72 different images of 100 different objects having a viewpoint separation of $5°$. We have used MATLAB as the comput-ing platform. Precision and recall values are chosen as the measures of retrieval per-formance [1]. Precision indicates the percentage of retrieved images that are relevant to the query. In contrast, recall measures the percentage of all the relevant images in the database which are retrieved. Precision versus recall curves are obtained by changing the thresholds $\theta_1, \theta_2,$ and θ_3 in connection with the three semantic simi-larity measures $sm1, sm2,$ and $sm3$, respectively. So, for obtaining the curve using measure $sm1$, we vary θ_1 from 5 to the total number of decision trees in the ran-dom forest and retrieve only those training image(s) m for which $sm1(m, q) > \theta_1$. Likewise, we obtain the curves using $sm2$ and $sm3$ by varying θ_2 and θ_3 from 0.05 to 1.0.

We experimentally determine the optimal number of trees in the random forest to be 100. Please see Fig. 2 where the best retrieval performance for Coil 1 is achieved with $T = 100$. We now compare our performance with a *fuzzy weighting scheme* [5], a *term frequency–inverse document frequency (tfx)*-based approach [14], a method which only uses *term frequency (txx)* [15], and a visual word assignment model (vwa) [6].

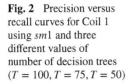

Fig. 2 Precision versus recall curves for Coil 1 using *sm*1 and three different values of number of decision trees ($T = 100, T = 75, T = 50$)

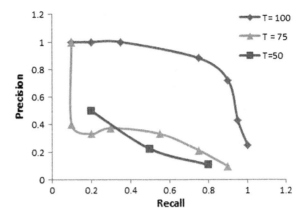

The precision versus recall curves are shown for two different query objects, namely Coil 3 and Coil 10 for each of the three semantic measures *sm*1, *sm*2, and *sm*3 are shown in Fig. 3 and Fig. 4, respectively. The curves clearly indicate that the retrieval performance using all three proposed semantic similarity measures yields superior results compared to the four competing methods. We now show the retrieved

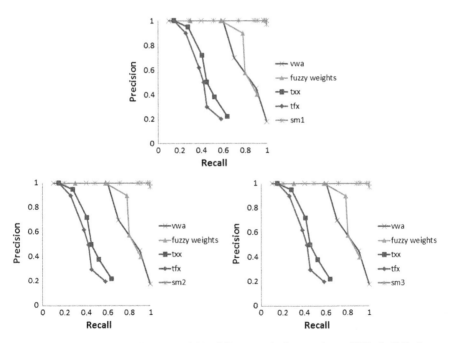

Fig. 3 Precision versus recall curves of five different methods, namely txx [15], tfx [14], fuzzy weighting [5], vwa [6], and current approach for Coil 3 with three different semantic similarity measures: *sm*1 (top), *sm*2 (bottom left), *sm*3 (bottom right)

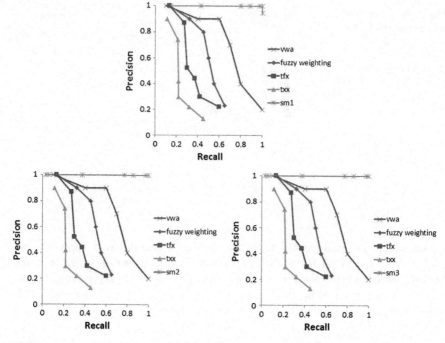

Fig. 4 Precision versus recall curves of five different methods, namely txx [15], tfx [14], fuzzy weighting [5], vwa [6], and current approach for Coil 10 with three different semantic similarity measures: *sm*1 (top), *sm*2 (bottom left), *sm*3 (bottom right)

images for Coil 3 and Coil 10 in Figs. 5 and 6. The retrieved results illustrate that the top five retrieved images for all three semantic measures are relevant. The rank and set of the relevant images are, however, different for different measures. Now, we include a failed case for the object Coil 9 in Fig. 7. This figure indicates that all three measures fail to retrieve only relevant images (images belong to the same class as that of the query image). The reason for failure is that there are quite a few extremely similar objects like Coil 9 in the database. Still, the measures *sm*2 and *sm*3 yield better results than *sm*1. This is because *sm*1 is only based on frequency of appearance of a training image in the SNS of a query image. In contrast, both *sm*2 and *sm*3 are derived from the characteristics (weights and levels of BoVW elements) of a query path.

We also present the recognition rate as an average precision for ten different objects, namely Coil 1 to Coil 10 of the COIL-100 database [5]. Please note that the recognition rate does not take into account any recall. In Table 1, we compare the recognition rates for the above objects of the proposed method with three different semantic similarity measures against four competing methods. The results clearly demonstrate that all three measures in our method have better performances than the competing methods. In nine out of ten cases, it turns out that the three measures become (single or joint) winners having achieved the highest recognition rate.

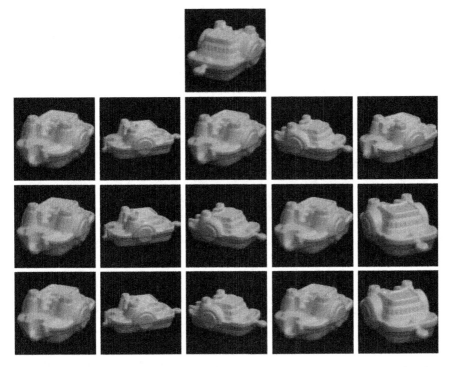

Fig. 5 Retrieval results for COIL-3: query image (first row), top five retrieved images based on *sm*1 (second row), *sm*2 (third row), and *sm*3 (fourth row). All five retrieved images for all three measures are relevant (belong to the same class as that of the query image)

Furthermore, all three average recognition rates, namely 89.7% from *sm*1, 93.2% from *sm*2, and 92.1% from *sm*3, clearly surpass the previously reported recognition rates. Once again, (and in fact, generally speaking), among the proposed three measures, *sm*2 and *sm*3, which carry more information, yield better results than *sm*1. Fourth best is [6] with an average recognition rate of 86%, followed by [5] with the reported average recognition rate of 80%. The other two methods [14, 15] are clearly quite behind with average recognition rates of 71.5% and 61.5%, respectively.

5 Conclusion

In this paper, we proposed a method of image retrieval using random forest-based new semantic similarity measures and SURF-based bag of visual words. The semantic similarity measures are derived from characterization of query paths and semantic neighbor sets in each decision tree of the random forest. Comparisons with some of the existing approaches on the COIL-100 database clearly show the merits of the proposed formulation. In future, we plan to perform more experiments with other similar

Fig. 6 Retrieval results for COIL-10: query image (first row), top five retrieved images based on *sm*1 (second row), *sm*2 (third row), and *sm*3 (fourth row). All five retrieved images for all three measures are relevant (belong to the same class as that of the query image)

Table 1 Recognition rate comparison among different competing methods: txx [15], tfx [14], fuzzy weighting [5], vwa [6], and current method with semantic similarity measures *sm*1, *sm*2, and *sm*3

Image	txx	tfx	Fuzzy weighting	vwa	sm1	sm2	sm3
Coil 1	0.5	0.4	0.65	0.8	0.98	0.98	0.98
Coil 2	0.4	0.1	0.45	0.6	0.88	0.89	0.89
Coil 3	0.9	0.95	1.0	1.0	1.0	1.0	1.0
Coil 4	1.0	0.9	1.0	1.0	1.0	1.0	1.0
Coil 5	0.25	0.1	0.75	0.75	0.97	0.98	0.98
Coil 6	1.0	1.0	1.0	1.0	1.0	1.0	1.0
Coil 7	1.0	0.85	0.95	1.0	0.94	0.93	0.91
Coil 8	0.55	0.5	0.7	0.75	0.94	0.94	0.94
Coil 9	0.7	0.6	0.6	0.7	0.26	0.6	0.51
Coil 10	0.85	0.75	0.9	1.0	1.0	1.0	1.0
Average	0.715	0.615	0.8	0.86	0.897	0.932	0.921

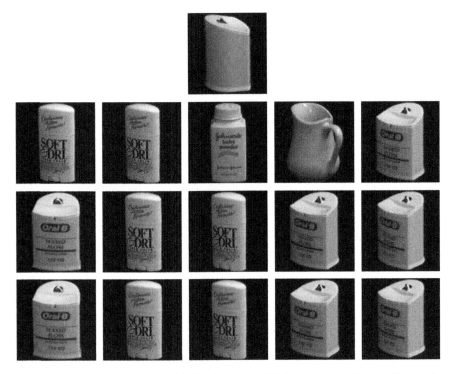

Fig. 7 Retrieval results for COIL-9 showing some failed cases: query image (first row), top five retrieved images based on sm1; only fifth image is from the relevant class (second row), top five retrieved images based on sm2; first, fourth, and fifth from relevant classes (third row), top five retrieved images based on sm3; first, fourth, and fifth from relevant classes (fourth row)

approaches on additional databases like Oxford buildings [11]. We will also exploit contextual and structural information in random forests [16] as well as explore deep learning-based approaches [17] to further improve the retrieval performance.

References

1. Datta, R., Joshi, D., Li, J., Wang, James Z., Image retrieval: Ideas, influences, and trends of the new age, ACM Computing Surveys, 40(2), 1–60, (2008).
2. Sivic, J., Zisserman, A.: Video Google: Efficient Visual Search of Videos, In Toward Category-Level Object Recognition, 127–144, (2006).
3. Bay, H., Ess, A., Tuytelaars, T., and Van Gool, L.: Speeded-up robust features (SURF), Computer Vision and Image Understanding, 110(3), 346–359, (2008).
4. Lowe D. G.: Distinctive Image Features from Scale-Invariant Keypoints, International Journal of Computer Vision, 60(2), 91–110, (2004).
5. Bouachir, W., Kardouchi, M., Belacel, N.: Improving Bag of Visual Words Image Retrieval: A Fuzzy Weighting Scheme for Efficient Indexation, Proc. SITIS, 215–220, (2009).

6. Mukherjee, A., Chakraborty, S., Sil, J., Chowdhury, A.S.: A Novel Visual Word Assignment Model for Content Based Image Retrieval, Proc. CVIP, Balasubramanian Raman et al. (eds.), Springer AISC, Vol. 459, 79–87, (2016).

7. Rahman, M.M., Bhattacharya, P., Kamel, M., Campilho A.: Probabilistic Similarity Measures in Image Databases with SVM Based Categorization and Relevance Feedback, Proc. ICIAR, Springer LNCS, Vol. 3656, 601–608, (2005).

8. Liu Y., Zhang D., Lu G., Ma W-Y.: A survey of content-based image retrieval with high-level semantics, Pattern Recognition, 40(1), 262–282, (2007).

9. Fu, H., Qiu G.: Fast Semantic Image Retrieval Based on Random Forest, Proc. ACM MM, 909–912, (2012).

10. Moosman, F., Triggs, B. and Jurie, F.: Fast Discriminative Visual Codebooks using Randomized Clustering Forests, Proc. NIPS, 985–992, (2006).

11. Dimitrovski, I., Kocev, D., Loskovska, S., Dzeroski, S.: Improving bag-of-visual-words image retrieval with predictive clustering trees, Information Science, 329(2), 851–865, (2016).

12. Nene, S. A., Nayar, S. K., Murase, H.: Columbia Object Image Library (COIL-100), Tech. Report, Department of Computer Science, Columbia University CUCS-006–96, (1996).

13. Breiman, L.: Random Forests, Machine Learning, 45, 5–32, (2001).

14. Sivic, J., Zisserman A.: Video Google: A Text Retrieval Approach to Object Matching in Videos, Proc. ICCV, 470–1477, (2003).

15. Newsam, S., Yang Y.: Comparing global and interest point descriptors for similarity retrieval in remote sensed imagery, Proc. ACM GIS, Article No. 9, (2007).

16. Kontschieder P., Rota Bulo S., Pelillo M.: Semantic Labeling and Object Detection, IEEE Transactions on Pattern Analysis and Machine Intelligence, 36(10), 2104–2116, (2014).

17. Wan J. et al.: Deep Learning for Content-Based Image Retrieval: A Comprehensive Study, Proc. ACM MM, 157–166, (2014).

Rotation Invariant Digit Recognition Using Convolutional Neural Network

Ayushi Jain, Gorthi R. K. Sai Subrahmanyam and Deepak Mishra

Abstract Deep learning architectures use a set of layers to learn hierarchical features from the input. The learnt features are discriminative, and thus can be used for classification tasks. Convolutional neural networks (CNNs) are one of the widely used deep learning architectures. CNN extracts prominent features from the input by passing it through the layers of convolution and nonlinear activation. These features are invariant to scaling and small amount of distortions in the input image, but they offer rotation invariance only for smaller degrees of rotation. We propose an idea of using multiple instance of CNN to enhance the overall rotation invariant capabilities of the architecture even for higher degrees of rotation in the input image. The architecture is then applied to handwritten digit classification and captcha recognition. The proposed method requires less number of images for training, and therefore reduces the training time. Moreover, our method offers an additional advantage of finding the approximate orientation of the object in an image, without any additional computational complexity.

Keywords Convolutional neural networks · Rotation invariance · Digit recognition · LeNet

A. Jain (✉) · D. Mishra
Department of Avionics, Indian Institute of Space Science & Technology,
Trivandrum 695547, Kerala, India
e-mail: ayushijain.168@gmail.com

D. Mishra
e-mail: deepak.mishra@iist.ac.in

G. R. K. Sai Subrahmanyam
Department of Electrical Engineering, Indian Institute of Technology,
Tirupati 517506, AP, India
e-mail: rkg@iittp.ac.in

© Springer Nature Singapore Pte Ltd. 2018
B. B. Chaudhuri et al. (eds.), *Proceedings of 2nd International Conference on Computer Vision & Image Processing*, Advances in Intelligent Systems and Computing 703, https://doi.org/10.1007/978-981-10-7895-8_8

1 Introduction

Neural Networks, as the name suggests, are the artificial structures inspired by the topological arrangement of neurons in human brain [1]. Fully connected neural networks require specific features of input for training and testing. The number of free parameters is also very high for such networks. They are not immune to scale changes and distortions in the input image. All such limitations of conventional neural networks are overcome in convolutional neural networks. CNN is a deep learning architecture which extracts the complex and discriminative features of the input using a series of convolutions, nonlinear activations, and subsampling. More details about the CNN architectures and functioning of each layer can be found in [1].

Recently, convolutional neural networks are becoming more popular and are being used in many areas including image processing, computer vision, speech analysis, natural language processing. They find application in digit recognition [2], image super-resolution [3], face detection [4], gesture segmentation [5], rotation invariant facial expression recognition [6].

CNNs use images directly as input. They learn a hierarchy of features which can then be used for classification purposes. Repeated convolution of input image with learnt kernels gives a hierarchy of features. The obtained feature maps are invariant to translation and distortions in the input image. Up to some extent, the features are invariant to rotation as well, provided that the degree of rotation is small.

Efforts have been made to employ rotation invariance in CNN by training the network with all possible rotations, i.e., data augmentation [4, 7]. Thus the size of training dataset becomes extremely large. Similar work that uses more than single unit of CNN to extract rotation invariant features has been done by [8], which has computationally complex training as they rotate the training input at multiple angles. Spatial transformer networks (STNs) [9] are also another work to achieve transformation invariant classification. A STN module generates the parameter of transformation from an input feature map. It spatially transforms the feature map with the obtained parameters to minimize the overall cost function. This STN module can be inserted after any layer in CNN. But, additional computations involved in the module makes CNN training complex.

The proposed architecture trains a CNN and uses it numerous times in parallel for testing. The network is trained with images having digits at single orientation and classifies the images with different orientations of the same digit. Thus, the training data size is smaller as compared to data augmentation, also the training time is reduced. The proposed approach springs from the intuition and observation that the CNN tested with roughly close to (need not be exactly same as) actual orientation of the digit/object with which it is trained, gives higher response than the CNNs tested with other orientations. As an additional application, our method can identify the approximate orientation of the digit in an image. We have applied the method on handwritten digit classification and rotated captcha recognition.

2 Proposed Method: RIMCNN

The orientation of an object plays an important role in tasks such as object classification, digit recognition, texture classification. The object can be oriented at any angle in an image. However, if the CNN is trained with single orientation, it will be unable to classify different orientation.

We propose a method of introducing rotational invariance using multiple instance of convolutional neural network (RIMCNN). We aim to construct a deep learning architecture which is robust to orientation of the object with respect to center of the image. The proposed idea involves one-time training of a single CNN unit, and later that trained unit can be used multiple number of times in parallel for testing, depending on the application.

The proposed method can be described in two phases:

- Training
- Testing

2.1 Training

This phase is same as training any CNN architecture with a version of backpropagation algorithm. A single unit of CNN is trained till it converges to give minimum training error. The training data includes images with only one orientation (say $0°$). No rotation is involved during training phase. Thus, the number of images in training data is lesser as compared to that of data augmentation.

We have used the LeNet-5 architecture [10] in our experiments. LeNet architecture is constructed using five layers: two convolution layers, two subsampling layers, and a fully connected layer. We choose six maps in each convolution layer and subsampling layer. Kernel size is selected as 5×5 for convolutional layers, and 2×2 window is used for subsampling. The architecture is as shown in Fig. 1.

2.2 Testing

Here is the main contribution of our work. From digit classification results discussed in the next section, we can observe that the network trained with particular orientation can give good classification accuracy for test images having rotations only up to $\pm 20°$ of the training orientation. Based on this observation, we rotate the test image by 'N' different angles (say $L_1, L_2, \ldots, L_N$) equally spaced in the range of rotations involved in the test data, such that one among the N resulting images gets the

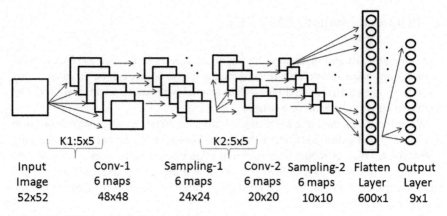

Input	Conv-1	Sampling-1	Conv-2	Sampling-2	Flatten	Output
Image	6 maps	6 maps	6 maps	6 maps	Layer	Layer
52x52	48x48	24x24	20x20	10x10	600x1	9x1

Fig. 1 RIMCNN training: LeNet-5 architecture

rotation $\pm 20°$ closer to the orientation with which the network is trained. The resulting rotated images are then fed to N different LeNets. The LeNet architecture trained in previous phase is used multiple (N) times in parallel. The number of LeNets (N) can be selected depending on the application.

Generally, in any classification task using CNN, the image is assigned to that class which gives the maximum response or activation at the output layer. We use a slightly different logic to decide the label. Suppose, there are 'C' classes, then we get the vector of length 'C $\times$ 1' at the output layer of a LeNet. And since we use 'N' parallel units of LeNet, we get N different vectors of length 'C $\times$ 1'. The image is assigned to the class which gives the highest response among all NC activations at output.

The reason behind deciding the label by such an approach is that if the image is rotated at different angles then the resulting image with net orientation which is close to the orientation that was included during the training of LeNet will give the maximum response for correct label at the output layer, which will be the highest among all NC responses obtained from 'N' resulting images. Closer the orientation of the image with training orientation, higher is the response at the output layer. Thus, the output label is decided by the LeNet that gets the image with orientation closer to $0°$, which comes out to be the correct label as the network is trained with the same orientation. This concept is demonstrated and verified with different experiments in following sections. The architecture used for testing is as shown in Fig. 2. The testing procedure is briefly described in Algorithm 1.

2.3 Selecting Number of LeNets and Rotation Angles

The number of LeNets 'N' is decided according to the application. If the application involves the images with rotations ranging between the angles $[-D_{max}, D_{max}]$, then

Algorithm 1: RIMCNN Test Phase

Input : Test Image, Trained model of *LeNet*, No. of LeNets(N), No. of classes(C), Range of
rotations in test data[D_{min},D_{max}]
Output: Label, Approximate Orientation
Initialization: *Outputs* = zeros(N,C)
begin
StepSize = (D_{max} - D_{min}) / N
angles = D_{min} : *StepSize* : D_{max}
for *i=1:N* **do**
    $\mid$ *RotatedImage* = Test image rotated with *angles*[i]
    $\mid$ *Outputs*[i] = *LeNet*'s output layer response for *RotatedImage* input
end
[*RowIndex,ColIndex*] = Index where *Outputs* is Maximum
Label = *ColIndex*
Approximate Orientation = angles[*RowIndex*]
end

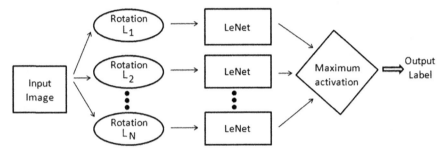

Fig. 2 RIMCNN test phase: a single unit of LeNet trained with 0° oriented images is used N times
in testing

the angles L_1, L_2, ..., L_N can be selected in the same range $[-D_{max}, D_{max}]$ at regular
intervals, but with opposite sign. Higher the number of CNN units 'N', flatter and
better is the accuracy plot, but higher is the testing time. So there is always a trade-off
between the number of CNN units and testing time.

2.4 Finding Approximate Orientation of the Digit

Once the classification label is obtained then the corresponding angle of rotation L_k
which gives the maximum response determines the orientation of object in the input
image. As the maximum activation is given by L_k which brings the object orientation
closer to 0°, the negative of angle L_k more or less gives the approximate orientation.
Thus, if the angles for rotation span the complete range of rotation from 0° to 360°,
the proposed architecture can be used to determine the orientation of the object with
respect to center of the image.

3 Experiments and Results

We applied RIMCNN for handwritten digit classification and rotated captcha recognition. We observed that RIMCNN performs better than single unit of LeNet in all applications. The results are discussed in detail below. All the experiments have been performed using a high-level neural networks library Keras (1.0.5) running on top of Theano (0.8.2). The computer used is Intel(R) Xeon(R) X5675 with 24 GB RAM and NVIDIA Quadro 6000 graphics card.

3.1 Handwritten Digit Classification

It is a known fact that the letters and digits made by people are not always similar. Also the documents may be scanned in any direction, yielding rotational digits/characters in it. Some people make erect digits, whereas others might have slanted handwriting. Thus, the proposed idea is applicable for handwritten digit recognition because the network trained with erect digits can be used to classify slanted digits as well, thereby making the architecture suitable for all users irrespective of their handwriting.

3.1.1 Dataset

We have used the MNIST dataset [11] for demonstrating our results on digit classification.

Pre-Requisite: To demonstrate the rotation invariant capabilities of RIMCNN, the test images are rotated at different angles (anticlockwise) and the network is tested for those rotated digits. Thus, the dataset requires the following modifications:

- The 28×28 image is centered in 52×52 image so that the digits do not suffer from any distortion after rotation.
- After rotation, the samples of class 6 and class 9 may resemble same and cause ambiguity in classification. Thus, all the samples belonging to class 9 in training and test data are discarded. Thus, the resulting problem contains nine classes.

  However, this step can be relaxed and all digits from 0 to 9 can be classified if the test data has rotations ranging within 90° of the orientation involved in the dataset, such that there is no possibility for the two classes (6 & 9) resembling each other under any rotation.

The details of original MNIST dataset and the modified dataset are shown in Table 1.

Table 1 MNIST dataset details

MNIST dataset	Actual	Modified
Number of training images	60000	54000
Number of test images	10000	9000
Image size	28×28	52×52
Number of classes	10	9

3.1.2 LeNet Training

A single unit of LeNet described in Fig. 1 is trained with only 54000 images in the modified MNIST dataset, with batch size of 50 images. Loss function is selected as 'categorical crossentropy', and 'adadelta' optimizer with default parameters is selected for compiling the model in Keras. The network converges to give training accuracy of 99.53% after 100 epochs. Training time for 100 epochs comes out to be 2000 s.

3.1.3 Behavior of LeNet for Rotated Digits

The trained LeNet is then tested with rotated test samples. Figure 6 shows the behavior of LeNet for rotated digits at different angles. The test error comes out to be symmetrical around 180°. It can be observed that LeNet can be considered robust up to ±20° of rotation, whereas test error greatly increases for higher degrees of rotations. This observation helps us to select the number of LeNets and their corresponding angles of rotation in RIMCNN.

3.1.4 RIMCNN for Digit Classification

RIMCNN is tested with digits at different orientations. The processing of the architecture can be understood from Fig. 3. Five images of digit '3' at different orientations are given to RIMCNN with LeNets having rotations at angles 0°, −30°, −60° and −90°. In the tables, it can be seen that the LeNet which gets the image with orientation closer to 0° classifies it correctly and gives the maximum response, whereas other LeNets may not give correct label and produce the response which is always lesser than the maximum response. Thus, the overall classification label comes out to be correct.

The selected angles $(L_1, L_2, ..., L_N)$ should be able to cover the entire range of rotations in the test data with tolerance of ±20°, i.e., the angles can be selected with maximum interval of 40° between two angles to get a fair classification. However, classification accuracy is improved further by making the interval even

N = 4, Angle of rotation (degrees) = 0,-30,-60,-90

Orientation	0°				30°				50°				60°				90°			
Class Labels	L_1 -0	L_2 -30	L_3 -60	L_4 -90	L_1 -0	L_2 -30	L_3 -60	L_4 -90	L_1 0	L_2 -30	L_3 -60	L_4 -90	L_1 0	L_2 -30	L_3 -60	L_4 -90	L_1 0	L_2 -30	L_3 -60	L_4 -90
0	0.000	0.000	0.006	0.000	0.000	0.000	0.000	0.006	0.000	0.000	0.000	0.012	0.000	0.000	0.000	0.000	0.031	0.001	0.000	0.000
1	0.000	0.000	0.000	0.000	0.000	0.000	0.000	0.000	0.000	0.000	0.000	0.000	0.000	0.000	0.000	0.000	0.000	0.000	0.000	0.000
2	0.000	0.000	0.037	0.000	0.000	0.000	0.000	0.037	0.004	0.000	0.000	0.122	0.004	0.000	0.000	0.992	0.000	0.000	0.001	0.000
3	1.000	0.992	0.000	0.000	0.995	1.000	0.992	0.000	0.029	0.998	1.000	0.701	0.029	0.995	1.000	0.000	0.000	0.010	0.994	1.000
4	0.000	0.000	0.939	0.000	0.000	0.000	0.000	0.939	0.000	0.002	0.000	0.083	0.000	0.000	0.000	0.000	0.347	0.944	0.000	0.000
5	0.000	0.000	0.000	0.000	0.002	0.000	0.000	0.000	0.783	0.000	0.000	0.000	0.783	0.002	0.000	0.000	0.000	0.000	0.005	0.000
6	0.000	0.000	0.000	0.998	0.000	0.000	0.000	0.000	0.000	0.000	0.000	0.000	0.000	0.000	0.000	0.000	0.622	0.041	0.000	0.000
7	0.000	0.000	0.010	0.002	0.000	0.000	0.000	0.010	0.183	0.000	0.000	0.000	0.183	0.000	0.000	0.000	0.000	0.003	0.000	0.000
8	0.000	0.000	0.007	0.000	0.000	0.000	0.000	0.007	0.001	0.000	0.000	0.082	0.001	0.000	0.000	0.000	0.000	0.003	0.000	0.000
Maximum activation	1.000				1.000				1.000				1.000				1.000			
Final classification label	3				3				3				3				3			

Fig. 3 Output responses of LeNet units in RIMCNN for images of digit '3' rotated at angles 0°, 30°, 50°, 60° and 90°(anticlockwise)

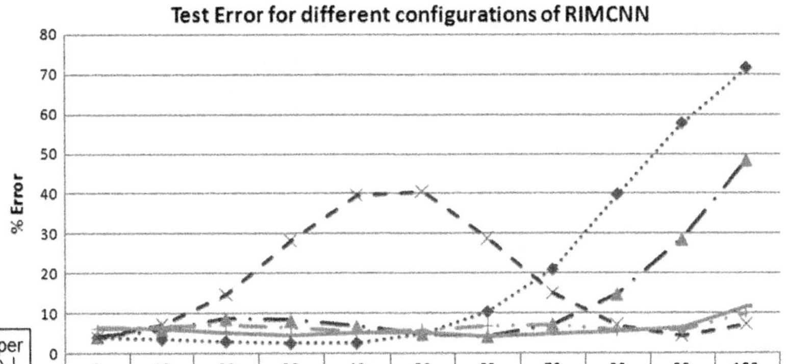

Rotation per LeNet (L_k)↓	0	10	20	30	40	50	60	70	80	90	100
0,-45	4.0267	3.6	3.037	2.58	2.66	4.93	10.4	20.98	39.81	57.7	71.55
0,-60	4.238	5.95	8.66	8.298	6.73	4.772	4.18	7.23	14.75	28.598	48.45
0,-90	4.004	7.27	14.67	28.26	39.61	40.523	28.76	14.87	6.99	4.483	7.2
0,-45,-90	6.08	6.75	7.13	6.59	5.48	5.93	6.92	7.25	5.83	5.78	10.156
0,-30,-60,-90	6.38	6.08	5.23	4.5	5.29	5.16	4.24	4.95	5.53	6.37	11.6

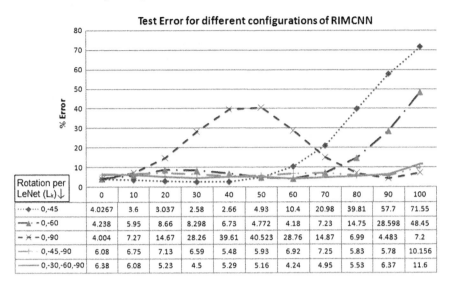

Fig. 4 Different configurations of RIMCNN for rotated test data

Table 2 Number of samples misclassified per class by RIMCNN (N = 4) for 9000 test images rotated at different angles

Angle of rotation	0	10	20	30	40	50	60	70	80	90	100
Classification label	No. of images misclassified										
0	13	19	11	11	13	7	8	10	17	12	43
1	26	44	22	18	14	16	12	18	16	15	37
2	65	38	56	42	54	77	50	86	102	120	216
3	55	60	41	40	49	45	39	49	64	75	143
4	73	67	49	30	48	35	24	33	45	35	77
5	45	36	30	26	42	70	86	113	107	146	183
6	99	78	67	54	53	48	31	31	33	37	77
7	136	142	136	118	134	104	68	46	58	64	145
8	62	63	58	66	69	62	63	59	55	69	122

smaller, which can be observed in Fig. 4. RIMCNN with N = 4 and rotations 0°, −30°, −60° and −90° is able to produce very good classification accuracy as compared to RIMCNN with N = 2 and N = 3. The no. of samples in error from each class for 9000 images rotated at different angles is shown in Table 2. It can be noticed that even at higher degrees of rotation, maximum images of most of the digits are correctly classified.

We give the comparison of proposed architecture with LeNet in Fig. 5. Our method yields results which are better than those obtained by LeNet.

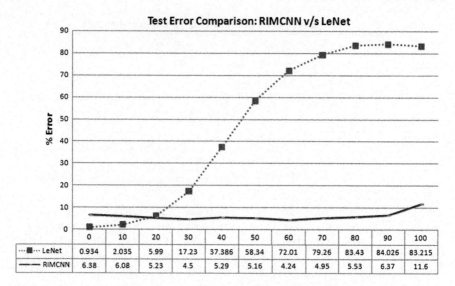

	0	10	20	30	40	50	60	70	80	90	100
LeNet	0.934	2.035	5.99	17.23	37.386	58.34	72.01	79.26	83.43	84.026	83.215
RIMCNN	6.38	6.08	5.23	4.5	5.29	5.16	4.24	4.95	5.53	6.37	11.6

Fig. 5 Comparison of RIMCNN with different architectures for test data rotated between 0° to 100°

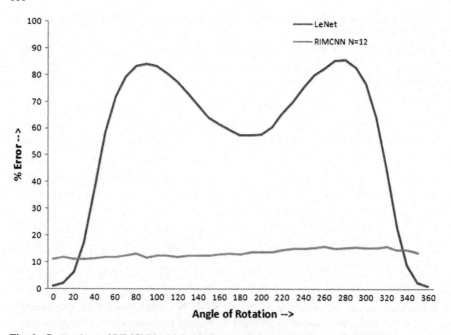

Fig. 6 Comparison of RIMCNN with LeNet for test data rotated between 0° to 360°

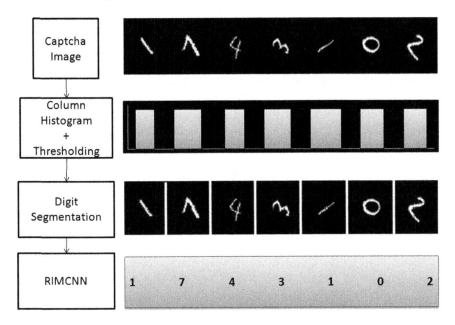

Fig. 7 Captcha recognition with RIMCNN

RIMCNN with N = 12 and rotations uniformly sampled between 0° to 360° is used for testing rotated data between 0° to 360°. Even for higher degrees of rotation in input image, the classification error is very low, as can be seen in Fig. 6.

3.2 Captcha Recognition Using RIMCNN

As RIMCNN is able to classify rotated digits, it can be applied to decode simple captcha images with non-overlapping and rotated digits. The steps involved along with the intermediate results are shown in Fig. 7. Initially, a histogram is formed by summing the nonzero pixels in each column of the binary captcha. The region which has a digit will produce higher-valued bins, as shown in the histogram, whereas the background columns will have negligible values. By thresholding the values in the histogram, each digit can be segmented. The segmented digits are then given to RIM-CNN for classification, which yields the labels for each digit. Here, RIMCNN with N = 4 and rotations 0°, 30°, −45° and 60° produces the correct labels.

4 Conclusion

The proposed architecture of RIMCNN is of great benefit for the problems which involve images with rotated objects. It works considerably well even for higher degrees of rotation and can be easily applied with limited training data. In all the tasks discussed above, it gives better classification as compared to LeNet without increasing the training time. The method of using multiple LeNets in testing phase is easier to implement. It is used for the first time as per our knowledge and has a lot of potential in terms of saving computational cost and gives satisfactorily results. We observed promising results in our experiments, and further this work can be extended for more complex data. Its ability to find the orientation of objects in an image can be utilized in many applications such as object tracking, robotics. Further, it can be applied to texture classification and robust object classification task.

References

1. Yann LeCun and Yoshua Bengio. Convolutional networks for images, speech, and time series. *The handbook of brain theory and neural networks*, 3361(10):1995, 1995.
2. Sajjad S Ahranjany, Farbod Razzazi, and Mohammad H Ghassemian. A very high accuracy handwritten character recognition system for farsi/arabic digits using convolutional neural networks. In *Bio-Inspired Computing: Theories and Applications (BIC-TA), 2010 IEEE Fifth International Conference on*, pages 1585–1592. IEEE, 2010.
3. Chao Dong, Chen Change Loy, Kaiming He, and Xiaoou Tang. Image super-resolution using deep convolutional networks. *IEEE transactions on pattern analysis and machine intelligence*, 38(2):295–307, 2016.
4. Fok Hing Chi Tivive and Abdesselam Bouzerdoum. Rotation invariant face detection using convolutional neural networks. In *International Conference on Neural Information Processing*, pages 260–269. Springer, 2006.
5. D. Wu, L. Pigou, P. J. Kindermans, N. D. H. Le, L. Shao, J. Dambre, and J. M. Odobez. Deep dynamic neural networks for multimodal gesture segmentation and recognition. *IEEE Transactions on Pattern Analysis and Machine Intelligence*, 38(8):1583–1597, 2016.
6. B. Fasel and D. Gatica-Perez. Rotation-invariant neoperceptron. In *18th International Conference on Pattern Recognition (ICPR'06)*, volume 3, pages 336–339, 2006.
7. Shih-Chung B Lo, Heang-Ping Chan, Jyh-Shyan Lin, Huai Li, Matthew T Freedman, and Seong K Mun. Artificial convolution neural network for medical image pattern recognition. *Neural networks*, 8(7):1201–1214, 1995.
8. Dmitry Laptev, Nikolay Savinov, Joachim M. Buhmann, and Marc Pollefeys. TI-POOLING: transformation-invariant pooling for feature learning in convolutional neural networks. *CoRR*, arXiv:1604.06318, 2016.
9. Max Jaderberg, Karen Simonyan, Andrew Zisserman, and Koray Kavukcuoglu. Spatial transformer networks. *CoRR*, arXiv:1506.02025, 2015.
10. Yann LeCun, Léon Bottou, Yoshua Bengio, and Patrick Haffner. Gradient-based learning applied to document recognition. *Proceedings of the IEEE*, 86(11):2278–2324, 1998.
11. Yann LeCun and Corinna Cortes. Mnist handwritten digit database. *AT&T Labs [Online]. Available:* https://yann.lecun.com/exdb/mnist, 2010.

Stochastic Assimilation Technique for Cloud Motion Analysis

Kalamraju Mounika, J. Sheeba Rani and Gorthi Sai Subrahmanyam

Abstract Cloud motion analysis plays a key role in analyzing the climatic changes. Recent works show that Classic-NL approach outperforms many other conventional motion analysis techniques. This paper presents an efficient approach for assimilation of satellite images using a recursive stochastic filter, Weighted Ensemble Transform Kalman Filter (WETKF), with appropriate dynamical model and image warping-based non-linear measurement model. Here, cloud motion against the occlusions, missing information, and unexpected merging and splitting of clouds has been analyzed. This will pave a way for automatic analysis of motion fields and to draw inferences about their local and global motion over several years. This paper also demonstrates efficacy and robustness of WETKF over Classic-Non-Local-based approach (Bibin Johnson J et al., International conference on computer vision and 11 image processing, 2016) [1].

1 Introduction

Cloud motion analysis is widely used in different domains like weather forecasting, meteorological applications, monitoring hazards. But, this has been a challenging task because of problems like non-rigid motion of clouds, splitting and merging of clouds, presence of many clouds of which each have a different shape, size, and also move in different directions. Traditional approaches for cloud motion analysis discussed in the literature include techniques using template matching [2],

K. Mounika (✉) · J. Sheeba Rani
Department of Avionics, Indian Institute of Space Science and Technology,
Trivandrum, India
e-mail: mouni.kalamraju@gmail.com

J. Sheeba Rani
e-mail: sheeba@iist.ac.in

G. S. Subrahmanyam
Electrical Engineering, Indian Institute of Technology, Tirupati, India
e-mail: rkg@iittp.ac.in

© Springer Nature Singapore Pte Ltd. 2018
B. B. Chaudhuri et al. (eds.), *Proceedings of 2nd International Conference on Computer Vision & Image Processing*, Advances in Intelligent Systems and Computing 703, https://doi.org/10.1007/978-981-10-7895-8_9

artificial neural network for contour shape matching [3], pixel-based motion estimation, i.e., optical flow estimation. The methods discussed in [2, 3] belong to category of feature-based motion estimation techniques, in which, objects to be tracked are segmented initially and then, motion is estimated over consecutive frames based on correlation or other matching techniques. In [4], clouds were identified at different thresholds and motion of clouds are then tracked using optical flow. The drawback of these methods is that they do not estimate or do not use intra-cloud motion and are not robust to occlusion and missing data cases [1].

In this method, a pixel-based motion estimation approach is used in order to get both intra-object and inter-object flow vectors. Basically, there are two approaches for estimating flow vectors. They are variational approaches [5, 6] and stochastic approaches [7]. The drawback of variational approaches is that they incorporate preconditioning stages in optimization or covariance specifications which include smoothing [8]. So, in this paper, a stochastic recursive filter called Weighted Ensemble Transform Kalman Filter (WETKF) [8] is used for optical flow estimation. It can also be considered as a particle filter extension of the Ensemble Transform Kalman filter (ETKF) [9, 10]. The only difference between Ensemble Kalman Filter (EnKF) [11] and ETKF is that ETKF do not use perturbed observations. In this paper, the generality and applicability of WETKF for cloud motion estimation is investigated. For experimentation, October month's Meteosat (HRI) [12] satellite data is considered. The main contribution of this paper is to demonstrate that WETKF is a very potential approach for cloud motion estimation, and it gives an automatic approach to access the cloud motion in spite of occlusion, missing information, splitting, merging, and fully non-rigid cloud motion.

In this paper, efficacy of WETKF approach over Classic-NL [1, 13] is demonstrated. Classic-NL is a Horn and Schunk optical flow estimation technique combined with modern optimization strategies. As discussed in [1], this method is supposed to handle large motion and occlusion cases. But, for missing data cases, this method fails. It also requires pre-processing and post-processing of data for complete analysis, and consistency of results is not observed. The proposed method, using WETKF, can overcome these drawbacks. Because of its recursive nature, it can automatically handle missing data cases, no pre-processing or post-processing of data is required, and results are consistent. The structure of this paper is organized as follows: Sect. 2 summarizes the steps of cloud motion estimation using WETKF; Sect. 3 demonstrates an insight into its effectiveness by presenting some experimental results and discusses its robustness; Sect. 4 draws some conclusions on this work.

2   Cloud Motion Estimation Using WETKF

Cloud motion estimation using WETKF involves a dynamical model and a measurement model. Dynamical model describes the formulation of state variable, which we have considered as vorticity (curl of velocity) for cloud analysis. Here, a stochastic filtered version of vorticity–velocity Navier–Stokes formulation

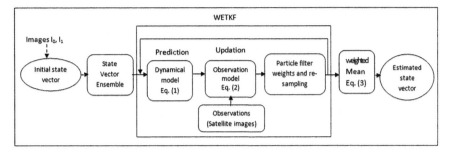

Fig. 1 Block diagram showing cloud motion analysis using WETKF

is considered as dynamical model, assuming cloud motion as a special case of fluid motion. Measurement model forms the relation between observations, i.e., cloud images and state variable. A non-linear image warping error is considered as the measurement model. Figure 1 shows the block diagram of cloud motion estimation using WETKF. The process of cloud image assimilation using WETKF is explained below.

2.1 Initialization of State Vector

In this cloud motion estimation analysis, vorticity is considered as state variable. Initially, at time $k = 0$, the N particles of state (i.e., ensemble size being N) $\left\{\xi_0^{a,(i)}, i = 1, 2, \ldots, N\right\}$ are initialized with vorticity, which is considered as curl of velocity obtained by performing optical flow estimation on images I_0 and I_1. Ensemble is generated by adding Gaussian noise to initial state vector. From time $k = 1$ to T, we perform the following three steps for the WETKF assimilation of state (vorticity) $\hat{\xi}_k$ using corresponding satellite images (observations).

2.2 Forecasting

Here, as discussed, a 2-D Navier–Stokes equation with a stochastic forcing function is considered as a dynamical model. At time k, the ensemble obtained from previous step are propagated through the model:

$$d\xi_t = -\nabla\xi_t.\vartheta dt + v\Delta\xi_t dt + \eta dB_t \tag{1}$$

to supply the forecast ensemble members $\xi_t^{f,(i)}$, expressed at time t. Here, the state vector represents the vorticity (ξ) at each image grid point coordinates, v represents viscosity constant, and ϑ represents velocity vector.

2.3 Correcting

In this step, we directly incorporate the available observations, i.e., satellite images and update forecast ensemble. Observational model is derived from brightness consistency assumption. In many optical flow estimation techniques like Horn and Schunk, Lukas kanade, brightness consistency equation is linearized. In this method, non-linear version of it is employed as:

$$I(x, k) = I(x + d(x), k + 1) + \gamma dB \quad \forall x \in \Omega_I \tag{2}$$

where $d(x) = \int_{k-1}^{k-\delta t} \vartheta(x_t) dt$ and Ω_I represents complete image plane. Using this model, forecast ensemble and inverse of covariance matrix, corrected ensemble is obtained from Kalman analysis equation [9].

2.4 Particle Filter Weights and Re-sampling

In this step, the ensemble members are weighted with a likelihood function of observations and then, re-sampling is performed. Finally, a weighted mean, w_i, of ensemble obtained is considered as the vorticity estimate at time k :

$$\xi_k = \sum_{i=1}^{N} w_i \xi_k^{a,(i)} \tag{3}$$

and the corresponding velocity field is deduced through the Biot-Savart kernel as $\vartheta = \nabla^\perp G * \xi$ where $G(.) = \frac{\ln(|.|)}{2\pi}$ is the green kernel associated to Laplacian operator [8].

Matrix inversion problem: The observations in our analysis, i.e., satellite images, will be usually of high resolution. In Kalman filter, Kalman gain calculation involves inversion of covariance term. This covariance term depends on state vector dimension, which is resolution of image. Thus, inversion of $n \times n$ dimensional covariance term (where n-state vector dimension) results in complex computation. In this method, an ensemble of state vector is considered instead of single state vector. Sample covariance, obtained from ensembles, is considered in place of original covariance term in Kalman gain calculation. This helps in reducing the computational complexity as the dimension of covariance matrix to be inverted is reduced from n to N, where N is the size of ensemble considered ($N \ll n$). This inversion can be performed efficiently by considering singular value decomposition of ensemble matrix [8].

3 Experimental Results

In this section, we present our WETKF-based direct image assimilation results on satellite images. In order to validate the accuracy of our method, we have initially tested with Particle Image Velocimetry (PIV) images for which ground truth data is available.

3.1 Particle Image Velocimetry (PIV) Images

In this section, comparison of WETKF and CNL on fluid images is established. This forms the motivation for further application of these methods in cloud motion estimation and analysis. Recent work [1] demonstrates that CNL (*Classic − NL*) outperforms many conventional optical flow approaches, and CNL can be successfully applied to cloud motion estimation. Hence, we compare results of WETKF and CNL in Figs. 2 and 3 to show efficacy of WETKF over CNL.

For particle images, following results are obtained with a simple 2-D velocity–vorticity formulation of Navier–Stokes equation as dynamic model. Ensemble size is considered as 200. Figure 2 compares estimated vorticity field obtained from WETKF with true vorticity field and CNL estimated vorticity field.

In Fig. 3, solid black line and dotted blue line indicate results from WETKF with ensemble sizes 200 and 50, respectively. From the Fig. 3, it is quantitatively shown that WETKF gives better results than CNL and Lucas Kanade (LK) [14] and also can be observed that as ensemble size increases, estimation error decreases.

As ground truth is available only for fluid images, the scenarios which occur in cloud images are simulated on fluid image sequence to test and compare the capability of WETKF over CNL. Here, the results of large occlusion cases are demonstrated. Figure 4 shows comparison of estimated fields in case of complete occlusion. The small error generated in estimating flow vectors is nullified over iterations, and this error is far less than the estimate obtained from CNL without any occlusion.

This ability of predicting flow vectors even in absence of data is practically useful when data from satellites like INSAT-3D, KALPANA etc., are used as they do not provide data during certain time interval per day.

3.2 Satellite Images

Here, WETKF is applied on satellite images. The analysis is done over both cloud images and cyclone images. Satellite images are taken from Meteosat-7 satellite (HRI).

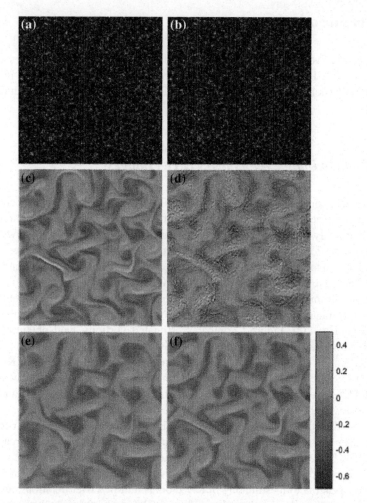

Fig. 2 Comparing true and estimated vorticity fields, **a** PIV image—K, **b** PIV image—$K + 1$, **c** true vorticity field, **d** estimated vorticity field with CNL, **e** estimated vorticity field with 50 ensemble WETKF, **f** estimated vorticity field with 200 ensemble WETKF

3.2.1   Cropping Cloud Region

A cropped region of 512×512 pixels in the satellite images during September, 2014, displaying normal clouds is considered, and WETKF is applied over these images. The following result is obtained when ensemble size $N = 200$. Figure 5 shows flow fields obtained from CNL and WETKF at 30th iteration. The obtained flow estimates are consistent and agree with visual observation. Vorticities are clearly visible, and splitting and merging of clouds is also properly shown by flow vectors.

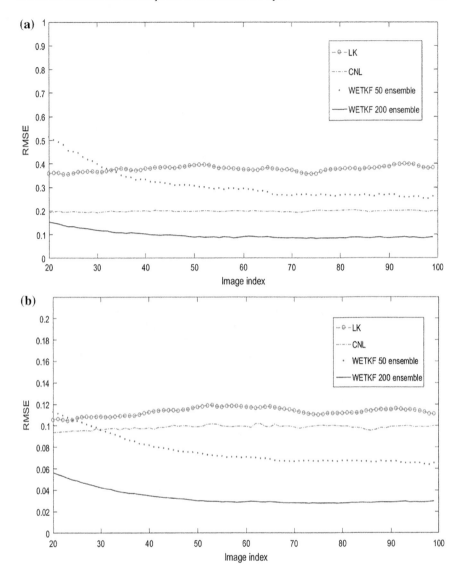

Fig. 3 RMSE plots. **a** RMSE in estimating velocity, **b** RMSE in estimating vorticity

WETKF performance is analyzed in cases of large occlusion and missing data. From the results, it is clear that WETKF is able to predict best estimate of flow vector even in complete occlusion and missing data cases. As satellite images are available for every 30 min interval, missing one frame causes significant change in motion estimation.

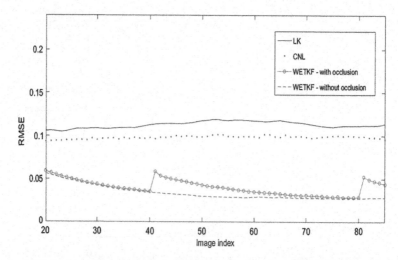

Fig. 4 RMSE plot showing results of WETKF with and without occlusion cases. At 40th and 80th iterations, there is occlusion

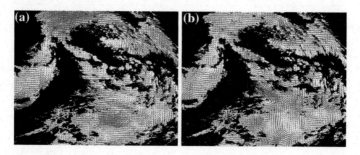

Fig. 5 Cloud images with flow vectors estimated by **a** CNL **b** WETKF

3.2.2 Cropping Cyclone Region

Now, we have applied WETKF on cyclone images taken from Meteosat satellite during October, 2014, over Bay of Bengal. These cyclone images are cropped into 340 × 340 pixels size.

From Fig. 6, WETKF gives better results than CNL because CNL does not consider non-rigid motion in its flow estimation. From the results, it is observed that CNL results are more uniform, which is not desired in case of cloud images. The expected motion vectors in the experimental images, i.e., in northern hemisphere region are to be in anticlockwise direction [15], and as cloud motion is similar to fluid motion, local rotations are possible. These two are clearly seen in WETKF results, whereas CNL results do not contain local rotations.

As flow vectors are obtained, tracking can be done by simply selecting vorticity center of cyclone and track this center using flow vectors. A random selected point

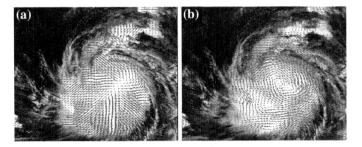

Fig. 6 Comparison of WETKF and CNL outputs of Hudhud cyclone at 0200 UTC on October 11, 2014: **a** CNL output, **b** WETKF output

Fig. 7 Track of intra-cloud motion obtained from flow vectors

tracking can also be done. Figure 7 shows a track of intra-cloud point using flow vectors estimated from WETKF. This intra-cloud track is used in splitting and merging cases of cloud motion analysis.

3.3 Validation

Validation of results can be done by considering model data as ground truth. For this, height or pressure level of the clouds should be known. Here, the pressure level of clouds is obtained from brightness temperature.

Figure 8 shows comparison between tracks obtained from WETKF (solid red color line in figure) and best track data (considered as model track for cyclones represented as solid black line in figure) for Hudhud cyclone.

Fig. 8 Cyclone track

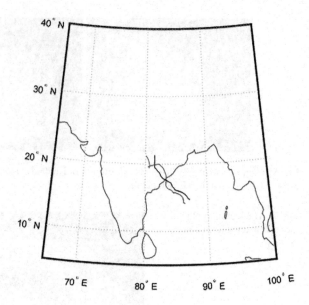

4 Conclusion

In this paper, a novel method for cloud motion estimation is proposed. Here, WETKF is used for data assimilation on real satellite images, and tracking of cyclones are done using flow vectors obtained. Results shown indicate that it has the ability to handle large data missing and occlusions. Future work includes obtaining the individual tracks for splitting and merging cases of cloud images and testing on different cyclones.

Acknowledgements We acknowledge Meteorological and Oceanographic Satellite Data Archival Centre (MOSDAC), Space Applications Centre, ISRO, European Organization for the Exploitation of Meteorological Satellites (EUMETSAT) for providing satellite images and INRIA for Particle Image Velocimetry (PIV) images.

References

1. Bibin Johnson, J. Sheeba Rani and G R K Sai Subrahmanyam: A novel visualization and tracking framework for analyzing inter/intra cloud pattern formulation to study their impact on climate. International conference on computer vision and Image processing. feb(2016)
2. J Schmetz, K Holmlund, J. Hoffman, B. Strauss, B. Mason, V. Gaertner, A. Koch and L. Van De Berg: Operational cloud-motion winds from Meteosat infrared images. Journal of Applied Materology, 132, 1206–1225 (1993)
3. S cote and A. R. L. Tatnall: A neural network-based method for tracking features from satellite sensor images. Remote Sensing, 16, 3695–3701(1995)

4. Harish Doraiswamy, Vivek Natarajan and Ravi S Nanjundiah: An exploration framework to identify and track movement of cloud systems, IEEE Transactions on Visualization and Computer Graphics, 19(12), 2896–2905(2013)
5. Nicolas Papadakis, Thomas Corpetti and Etienne Memin: Dynamically consistent optical flow estimation. International conference on computer vision (2007)
6. Thomas C, Corpetti T and Memin E: Data assimilation for convective cells tracking on meteorological image sequences, IEEE Trans. Geosci Remote Sens, 48(8), 3162–3177(2010)
7. N. Papadakis, E. Memin, A. Cuzol and N. Gengembre: Data assimilation with the weighted ensemble Kalman filter, Tellus A, 62, 673–697(2010)
8. S. Beyou, A. Cuzol, Gorthi Sai Subrahmanyam and E. Memin: Weighted ensemble transform Kalman filter for image assimilation, Tellus A, 65(2013)
9. C. H. Bishop, B. J. Etherton and S. J. Majumdar: Adaptive Sampling with the Ensemble Transform Kalman Filter. Part I: Theoretical Aspects, Monthly Weather Review, 129, 420–436(2001)
10. M K Tippett, J L Anderson, C. H. Bishop, T. M. Hamill and J. S. Whitaker: Ensemble Square Root Filters, Monthly Weather Review, 131, 1485–1490(2003)
11. G. Evensen: The Ensemble Kalman Filter: theoretical formulation and practical implementation, Ocean Dynamics, 53, 343–367(2003)
12. HRI:Image data in the form of High Rate Transmissions available at 30-minute intervals with coverage over the Indian Ocean (https://www.eumetsat.int)
13. Deqing Sun, Stefan Roth and Michael J Black: A quantitative analysis of current practices in optical flow estimation and the principles behind them., International Journal of Computer Vision, 106(2), 115–137(2014)
14. Dhara Patel, Saurabh Upadhyay: "Optical Flow Measurement using Lucas kanade Method", International Journal of Computer Applications, 61, 6–10(2013)
15. Nelson, Stephen (Fall 2014): "Tropical Cyclones (Hurricanes)", Wind Systems: Low Pressure Centers, Tulane University

Image Contrast Enhancement Using Hybrid Elitist Ant System, Elitism-Based Immigrants Genetic Algorithm and Simulated Annealing

Rajeev Kumar, Anand Gupta, Apoorv Gupta and Aman Bansal

Abstract Contrast enhancement is a technique which is used to expand the range of intensities within the image to make its features more distinct and easily perceptible to the human eye. It has found many applications ranging from medical to satellite imagery where the primary aim is to find hidden or minute details within an image. Through literary research, the authors have realised that the existing approaches lag behind in enhancing the contrast of an image. Hence in the present paper, an improved contrast enhancement technique is proposed which is based on the hybrid combination of nature-based metaheuristics: Elitist Ant System (EAS), Elitism-based Genetic Algorithm (EIGA) and Simulated Annealing (SA). EAS and EIGA work together to search globally for the optimum solution which is then refined by SA locally. Through experiment, it is observed that the proposed algorithm is efficiently improving the contrast of an image when compared with existing algorithms.

Keywords Contrast enhancement · Elitist ant system · Elitism-based immigrants genetic algorithm · Simulated annealing

The authors contributed equally to this paper.

R. Kumar · A. Gupta
Department of Computer Engineering, NSIT, University of Delhi, New Delhi, India
e-mail: rajeevrai123@yahoo.com

A. Gupta
e-mail: omaranand@nsitonline.in

A. Gupta (✉) · A. Bansal
Department of Information Technology, NSIT, University of Delhi, New Delhi, India
e-mail: apoorv2711@gmail.com

A. Bansal
e-mail: bansalaman2905@gmail.com

© Springer Nature Singapore Pte Ltd. 2018
B. B. Chaudhuri et al. (eds.), *Proceedings of 2nd International Conference on Computer Vision & Image Processing*, Advances in Intelligent Systems and Computing 703, https://doi.org/10.1007/978-981-10-7895-8_10

1 Introduction

The petabytes of digital images in the world reinforce the need for better quality and enhanced imagery. Hence, digital image enhancement techniques are used to improve upon how we perceive these images and the information they provide. These images can then provide 'better' or 'improved' inputs to consequent digital image processing systems.

Contrast enhancement is an important aspect of image enhancement and is concerned with the difference between the bright and the dark areas of an image [1]. The effect of shadows or highlights in an image makes contrast enhancement an important aspect of image enhancement, as the colour variations are not clearly discernible in such an image owing to an eye's relative insensitivity to variations in dark and bright colours [1]. A considerable percentage of the images we deal with regularly, ranging from the images of our own cameras to remote sensing and medical imagery, suffer from poor contrast. This makes necessary, the need for contrast enhancement. For the past few years, *global intensity transformation* techniques are being employed to increase the contrast [2]. In these techniques, an optimal transfer function is formed which maps the grey levels in an image to the corresponding new intensity values. Hence, optimisation techniques such as nature-based metaheuristics are found many applications in the field of contrast enhancement. In the next subsection, previous applications of these metaheuristics in the field of image contrast enhancement are discussed in detail.

1.1 *Related Work*

In recent years, nature-based metaheuristics like Ant Colony Optimisation (ACO) [3], Genetic Algorithm (GA) [4], Simulated Annealing (SA) [5], or Artificial Bee Colony (ABC) [6] have found a plethora of applications in the real-world image processing. For improving the contrast of an image, nature-based meta-heuristics are employed to find the optimal transformation functions. ACO, which is used in contrast enhancement [7, 8], is a nature-based metaheuristic that is inspired by the movement pattern of the ants. Similarly, GA which is used in [9, 10] is another nature-based metaheuristic algorithm which uses genetic operators such as selection, crossover and mutation to find the most optimal transfer function. Even though the above nature-based metaheuristics provide adequately enhanced images, sometimes the best possible solutions are left out because the guiding process of the metaheuristic is random in nature [11]. Hence, to improve the guiding process of the metaheuristics, hybridisation of these algorithms is carried out. Like in [12], a combination of ACO and GA is used, where GA acts on the solution space created by the ACO, thus improving the transfer function. Similarly, in [13], a hybrid of ACO, GA and SA is used. ACO and GA work together to search globally for the best solution, whereas SA is used to optimise the result locally.

1.2 Motivation

After implementing the hybrid algorithm [13], the authors have observed that the above hybrid algorithm is an improvement over previously found contrast enhancement algorithms. However, it is felt that [13] has not fully exploited the capabilities of the metaheuristics involved. The authors have observed the following limitations:

A. The initial form of ACO used in [13] lacks the exploitation of global best solution to the fullest.
B. GA used in [13] lacks adequate exploration powers.

The initial form of ACO algorithm used in [13] is subjected to several improvements, which have a considerable impact on the results. Primarily, these improvements have laid their focus on two areas [14]. The solutions obtained by the ants can be improved radically by the use of a local search algorithm. Secondly, these improvements focus on exploiting the global best solution to the fullest. Though [13] takes advantage of using a local search algorithm, i.e. SA, to improve the solution found by ants, it does not exploit the global solution as well as it can. Exploitation and exploration are the two predominant concepts that help in characterising every heuristic search. When exploration of a search is enhanced, the exploitation reduces and when exploitation is strengthened, the search gets diluted making them two conflicting concepts [15]. GA tends to give preference to the best individual in the population so as to direct the algorithm to an optimum, but high selective pressure can lead to reduced exploration and premature convergence of the algorithm which leads to poor results [16]. Moreover, the search is not highly explorative when we deal with a fixed set of individuals initialised in the beginning of the execution of the algorithm as in [13]. Though [13] tries strengthening the exploration with the help of mutation, convergence might not happen at all if the exploration is increased by a huge factor using mutation [17]. Therefore, GA used in [13] lacked a method wherein the explorative power of search is increased for individuals in the bad region of the state space and at the same time exploitation power of search is not affected for individuals in the high fitness region of the state space. In the next subsection, the above-mentioned limitations have been addressed by the authors.

1.3 Contribution

To overcome the above-mentioned drawbacks, the authors propose an optimised contrast enhancement algorithm. Through literary research, it has been observed that original form of ACO did not incorporate elitist ants, resulting in less optimum solutions [14]. The inclusion of these elitist ants improved upon the ant system's ability to find better tours, and that too in fewer iterations [14]. This way, a better tour is obtained by taking advantage of the increased exploitation of the global search space. Therefore, to enhance the capabilities of Ant Colony Optimisation (ACO) algorithm, the use of Elitist Ant System (EAS) is proposed. The GA algorithm used in [13] did not have sufficient explorative power as after few generations due to selective pressure the population starts lacking diversity. In order to efficiently search the solution space, we need to maintain the variations and diversities in the population and for this reason; Elitism-based Immigrants Genetic Algorithm (EIGA) is employed in the proposed hybrid algorithm [18, 19]. After observing the flux of immigrant population that moved in and out with respect to the main population between two generations, this algorithm had been proposed in [18]. The algorithm replaces the worst performing individuals in the population by new immigrants thereby maintaining the diversity [18]. The immigrants can be generated randomly, but these can divert the search process by an unacceptable degree. Thus, the proposed algorithm uses an *Elitism-based Immigrants Genetic Algorithm* (denoted EIGA) where the immigrants are generated on the basis of the best individual in the population. Elitism-based immigrants maintain the delicate balance between the exploration and exploitive power of the algorithm [18]. The enhanced diversity of the population achieved by immigration helps the algorithm tackle multimodal functions better and can enable it to search for multiple global and local optima [16].

The contributions proposed by the authors are summarised in Fig. 1 where Fig. 1a represents the basic framework of the hybrid algorithm as proposed in [13], whereas Fig. 1b shows the basic framework of proposed algorithm. From Fig. 1, it can be observed that following modifications have been proposed:

A. ACO is replaced by EAS as it enhances the exploitation property of the ACO.
B. GA is replaced by EIGA as it enhances the exploration property of GA without affecting its exploitation prowess.

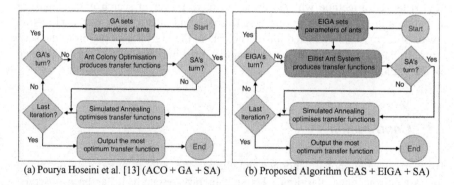

(a) Pourya Hoseini et al. [13] (ACO + GA + SA)   (b) Proposed Algorithm (EAS + EIGA + SA)

Fig. 1 Comparison of framework of [13] and proposed algorithm

Rest of this paper is organised in the following manner. The proposed work is explained in detail in Sect. 2. Section 3 compares the performance of the proposed method with other contrast enhancement approaches through experimentation and, Sect. 4 concludes our work while discussing the prospects of the future work.

2 Proposed Algorithm

The detailed flowchart of proposed algorithm is shown in Fig. 3. The algorithm consists of three components, i.e. Elitist Ant System (EAS), Elitism-based Immigrants Genetic Algorithm (EIGA) and Simulated Annealing (SA). The proposed algorithm considers greyscale 8-bit images whose pixel intensities range from 0 to 255. The algorithm can also enhance the contrast of colour images by operating upon the *Intensity* component of HSI colour model of images [20]. To enhance the contrast of an image, a mapping function called transfer function is required which transforms the pixel intensities to the desired intensities. Thus, the search space defined for the transfer function is a plane of size 256×256. To find the optimum transfer function in the search space, EAS first generates a set of ants. These ants explore the search space for the optimal transfer function. The trace of an ant's movement generates the required transfer function. The movement of these ants is controlled by their chromosomes which are initialised by EIGA. Hence, the chromosomes set by EIGA are responsible for controlling the movement parameters of ants. SA is used to optimise the transfer function generated by EAS. In SA, a set of recently generated transfer functions are selected randomly and are locally optimised. To terminate the search process, the proposed algorithm is run for 100 iterations. The frequency of execution of EIGA and SA is varied as the algorithm proceeds and is divided into three equal phases which are detailed in Table 1. As EIGA is responsible for the setting the movement parameters of the ants, the frequency of EIGA is high during the first phase of the algorithm to enhance the algorithm's exploration powers. Similarly, the number of immigrants is higher in the first phase to support exploration. Both the frequency and the number of immigrants are decreased as the algorithm proceeds to second and third phase to aid

Table 1 Parameters of EIGA and SA

Phase	EIGA		SA			
	Execution interval (iteration)	Number of elitism-based immigrants	Execution interval (iteration)	Number of transfer func. selected	Number of points on transfer func.	Number of SA iterations
1	5	2	10	1	2	3
2	6	2	6	2	4	6
3	7	1	3	4	6	12

convergence. The frequency of SA is kept low in the first phase and is increased as the algorithm proceeds to second and third phase to further optimise the generated transfer function and their corresponding pheromone trail. The various important components of the proposed algorithm are now discussed in detail.

2.1 Elitist Ant System

Elitist Ant System which is a nature-based metaheuristic inspired by the movement pattern of ants [14]. Ants deposit pheromone as they move, which becomes a measure of their performance evaluation. In EAS, elitist ants are selected based on their fitness values and are made to deposit much higher level of pheromone along their paths. A problem might arise in deciding the number of elitist ants to be used as too many of these elitist ants can decrease the explorative power of the algorithm, resulting in the algorithm getting trapped in local minima [14]. By carrying out experiments repeatedly, it has been found that a single elitist ant provides the most optimum solution and maintains the balance between exploration and exploitation.

In Fig. 2, EAS has been shown in red. The movement of a population of 20 ants defines the transfer function for the given image. This movement of the ants through the search space is guided through a probability function defined as (1),

$$P = \frac{(1+\tau_i)^{\alpha} * \left[\left(1 + (K_i/\gamma)^{10}\right) * \eta_i\right]^{\beta}}{\sum_{i \in G(i)} (1+\tau)^{\alpha} * \left[\left(1 + (K_i/\gamma)^{10}\right) * \eta_i\right]^{\beta}} \tag{1}$$

where α and β control the relative importance of pheromone trail against heuristic value. $G(i)$ is the set of neighbourhood points around ant and τ_i is the amount of pheromone deposited on these points [13]. The value of η_i is defined as C_{up} for the up neighbour, 1 for the up-right neighbour, C_{right} for the right neighbour and for other neighbours it is set as zero. For up and right neighbours, K_i is set to $Intensity_{input} - Intensity_{input-min}$ in the horizontal axis and $Intensity_{output}$ in the vertical axis for the current position of ant, respectively, and for the other neighbours it is set to zero. γ helps guide the ants towards the upper right corner. When ants move across the search space, they leave a pheromone trail behind them to guide other ants so that they can reach the destination efficiently. Also, this pheromone keeps on evaporating so that the search space is properly explored. As discussed before, utilisation of an elitist ant in the algorithm provides better results by depositing higher levels of pheromones on the path of the best ant. The mathematical function of pheroas (2),

$$\tau_{ij}(t+1) = (1-\rho) * \tau_{ij}(t) + \left(\sum_{k=1}^{20} \Delta\tau_{ij}^{k}(t)\right) + \Delta\tau_{ij}^{BA} \tag{2}$$

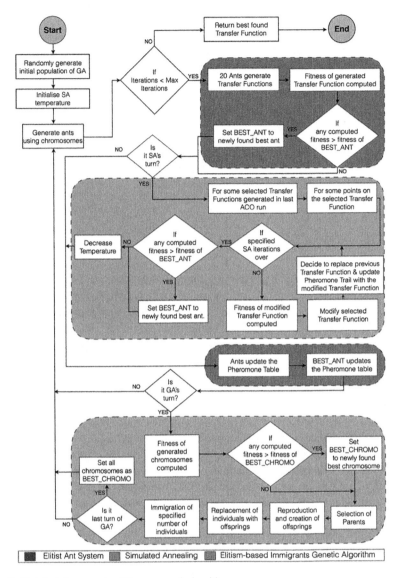

Fig. 2 Detailed framework of the proposed algorithm

where τ_{ij}^k is the quantity of pheromone deposited by the kth ant in between points i and j and is equal to $F_k/(30 \times F_{BA})$. F_k is the fitness of the kth ant and F_{BA} is the fitness of the best ant. τ_{ij}^{BA} is the quantity of pheromone deposited by the best ant, i.e. elitist ant and is equal to F_{BA}. ρ is the pheromone evaporation rate and is set to 0.4.

To compute the quality of the produced images a fitness function needs to be defined. The proposed algorithm employs the fitness function from [13] where it has been studied in detail. The fitness function is found to provide higher contrast images while preserving their natural look [13]. The fitness function is defined as (3),

$$F = \sqrt[3]{STD * ENTROPY * SOBEL} \tag{3}$$

where STD is the global standard deviation of intensities and is a measure of contrast of an image, $ENTROPY$ is the global entropy that measures the randomness in the image and $SOBEL$ is defined as $mean(|SOBEL_{vertical}| + |SOBEL_{horizontal}|)$.

2.2 Elitism-Based Genetic Algorithm

It is an evolutionary metaheuristic algorithm which searches for the optimised solutions in the search space using genetic operators including selection, recombination (i.e. crossover and mutation), replacement and immigration operators. In Fig. 2, EIGA is shown in blue colour. In the beginning of the algorithm execution, EIGA is used to initialise the parameters of EAS. A set of ten chromosomes is initialised randomly in the beginning of the algorithm. Each chromosome is then used to set parameters of two ants. Every chromosome consists of five genes, i.e. α, β, γ, C_{up}, C_{right}, which define the movement parameters of the [13]. The genes α and β are in between 0 and 5. C_{up} and C_{right} can vary between 0 and 3, while γ can vary from 100 to 250. During parent selection phase, two individuals are selected on the basis of roulette wheel technique. These are then allowed to perform uniform crossover operation with a probability of 0.85, which generates two offspring. The fitness of the chromosome is equal to the sum of the average fitness values of the transfer functions produced in between two EIGA iterations by each of the two ants created using the specified chromosome and the maximum fitness value produced by either of the two ants. The two produced offspring then replace the worst ant in the population along with the worst ant out of the two parents. If both represent the same ant, then two worst individual ants are replaced by these offspring. Mutation probability is set to 5% where one of the gene values can change with ±10% of the original value.

Immigration is applied after replacement of the ants by the new offspring, and it creates elitism-based immigrants for the GA population. Every gene in the generated individual has a 25% probability of being similar to the corresponding gene of the best chromosome. This immigrated individual then replaces the individual having the least fitness. As described earlier, this elitism trait along with the replacement strategy ensures that the generated migrants while enhancing the diversity do not hamper the exploitation property of the algorithm. The number of immigrants is carefully decided and kept low, at two, so that the exploitation power of the algorithm is not affected and is reduced further to one in the later iterations in order to enhance the convergence of the algorithm.

2.3 Simulated Annealing

Simulated Annealing is an optimisation technique that is inspired by the annealing process in metallurgy, which involves controlled cooling of a heated metal [5]. In Fig. 2, SA has been shown in green. In the proposed work, SA is used to further optimise the transfer function generated by EAS. In SA, the best transfer function along with some randomly selected transfer function generated in the last run of EAS are selected. Then random points are selected on the transfer functions which are then optimised using the SA algorithm. A detailed explanation of neighbour selection is provided in [13]. The probability function to select the neighbour is defined as (4),

$$P = \begin{cases} e^{\frac{F^{new} - F^{old}}{0.05*T*F^{old}}} & if\ F^{new} < F^{old} \\ 1 & if\ F^{new} > F^{old} \end{cases} \tag{4}$$

where F^{old} and F^{new} are the value of the current fitness and the value of the resulting fitness, respectively. The temperature which is initially set to 200 and is cooled by 50% after every SA phase. After optimising the transfer functions, ant's pheromone trails are also adjusted as per the changes in the new transfer functions.

3 Experimental Results and Analysis

The proposed algorithm was tested on a system with Intel i7-4700MQ processor (clock speed of 2.40 GHz), 8 GB RAM and was implemented on MATLAB. To benchmark the performance of the algorithm, commonly used images available in MATLAB and on the internet are used. The prowess of the algorithms to enhance the contrast is quantified on the basis of the fitness values of the output images, defined in (3).

The proposed algorithm is compared with the method implemented in [13], to show the overall increase in the capabilities of enhancing the contrast of the image as compared to [13]. Further, the algorithm is compared to histogram equalisation [20] to show its supremacy over traditional contrast enhancement techniques. Lastly, to show the effects of SA algorithm, the proposed algorithm with SA turned off is also tested. In Table 2, the results of the fitness values of the enhanced images by using the above-mentioned algorithms are stated. In Fig. 3, the images enhanced by using the proposed algorithm are shown.

Table 2 Fitness values of images

Image name	Original	Histogram equalisation	Hoseini and Shayesteh [13] (ACO + GA + SA)	EAS + EIGA	Proposed algorithm (EAS + EIGA + SA)
Cameraman	24.96	27.69	29.77	30.03	**30.95**
Tire	25.89	28.89	31.63	32.13	**33.15**
Pout	13.40	27.72	28.89	29.73	**30.43**
Beach	21.00	20.75	23.69	24.43	**24.68**
Factory	26.76	33.13	34.17	34.65	**35.71**
House	19.17	24.11	24.57	24.35	**25.14**
Chang	8.74	25.87	26.78	27.83	**28.34**
Crowd	15.98	27.99	28.94	29.57	**30.62**
X-ray	11.90	16.34	24.08	24.58	**24.86**

(a) Cameraman   (b) Tire   (c) Pout

(d) Factory   (e) House   (f) Chang

Original   Enhanced
(g) Crowd

Fig. 3 Contrast-enhanced images using the proposed algorithm

By comparing the fitness values of the images produced by Hoseini and Shayesteh [13] with the fitness value of the images produced by the proposed algorithm, it is seen that the proposed algorithm outperforms [13] in enhancing the contrast of the images. In all of the test images, the proposed algorithm provides higher fitness as compared to [13], especially in the case of images *Pout* and *Tire* where there is a significant improvement in the fitness value. The proposed method achieves these better results in almost similar time as the previous method, i.e. [13]. In Fig. 4, the output image produced by Hoseini and Shayesteh [13] is compared with the image produced by the proposed algorithm. It can be seen that the output image produced by the proposed algorithm has better contrast as compared to the image produced by Hoseini and Shayesteh [13].

To show the effects of Simulated Annealing, it is turned off and is then compared to the proposed algorithm. From Table 2, by comparing the values of EAS + EIGA and the proposed algorithm, the effect of SA can be seen at refining the global solutions produced by EAS and EIGA. Moreover, EAS + EIGA tends to perform slightly better than [13]. Though the difference in fitness value is not much, turning SA optimisation off greatly increases the computation speed of the algorithm as verified in [13]. Therefore, EAS + EIGA provides slightly better results than [13] in considerably less time and is suited for applications where computational time is of great importance.

The proposed algorithm is further compared to histogram equalisation in Fig. 4 from where it is seen performing significantly better in enhancing the contrast of the

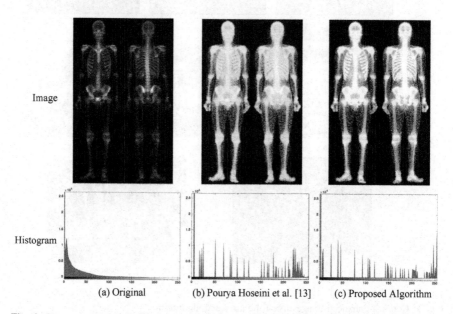

Image

Histogram

(a) Original   (b) Pourya Hoseini et al. [13]   (c) Proposed Algorithm

Fig. 4 Comparison of performance of proposed algorithm versus [13] on image *X-ray*

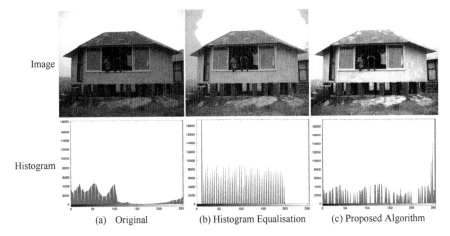

(a) Original   (b) Histogram Equalisation   (c) Proposed Algorithm

Fig. 5 Comparison of performance of proposed algorithm versus histogram equalisation on the image *Beach*

images. Though histogram equalisation takes much lesser time in generating the output images as compared to any metaheuristic-based contrast enhancement algorithm including the proposed algorithm [13], it is unable to enhance the contrast of the image adequately. Moreover, the image sometimes lacks the natural look. In Fig. 5, the images produced by histogram equalisation are compared to the image produced by the proposed method. It is evident that the image produced by the proposed algorithm has higher contrast as compared to histogram equalisation. Further, the image produced by histogram equalisation has been distorted (In Fig. 5b, sky above the hut) and lacks the natural look.

4   Conclusion and Future Work

In this work, the authors have proposed an improved contrast enhancement algorithm. The proposed algorithm uses a hybrid of Elitist Ant System (EAS), Elitism-based Genetic Algorithm (EIGA) and Simulated Annealing (SA) metaheuristics. The experimental results have shown that the proposed algorithm performs better at enhancing the contrast of the images. The algorithm is also found to preserve the natural look of the images while enhancing the contrast. However, the better contrast of an image is susceptible to human perception. Hence, the requirement of better and advanced contrast enhancement technique has always been felt. The proposed algorithm can be improved by using a hybrid algorithm with newer and advanced nature-based metaheuristics such as grey wolf optimisation (GWO) [21] and fireworks optimisation [22], which generally outperform global optimisation algorithms such as ACO and GA. Moreover, more advance parameters such as feature similarity index (FSIM) [23] and edge-based contrast measurement (EBCM) [24] can be used to further evaluate and compare the proposed contrast enhancement algorithm.

References

1. Shefali Gupta, Yadwinder Kaur: Review of Different Local and Global Contrast Enhancement Techniques for Digital Image. International Journal of Computer Applications, Vol. 100, No.18 (August 2014).
2. Md. Hasanul Kabir, M. Abdullah-Al-Wadud, Oksam Chae: Global and Local Transformation Function Mixture for Image Contrast Enhancement. In: Proceedings of Digest of Technical Papers International conference on Consumer Electronics 2009, Las Vegas, NV, 2009, pp. 1–2.
3. M. Dorigo and L. Gambardella: Ant colony system: A cooperative learning approach to the traveling salesman problem. IEEE Transactions on Evolutionary Computation, Vol. 1 (1997), pp. 53–66.
4. Melanie M: An introduction to genetic algorithms. First MIT Press edition, 1998, Cambridge.
5. S. Kirkpatrick, C. D. Gelatt Jr., M. P. Vecchi: Optimization by Simulated Annealing. Science, Vol. 220 (13 May 1983) pp. 671–680.
6. D. Karaboga: An idea based on honey bee swarm for numerical optimization. Technical Report-TR06, Erciyes University, Computer Engineering Department, 2005.
7. Kanika Gupta, Akshu Gupta: Image Enhancement using Ant Colony Optimization. IOSR Journal of VLSI and Signal Processing, Vol. 1 Issue 3 (Nov–Dec 2012) pp. 38–45.
8. Davinder Kumar, Satnam Singh, Vikas Saini: Ant Colony Optimization based Medical Image Enhancement. International Journal of Advanced Research in Computer Science and Software Engineering, Vol. 6 Issue 7 (July 2016) pp. 425–433.
9. F. Saitoh: Image contrast enhancement using genetic algorithm. In: Proceedings of 1999 IEEE International Conference on Systems, Man, Cybernetics, Tokyo, Vol. 4 (1999) pp. 899–904.
10. C. Munteanu and A. Rosa: Gray-scale image enhancement as an automatic process driven by evolution. Proceedings of IEEE Transactions on Systems, Man, and Cybernetics, Part B (Cybernetics), Vol. 34, no. 2 (April 2004) pp. 1292–1298.
11. Xin-She Yang: Nature Inspired Metaheuristic Algorithms, Second Edition. Luniver Press, University of Cambridge, United Kingdom, 2010.
12. Biao Pan: Application of Ant Colony Mixed Algorithm in Image Enhancement. Computer Modelling and New Technologies, Vol. 18 Issue 12B (2014) pp. 529–534.
13. Pourya Hoseini, Mohrokh G. Shayesteh: Efficient contrast enhancement of images using hybrid ant colony optimisation, genetic algorithm and simulated annealing. Digital Signal Processing, Vol. 23 (2013) pp. 879–893.
14. T. White, S. Kaegi, T. Oda: Revisiting Elitism in Ant Colony Optimization. In: proceedings of Genetic and Evolutionary Computation Conference, Chicago, USA, (2003) pp. 122–133.
15. K.G. Srinivasa, Venugopal K R, Lalit M Patnaik: A self-adaptive migration model genetic algorithm for data mining, Information Science, Vol. 177 Issue 20 (2005) pp. 4295–4313.
16. Deepti Gupta, Shabina Ghafir: An Overview of methods maintaining Diversity in Genetic Algorithms. International Journal of Emerging Technology and Advanced Engineering, Vol. 2 Issue 5 (May 2012) pp. 56–60.
17. W.Y. Lin, W.Y. Lee and T.P. Hong: Adapting Crossover and Mutation Rates in Genetic Algorithms. Journal of Information Science and Engineering, Vol. 19 (2003) pp. 889–903.
18. H. Cheng, S. Yang: Genetic Algorithms with Immigrants Schemes for Dynamic Multicast Problems in Mobile Ad Hoc Networks. Engineering Applications to A.I. (2009) pp. 1–35.

19. J. Grefenstette: Genetic algorithms for changing environments. In: Proceedings of the Second International Conference on Parallel Problem Solving from Nature (1992) pp. 137–144.
20. R. C. Gonzalez and R. E. Woods: Digital Image Processing, Third Edition, 2008.
21. S. Mirjalili, S. M. Mirjalili and A. Lewis: Grey wolf optimizer. Advances in Engineering Software, Vol. 69 (2014) pp. 46–61.
22. Tan and Y. Zhu: Fireworks algorithm for optimization. Advances in Swarm Intelligence: Lecture Notes in Computer Science, Vol. 6145 (2014) pp. 355–364.
23. L. Zhang, L. Zhang, X. Mou and D. Zhang: FSIM: A Feature Similarity Index for Image Quality Assessment. IEEE Transactions on Image Processing, Vol. 20 (2011) pp. 2378–2386.
24. T. Celik, T. Tjahjadi: Automatic Image Equalization and Contrast Enhancement Using Gaussian Mixture Modeling. IEEE Transactions on Image Processing, Vol. 21 (2012) pp. 145–156.

A Novel Robust Reversible Watermarking Technique Based on Prediction Error Expansion for Medical Images

Vishakha Kelkar, Jinal H. Mehta and Kushal Tuckley

Abstract Degradation of the host image by noise due to errors during data transmission is a major concern in telemedicine, especially with respect to reversible watermarking. This paper presents the effect of salt and pepper noise on prototypical prediction error expansion-based reversible watermarking and proposed prediction error expansion scheme using border embedding for gray scale medical images. In prototypical prediction error expansion, the accretion of the predicted error values is used for data insertion while in the proposed scheme, prediction error expansion using border embedding is used and aftermath of noise is demonstrated, respectively. A performance assessment based on peak signal-to-noise ratio (PSNR), total payload capacity, noise effect is conducted. Additional capacity and less mutilation of the host image in contrast to the pristine method in the presence of noise is obtained through the results.

Keywords Telemedicine · Reversible watermarking · Salt and pepper noise
Prediction error expansion · Peak signal-to-noise ratio · Capacity

V. Kelkar (✉)
UMIT, SNDT University, Mumbai, India
e-mail: kelkar.vishakha@gmail.com

J. H. Mehta
Department of Electronics and Telecommunication,
D.J. Sanghvi College of Engineering, Mumbai, India
e-mail: jinalmehta94@gmail.com

K. Tuckley
AGV Systems Pvt. Ltd. India, Mumbai, Maharashtra, India
e-mail: kushal.tuckley@agv-systems.in

© Springer Nature Singapore Pte Ltd. 2018
B. B. Chaudhuri et al. (eds.), *Proceedings of 2nd International Conference on Computer Vision & Image Processing*, Advances in Intelligent Systems and Computing 703, https://doi.org/10.1007/978-981-10-7895-8_11

1 Introduction

Recent developments in information and communication technologies have made digital information transfer easier and encouraged the quality of healthcare services. With the help of telemedicine, it is possible to exchange electronic health records and medical images between different healthcare entities. Potency of the tele-medicine application depends on the effectiveness of the data transmission channel and its auxiliary processing's while transmission over public networks [1]. For secure data transmission, digital watermarking is currently a promising technology in the digital intellectual property and information security domain [2]. Digital watermarking is the process of sheltering secret information in the primary image for content protection and verification. Based on its behavior, watermarking can be divided into two types, namely reversible and irreversible [3]. In reversible watermarking, the host image after watermark extraction is similar to the original elementary image [4]. Over the years, a lot of importance is been given to pre-diction error expansion (PEE)-based reversible watermarking techniques. In clas-sical PEE, correlation among the neighboring pixels is used to obtain prediction error which is realized with the help of median edge detector (MED). The selected pixels are customized in order to augment two times the prediction error. Multi-plication by two creates a void in the least significant bit (LSB) which is used for embedding one bit of data. The pixels to be infused are selected considering the overflow/underflow problem and by defining an optimum threshold. Prediction error expansion technique reaps embedding rates up to 1 bit per pixel (bpp) without multiple embedding [5].

Mutilation of the host image due to errors during data transmission is an important issue, especially with respect to reversible watermarking, as most of the existing reversible watermarking schemes are fragile in nature [2]. A watermarked image may get tampered deliberately or accidently during transmission. The implemented watermarking mechanism should be robust enough to detect and extract the watermark as well as the original primary image. Different types of attacks may mutilate the transmitted image [6]. Salt and pepper noise is one such attack where the intensity values of certain pixels are altered resulting in the dis-tortion of the original transmitted image [7].

This paper demonstrates how the distortion in the host image and the watermark can be scaled down and robustness against salt and pepper noise can be improved with the help of border embedding for gray scale medical images. Healthcare and medical images like X-ray, magnetic resonance imaging (MRI), computerized tomography (CT) scan are taken into account.

The paper is assorted as follows: The vintage PEE scheme is illustrated in Sect. 2. The proposed PEE technique using border embedding for gray scale medical images is explained in Sect. 3. Comparison of the results obtained by executing the two methods for 8-bit gray scale medical images in the presence of predefined noise density is discussed in Sect. 4, and Sect. 5 is the conclusion.

Fig. 1 Pixel plot

2 Prediction Error Expansion [5, 8]

In PEE, the correlation among the neighboring pixels is exploited for predicting the intensity of the pixel under consideration. Consider a pixel with intensity p in a gray scale image. The pictorial representation of pixel under consideration and the neighboring pixels u, v, w are as shown in Fig. 1. The predicted value $\hat{p}$ of a pixel p is acquired with the help of Median Edge Detector as follows:

$$\hat{p} = \begin{cases} \max(u, v) & \text{if } w \leq \min(u, v) \\ \min(u, v) & \text{if } w \geq \max(u, v) \\ u + v - w & \text{otherwise} \end{cases}$$

In case of border pixels, we can infer that the '$u + v - w$' condition is applicable. The prediction error $e_r = p - \hat{p}$ is augmented and message bit m is inserted in the primary image. Thus, the watermarked prediction error is given as,

$$e'_r = 2e_r + m. \tag{1}$$

Pixel intensity values of a gray scale image lie between [0, 255]. Embedding of the message bits can cause overflow or underflow; hence, only expandable locations are selected for insertion. A location of the primary image that satisfies (2) is an expandable location. The respective pixel intensities of the modified image should lie between 0 and 255. Such that, $0 \leq p' \leq 255$.

$$p + e_r \begin{cases} \leq 254 & \text{if } e_r \geq 0 \\ \geq 0 & \text{if } e_r < 0 \end{cases} \tag{2}$$

A portion of the expandable locations is chosen for embedding by employing a threshold value T'. Initially, the expandable locations are embedded by the payload. After the completion of payload insertion, the remaining locations are fed with original LSBs of the locations that would be employed for LSB replacement in the second stage (location map embedding). This serves as the subsidiary data for extraction of the original primary image at the receiver end. The length of subsidiary data is equal to that of the compressed location map. The threshold T' is selected depending on the size of the payload and the subsidiary data. The location map of the selected expandable locations is compressed and lodged into the primary image by LSB replacement. The final image at the transmitter end is given as,

$$p' = \hat{p} + e'_r = p + e_r + m. \tag{3}$$

For analyzing the effect of salt and pepper noise, noise of a particular density 'd' is added to watermarked image and then the image is sent for further processing.

At the receiver end, foremostly, the location map is redeemed. The secret data can be extracted and the original image can be revived by using the same predictor as the one deployed at the transmitter end. The received image is processed in the reverse raster scan order. Initially, the subsidiary data is redeemed and reimbursed into received image, later the payload is extracted. The prediction error at the receiver end is calculated for all the expandable locations. It is given as,

$$e'_r = p' - \hat{p} \tag{4}$$

The LSB of e'_r is the embedded message bit. The required pixel intensity can be computed with the help of original prediction error e_r as,

$$e_r = \frac{e'_r}{2} \tag{5}$$

Thus, the primary pixel intensities of the basic image can be retrieved as,

$$p = p' - e_r - m. \tag{6}$$

3   Prediction Error Expansion Using Border Embedding

3.1   Encoder First Stage

In this technique, application of basic PEE scheme in the first stage is followed by the location map embedding at the border. The first stage of embedding is analogous to that of basic PEE scheme. Consider a pixel of intensity a, having estimated intensity $\hat{a}$. Let m_a be the private message to be infused. A median edge detector is employed where the predicted value of the pixel under consideration is deduced. The difference between the original pixel intensity and the predicted intensity is augmented and is used for the insertion of message bits given as, $e_{ra} = a - \hat{a}$. The watermarked prediction error is given as,

$$e'_{ra} = 2e_{ra} + m_a \tag{7}$$

A location map is developed by taking overflow and underflow into account and by applying suitable threshold T'_a, depending on the size of the payload. The message bits m_a are inserted into the primary image; thus, the reworked pixel intensity is given as,

$$a' = \hat{a} + e'_{ra} = a + e_{ra} + m_a \qquad (8)$$

3.2 Encoder Second Stage

The location map computed in the first stage is compressed using lossless compression technique and embedded along the border of the image in concentric fashion starting from the edge toward the center in 'U'-shaped fashion as shown in Fig. 2. It is assumed that the payload to be inserted occupies one-fourth part of a specified image approximately. As the embedded pixel values at the border do not carry significant information with respect to medical images, their initial values can be neglected. Hence, it is not necessary to preserve the LSBs of the primary image, as like in the primitive PEE scheme, hence vanishing the need of subsidiary data. The location map is ingrained with the help of basic LSB replacement scheme.

3.3 Salt and Pepper Noise Attack

For understanding the effect of salt and pepper noise, noise of a particular density 'd,' having value equal to the one employed in classical PEE scheme is added to the final watermarked image at the encoder section.

3.4 Decoder First Stage

At the receiving end, initially, the compressed location map is extracted from the concentric 'U'-shaped border and decompressed to obtain the actual location map consisting the set of expandable locations.

Fig. 2 Border embedding

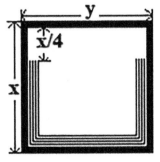

3.5 Decoder Second Stage

The prime aim at this instance is simply the payload extraction. This can be implemented with the help of classical PEE technique where the predicted values are calculated for the locations that were embedded with the help of median edge detector and decompressed location map. The prediction error at the receiver end is given as,

$$e'_{ra} = a' - \hat{a} \tag{9}$$

The LSB of e'_{ra} is the required message bit. The original pixel intensities of the host image can be retained as,

$$a = a' - e_{ra} - m_a \tag{10}$$

4 Results and Discussion

A performance assessment of the primitive PEE technique and proposed PEE using border embedding for gray scale medical images in the presence of noise is conducted. The parameters employed for judging the imperceptibility and visual quality are mean square error (MSE), PSNR [9, 10]. Another parameter determining the ability of the host image to embed larger payload by a particular technique is the total capacity or the maximum number of locations that could be embedded [8].

The MSE between a primary image I and watermarked image I_w having M rows and N columns can be calculated as,

$$MSE = \frac{1}{M * N} \sum_{M, N} [I(M, N) - I_w(M, N)]^2 \tag{11}$$

Mean square error determines the average of the squares of the errors. MSE for gray scale image can be used to calculate PSNR as follows,

$$PSNR = 10 \log_{10} \left[\frac{255^2}{MSE} \right] \tag{12}$$

Higher value of PSNR of the acquired image proves that the error between the images being compared is minimum. High PSNR also implies higher transparency and better visual quality of the image under consideration [8].

One other important parameter that determines the robustness of the technique against an attack is the effectiveness of the technique in the presence of noise. It is made sure that the distortions are limited to those factors which do not lead to excess degradation of the original image, and the integrity of the watermarking algorithm is conserved [11, 12]. Since most of the existing reversible watermarking techniques are feeble in nature, they can sustain only a small amount of noise [2]. In this paper, salt and pepper noise is taken into consideration. In this type of noise, black-and-white dots appear in the image which corresponds to the corrupted pixels which have respective intensities set to maximum value or has single bits flipped over. It occurs due to the errors during data transmission. Unaffected pixel values remain unaltered [13]. Generally, adulteration of the image due to salt and pepper noise is to a small extent [7].

The comparison between the two discussed techniques in the presence of salt and pepper noise of density d = 0.01 is carried out with the help of MATLAB. The techniques explained above were executed and tested for a set of 8 bit gray scale medical images. The payload is a binary test authorization image of the patient's personal details. A preconfigured threshold is selected and deployed for both the techniques. The compression technique used for both the methods is arithmetic coding. Detailed discussion regarding arithmetic coding can be obtained from [14]. The cover image considered for data hiding in this observation is Fig. 3 which is a 255 × 197, Kidney CT scan image, Fig. 4 is a binary payload of dimensions 132 × 59 and size 2.30 Kilo bytes (KB). Figure 5 is the watermarked image by basic PEE (Method 1, M1) while Fig. 6 is the watermarked image by PEE using border embedding (Method 2, M2). Figure 7 is the watermarked image by basic PEE (Method 1) while Fig. 8 is the watermarked image by PEE using border embedding (Method 2), both considering the effect of salt and pepper noise,

Fig. 3 Host image to be watermarked

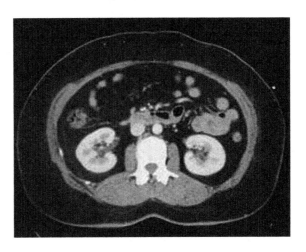

Fig. 4 Payload to be embedded

ABCD HEALTHCARE
NAME :XYZ
AGE:30 YEARS
DOCTOR: MR. PQR

Fig. 5 Watermarked image by M1 (no noise)

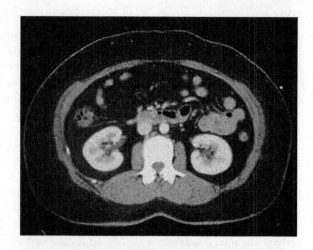

Fig. 6 Watermarked image by M2 (no noise)

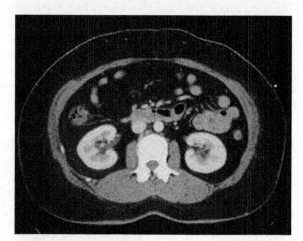

Fig. 7 Watermarked image by M1 + noise

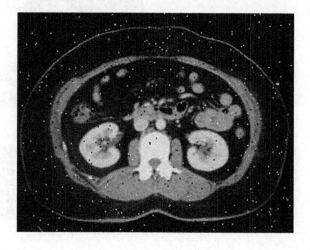

Fig. 8 Watermarked image by M2 + noise

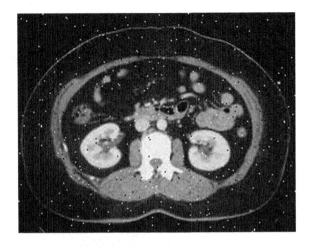

Fig. 9 Retrieved image by M1

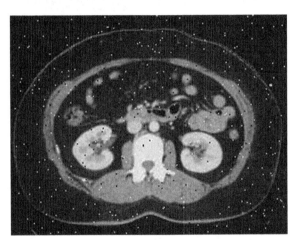

Fig. 10 Retrieved image by M2

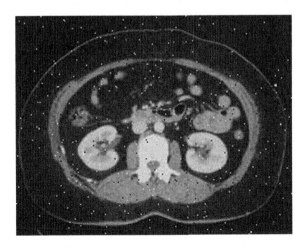

Fig. 11 Retrieved watermark
by M1

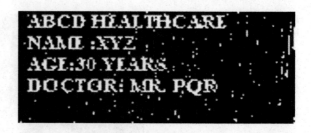

Fig. 12 Retrieved watermark
by M2

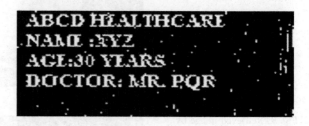

d = 0.01. The retrieved image obtained by method 1 and method 2 are Fig. 9 and
Fig. 10, respectively, while the retrieved payload for method 1 and method 2 is
Fig. 11 and Fig. 12, respectively. Table 1 shows the comparative results of method
1 and method 2 for Fig. 3 in terms of PSNR and total capacity.

It is observed that method 2 implementing PEE using border embedding has
better PSNR of the watermarked noisy image as well as that of retrieved noisy
payload and has an additional capacity in contrast to the primitive PEE method.
Also, the PSNR of the restored image by method 2 is greater than that of method 1
considerably. The comparative results of the image metrics for a set of 5 similar 8
bit gray scale medical images are listed in Table 2.

From Table 2, it is apparent that proposed PEE using border embedding outperforms the classical PEE scheme in terms of PSNR of the retrieved host image as
well as that of retrieved payload in the presence of salt and pepper noise (d = 0.01).
The elementary reason for this is that the probability of occurrence of the random
salt and pepper noise is higher in the upper section of an image under consideration,
in comparison with the mentioned border embedding in method 2. In addition to
this, the unsatisfactory performance of method 1 in the presence of noise is due to

Table 1 Parameter comparison of method 1 and method 2

Parameter	Method 1	Method 2
Total capacity (bits)	45655	49734
PSNR of watermarked image without noise	36.5891	39.4012
PSNR of watermarked image with noise	24.3557	24.3786
PSNR of retrieved image with noise	18.9851	23.9787
PSNR of the retrieved payload with noise	16.5035	18.0151

Table 2 Parameter comparison of method 1 and method 2

Metric image	Method 1 (Basic PEE)				Method 2 (PEE using border embedding)			
	Total capacity (bits)	PSNR of marked image	PSNR of retrieved image	PSNR of the retrieved payload	Total capacity (bits)	PSNR of marked image	PSNR of retrieved image	PSNR of the retrieved payload
1	29806	23.4714	18.8718	14.6158	41063	23.9560	20.1123	18.7997
2	29070	23.3070	18.8554	13.5374	47010	24.3861	22.0994	18.6630
3	24655	23.0164	18.4063	13.7171	36009	23.4763	20.9181	18.7022
4	33921	23.7218	19.4269	14.5498	48157	23.9726	21.8565	18.6575
5	32971	23.6658	20.7593	17.8116	47768	23.8938	22.3436	21.0852

the destruction of the pixels values of the compressed location map residing in the upper section of the image causing deterioration in the performance of the watermarking algorithm. The probability of this distortion is less along the proposed border. Also, the replaced original LSBs (in the decoded image) of the host image which forms the auxiliary data consist of pixels that were already corrupted by noise before reaching the decoder section. All this makes method 1 more sensitive to noise in comparison with proposed method 2 and leads to its strategy degradation.

The PSNR of the watermarked image is greater for method 2 in comparison with method 1. This is because the need to save the original LSBs of the host image that forms the overhead subordinate data having length equal to that of compressed location map is averted. This is true for the basic border embedding method as well as in the presence of noise. Also, due to this, an improvement in capacity is obtained.

It can also be concluded that the effectiveness of PEE using border embedding depends on the performance of the coding technique used for location map compression. Smaller the length of the compressed location map, less space it occupies, lesser is the effect of noise, greater is the PSNR at the encoder as well as that at decoder end.

With the execution and testing of the above methods, it is understood that the decoder deployed in PEE using border embedding can be realized with much ease in comparison with basic PEE scheme. The requirement for selection of the optimized perfect threshold to make room for the payload as well as the subsidiary data can be averted, since only the payload embedding is done in the expandable locations.

Thus, the only foreseen drawback of this method at this point is that if the size of the compressed location map is large, there is a possibility of its embedding area interfering with the region of interest which may lead to errors.

5   Conclusion

This paper presents a comparatively robust reversible watermarking technique in contrast to the basic PEE method. It is executed and tested for small amount of salt and pepper noise. For fixed payload size, proposed technique endures acceptable results in comparison with the basic PEE technique. It can withstand salt and pepper noise up till 5% (d = 0.05) corrupted image pixels. The host image as well as the infused payload can be obtained at the receiving end with some distortion, but still having a better PSNR and improved capacity in comparison with the original method.

References

1. Al-Haj, Ali. "Secured telemedicine using region-based watermarking with tamper localization." *Journal of digital imaging* 27.6 (2014): 737–750.
2. Khan, Asifullah, Ayesha Siddiqa, Summuyya Munib, and Sana Ambreen Malik. "A recent survey of reversible watermarking techniques." Information sciences 279 (2014): 251–272.
3. Coatrieux G, Maitre H, Sankur B: Strict integrity control of biomedical images. In: Proc. SPIE Security Watermarking Multimedia Contents III, SPIE 2001, vol. 4314, San Jose, CA January 2001, pp 229–240.
4. Nedelcu, Tudor, and Dinu Coltuc. "Alternate embedding method for difference expansion reversible watermarking." *Signals, Circuits and Systems (ISSCS), 2015 International Symposium on*. IEEE, 2015.
5. Thodi, Diljith M, and Jeffrey J. Rodriguez. "Prediction-error based reversible watermarking." *Image Processing, 2004. ICIP'04. 2004 International Conference on Image Processing*. Vol. 3. IEEE, 2004.
6. Pal, Koushik, Goutam Ghosh, and Mahua Bhattacharya. "Reversible digital image watermarking scheme using bit replacement and majority algorithm technique." *Journal of Intelligent Learning Systems and Applications* 4.03 (2012): 199.
7. Verma, Rohit, and Jahid Ali. "A comparative study of various types of image noise and efficient noise removal techniques." *International Journal of advanced research in computer science and software engineering* 3.10 (2013): 617–622.
8. Jinal H. Mehta, Vishakha Kelkar. "Comparison of Reversible watermarking using Prediction error expansion and Prediction error expansion considering Region of interest for Medical images." *2017 2nd International Conference for Convergence in Technology (I2CT)*, 2017.
9. Fallahpour, Mehdi, David Megias, and Mohammed Ghanbari. "High capacity, reversible data hiding in medical images." *Image Processing (ICIP), 2009 16th IEEE International Conference on*. IEEE, 2009.
10. Coatrieux, Gouenou, et al. "A review of image watermarking applications in healthcare." *Engineering in Medicine and Biology Society, 2006. EMBS'06. 28th Annual International Conference of the IEEE*. IEEE, 2006.
11. Shi, Yun Q. "Reversible data hiding." *International workshop on digital watermarking*. Springer Berlin Heidelberg, 2004.
12. Lin, Eugene T, and Edward J. Delp. "A review of fragile image watermarks." *Proceedings of the Multimedia and Security Workshop (ACM Multimedia'99) Multimedia Contents*. 1999.
13. Serener, A., and C. E. M. A. L. Kavalcioglu. "Teledermatology based medical images with AWGN Channel in Wireless Telemedicine System." *Proceedings of the 1st WSEAS International Conference on Manufacturing Engineering, Quality and Production Systems*.
14. Osorio, Roberto R., and Javier D. Bruguera. "Architectures for arithmetic coding in image compression." *Signal Processing Conference, 2000 10th European*. IEEE, 2000.

Integrated Feature Exploration for Handwritten Devanagari Numeral Recognition

Shraddha Arya, Indu Chhabra and G. S. Lehal

Abstract In this paper, the statistical feature extraction techniques are explored, incrementally combined using different methods and analyzed for the recognition of isolated offline handwritten Devanagari numerals. The techniques selected are zoning, directional distance distribution, Zernike moments, discrete cosine transform, and Gabor filter that encapsulate the mutually exclusive statistical features like average pixel densities, directional distribution, orthogonal invariant moments, elementary frequency components, and space frequency component, respectively. The standard benchmark handwritten Devanagari numeral database provided by ISI, Kolkata, is used for the experimentation and 1-nearest neighbor and support vector machine for classification. The accuracy achieved with individual feature extraction techniques ranges from 86.87% to 98.96%. Further, features are integrated with methods like feature concatenation, majority voting, and a new proposed methodology by us named winners pooling. The maximum recognition obtained through feature integration is 99.14%.

Keywords Handwritten Devanagari recognition · Zoning · Directional distance distribution · Zernike moments · Discrete cosine transform · Gabor filter

S. Arya (✉)
Department of Computer Science, Sri Guru Gobind Singh College, Chandigarh, India
e-mail: shraddhaarya@rediffmail.com

I. Chhabra
Department of Computer Science and Applications, Panjab University, Chandigarh, India
e-mail: chhabra_i@rediffmail.com

G. S. Lehal
Computer Science Department, Punjabi University, Patiala, Punjab, India
e-mail: gslehal@gmail.com

© Springer Nature Singapore Pte Ltd. 2018
B. B. Chaudhuri et al. (eds.), *Proceedings of 2nd International Conference on Computer Vision & Image Processing*, Advances in Intelligent Systems and Computing 703, https://doi.org/10.1007/978-981-10-7895-8_12

1 Introduction

Efficient feature extraction and correct classification are the pillars of any good handwritten character recognition system. A feature extraction technique despite being proficient has a limited scope in terms of overall optimization. This paper experiments with statistical feature extraction techniques and their integration for the recognition of offline handwritten Devanagari numerals. Numerals are the basic entities of any script, and their recognition is a prime concern. Devanagari is the script for many languages like Sanskrit, Hindi, Bhojpuri, Marathi, and Nepali. Hindi is officially the national language of India and the third most popular language of the world. Handwritten character recognition imposes challenge as it permits the infinite number of possible rendering of the character shapes and offline adds to the complexity as there is total lack of the real-time information as available in online recognition. A vivid description on feature extraction methods for character recognition is presented by Trier et al. [18]. The directional distance distribution (DDD)-based feature was introduced by Oh and Suen [11] which was applied to recognize CENPARMI handwritten numeral database through modular neural network classifier. Liu et al. [10] used Gabor features to recognize Roman handwritten digits using MNIST and CENPARMI database reporting 99.47% and 98.95% accuracy, respectively. Handwritten numerals of MNIST database were recognized by Wang et al. [19] applying Gabor filters and MQDF classifier giving 0.87% error rate. Bhattacharya and Chaudhuri [2, 4] developed the handwritten databases for Devanagari and Bangla thus providing the standard benchmark in the native script. Handwritten Devanagari numeral database of the same is used for experimentation and comparison purpose in our work and referenced as ISI database in the text. Following is the work reported using ISI database. Dimensional features with MQD-F classifier were used by Pal et al. [12] attaining 99.56% accuracy. Wavelet-based technique and Chain code histogram using MLP classifier were applied by Bhattacharya and Chaudhuri [2] reporting 99.27% accuracy. Jangid et al. [7] performed recursive subdivision of character image with support vector machine (SVM) giving 98.98% recognition rate. Experiments on feature combination have been performed by the researchers in order to compliment rather than supplement the feature vector potential. Hybrid techniques are likely to enhance performance as multivariate properties of character are captured. Different methods have been proposed in the literature for the feature combination ranging from simple concatenation, weighted combination, combination through feature selection using intelligent techniques like ant colony optimization, bee swarming, genetic algorithm, fuzzy logic, and combination at classifier level as in neural networks. The following work is reported using feature combination and feature selection using ISI database. Prabhanjan and Dinesh [13] used Fourier descriptor and pixel density from zones for feature extraction and Naive Bayes, instance-based learner, random forest and sequential minimum optimization for classification reporting 99.68% recognition. Singh et al. [16] applied zoning and direction feature with 1-nearest neighbor (1-NN) and quadratic Bayes classifiers giving 99.73% accuracy at 6% rejection rate. The same authors also utilized gradient-based feature with MLP classifier reporting 99.37% recognition rate [15] and applied pixel density values and zone-based gradient features with MLP classifier resulting in 98.17% recognition rate [17]. Bhattacharya et al. [3] used shape features with ANN and HMM classifiers reporting 92.83% recognition accuracy.

In this work, the features are integrated using different methods through 1-NN and SVM classifiers.

2 Feature Extraction and Classification

2.1 Feature Extraction

Zoning: In this technique, the character image is segmented into zones of equal size. The density values (Number_of_foreground_pixels/Total_number_of_pixels) are computed for each zone and used as the feature values. The zone size opted for our experimentation is 5×5, and hence the feature size is 25.

Directional Distance Distribution (DDD): It is a distance-based feature proposed by Oh and Suen [11]. Two sets of eight bytes say W (white) set and B (black) set are assigned to each pixel of the binarized numeral image. If the pixel is white, the set W stores the distance of nearest black pixel in each of eight directions $(0°, 45°, 90°, 135°, 180°, 225°, 270°, 315°)$ and the corresponding set B is filled with zero values. Similarly, for a black pixel, the set B stores the distance of nearest white pixel in each of eight directions and the corresponding set W is filled with zero values. The eight direction codes assumed are 0(E), 1(NE), 2(N), 3(NW), 4(W), 5(SW), 6(S), 7(SE). Two variations have been proposed for the case when the desired black or white pixel is not found on reaching the array boundary [11]: either the array is assumed circular or the search continues in the reflected direction. We have used the latter one and the search halts after the ray is reflected by two boundaries. The character image is divided into 5×5 zones. 16 DDD values are computed for each image pixel, and average DDD value is calculated for each zone. Thus, feature size is 400 $(5 \times 5 \times 16)$. The computation of DDD feature is illustrated in Fig. 1.

Zernike Moments: These are a set of complex polynomials which form a complete orthogonal set over a unit disk of $(x^2 + y^2 \leq 1)$ in polar coordinates. The Zernike polynomials are defined by (1) with constraints $(n \geq 0)$ and $n - |m|$ is even, $|m| \leq n$.

$$R_{nm}(\rho) = \sum_{s=0}^{(\frac{n-|m|}{2})} \frac{(-1)^s (n-s)! \rho^{(n-2s)}}{s! (\frac{n+|m|}{2} - s)! (\frac{n-|m|}{2} - s)!}. \tag{1}$$

The orthogonal rotational invariant moments (ORIMs) of order n and repetition m of a continuous signal $f(x, y)$ over a unit disk are defined by A_{nm} [14] as given in (2). The moment basis function $V_{nm}(x, y)$ is given by (3) where ρ is the length of vector from origin to pixel (x, y), i.e., $\rho = \sqrt{x^2 + y^2}$, θ is the angle between vector ρ

Fig. 1 Calculation of DDD feature

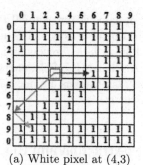

(a) White pixel at (4,3)

w0	w1	w2	w3	w4	w5	w6	w7	b0	b1	b2	b3	b4	b5	b6	b7
3	3	3	3	6	5	2	2	0	0	0	0	0	0	0	0

(b) WB encoding for white pixel at (4,3)

(c) Black pixel at (6,3)

w0	w1	w2	w3	w4	w5	w6	w7	b0	b1	b2	b3	b4	b5	b6	b7
0	0	0	0	0	0	0	0	3	1	1	1	1	4	8	2

(d) WB encoding for black pixel at (6,3)

(e) Directional distances of the pixels (4,3) and (6,3)

Direction	White Pixel [1]	Black Pixel[2]
0^0	3	3
45^0	3	1
90^0	3	1
135^0	3	1
180^0	6	1
225^0	5	4
270^0	2	8
315^0	2	2

[1]Distance of nearest black pixel for white pixel (4,3)
[2]Distance of nearest white pixel for black pixel (6,3)

and x-axis in counter clockwise direction, i.e., $\theta = \tan^{-1}(y/x)$, $i = \sqrt{-1}$ and $R_{m,n}(\rho)$ is the radial polynomial, $V^*_{nm}(x, y)$ is the complex conjugate of $V_{nm}(x, y)$. The basis function $V_{nm}(x, y)$ is orthogonal to each other. The image function $f(x, y)$ can be represented by (4).

$$A_{nm} = \frac{n+1}{\pi} \int \int_{x^2+y^2\leq 1} f(x, y)\, V^*_{nm}(x, y)\, dxdy \,. \tag{2}$$

$$V_{nm}(x, y) = V_{nm}(\rho, \theta) = R_{nm}(\rho)\, e^{im\theta} \,. \tag{3}$$

$$f(x, y) = \sum_n \sum_m A_{nm} V_{nm}(x, y) \,. \tag{4}$$

Discrete Cosine Transform (DCT): This method converts image data into its elementary frequency components. The DCT of a 2D array A[M, N] is a matrix say B[M, N] of same size as A whose each element f(u, v) is calculated by the given formula for corresponding element f(m, n) in A where M and N are the width and height of the image, respectively [8]. Its coefficients f(u, v) for f(m, n) are computed by (5).

$$f(u, v) = a(u)\, a(v) \sum_{m=0}^{M-1}\sum_{n=0}^{N-1} f(m, n) \cos\left[\frac{(2m+1)\pi u}{2M}\right] \cos\left[\frac{(2n+1)\pi v}{2N}\right] \,. \tag{5}$$

$$\text{where}\quad a(u) = \left\{ \begin{array}{l} \frac{1}{\sqrt{M}}, \ u = 0 \\ \sqrt{\frac{2}{M}}, \ 1 \leq u \leq M-1 \end{array} \right\}, \ a(v) = \left\{ \begin{array}{l} \frac{1}{\sqrt{N}}, \ v = 0 \\ \sqrt{\frac{2}{N}}, \ 1 \leq v \leq N-1 \end{array} \right\}.$$

The resultant matrix B[M, N] is the matrix with DCT coefficients f(u, v) corresponding to each pixel f(m, n), and the feature size is M × N. DCT cluster high-value coefficients in the upper left corner and low-value coefficients in the lower right corner of matrix B[M, N]. These high-value coefficients are obtained in zigzag pattern from top left corner of matrix and used as features.

Gabor Filter: A Gabor filter is two-dimensional linear filter defined as a complex sinusoidal plane modulated by a Gaussian envelope of an elliptical shape which can be tuned to a specific orientation and spatial frequency. Its computation in spatial domain is given by (6).

$$g\left(x, y\, ; f_0\, , \theta\right) = \frac{1}{2\pi\sigma_x\sigma_y} \exp^{-\left[\frac{x'^2}{2\sigma_x^2}+\frac{y'^2}{2\sigma_y^2}\right]} \cdot \exp^{i2\pi\left(u_0 x + v_0 y\right)} \,. \tag{6}$$

$$x' = x\cos\theta + y\sin\theta, \ y' = -x\sin\theta + y\cos\theta, \ u_0 = f_0\cos\theta, \ v_0 = f_0\sin\theta$$

where (x, y): spatial co-ordinates, f_0: center frequency, θ: sinusoidal plane wave orientation, (x', y'): rotation co-ordinates, (σ_x, σ_y): spread of the elliptical Gaussian envelope along x-axis and y-axis, respectively, (u_0, v_0): center spatial frequencies of the sinusoidal wave in Cartesian coordinates, (f_0, θ): the corresponding counterpart frequency magnitude $(f_0 = \sqrt{u_0^2 + v_0^2})$ and direction $\left(\theta = \tan^{-1}\left(u_0/v_0\right)\right)$ in polar coordinates [1].

2.2 Classification

The 1-NN and SVM with linear and radial basis function (RBF) kernel are used for the classification purpose. The minimum Euclidean distance metric is utilized as measure for classification through 1-NN. SVM is a discriminative classifier defined by a separating hyperplane. Given labeled training data, the algorithm returns an optimal hyperplane that classifies the unknown data. We have used LIB-SVM 3.21 classifier tool [5] with linear and RBF kernel. The cost and gamma parameters required for RBF kernel computation are computed using grid.py tool using fivefold cross validation [6].

3 Feature Integration

The features are combined selectively after feature extraction phase either before or after classification as per the chronological order of their individual recognition performance scores. The testing is done by combining features using either 1-NN or SVM classifiers. Three methods are designed for the experimentation.

3.1 Combining Features Using 1-NN Classifier

The features extracted through different feature extraction techniques have incompatible range and values. Hence, for justified and unbiased combination through 1-NN, the Euclidean distance metric obtained for each of the feature extraction technique is normalized to range (0, 1) using formula (7) where M': normalized distance metric, M: distance metric, (max, min): maximum and minimum feature value for combined training and testing data. Although features can be normalized before calculating distance metric, it is computationally expensive to first normalize each feature before metric computation as compared to normalizing distance metric directly. The sum total of the normalized metric of the selected features as per their individual recognition rate order of 1-NN performance is used for the classification.

$$M' = \frac{M}{(max - min)}.$$

(7)

3.2 Combining Features Before Classification Through SVM

Features extracted using different feature extraction techniques are first concatenated irrespective of their range of values and then scaled using the svmscale application in LIBSVM in the range $(-1, 1)$. The svmscale does scaling longitudinally with respect to values within a feature column and irrespective of the feature values in the same feature group or the adjacent feature group columns. This characteristic of svmscale permits us to concatenate feature vectors without the need to first scale them. The SVM classification is then done for the concatenated scaled feature vector using linear and RBF kernel.

3.3 Combining Features Using SVM Classifier Ensemble

In the present study, the SVM classifier with RBF kernel gives the optimum recognition performance in comparison to the 1-NN and SVM with linear kernel. Hence, it is used for feature combination through the classifier ensemble. The optimal classifier is selected to maximize the ensemble accuracy by minimizing the entry of wrong classification results which pool in by classifier ensemble. In this method, first, the features extracted using individual feature extraction techniques are classified using SVM with RBF kernel independently and then the classifiers result is combined using either of the following methods.

Majority Voting The majority voting is simple and effective technique that combines the output of different classifiers [9]. Each classifier assigns a class label to test-data which acts as a vote for the class. A class casted k number of votes is selected through majority voting technique if condition (8) is satisfied.

$$k \geq \frac{(n+1)}{2} \quad \text{if n is odd or} \quad k \geq \left(\frac{n}{2} + 1\right) \quad \text{if n is even .} \tag{8}$$

where k: number of votes and n: number of classifiers. For 'k' less than the given value, it is considered as rejection or neutral case and not counted under misclassification as per the technique [9].

Winners Pooling The principle used is to give more weight or priority to the class labels assigned by the feature giving higher recognition rate. The features are arranged in the decreasing order of their recognition rate as in (10). The algorithm given below explains the method.

Devnagari Numerals	०	१	२	३	४	५	६	७	८	९	Total
Training Data	1843	1891	1891	1882	1876	1889	1869	1869	1887	1886	18783
Testing Data	369	378	378	377	376	378	374	378	377	378	3763
Total	2212	2269	2269	2259	2252	2267	2243	2247	2264	2264	22546

Fig. 2 The size of ISI database

Steps: Algorithm for Winners Pooling

1. Assuming the order of recognition rate as given in (10), let a, b, c and d be the class labels for test-data assigned by DDD, DCT, Zoning and Gabor features using SVM with RBF kernel.
2. int final_class=0;
3. if ((a==b) OR (a==c) OR (a==d)) then final_class = a;

   elseif ((b==c) OR (b==d)) then final_class = b;
     elseif (c==d) then final_class = c;
       else final_class = -1;

   endif
4. if (final_class == -1)

   if (Rejection is permitted) then 'Misclassification' case;
     elseif (Rejection is not permitted) then final_class = a;
       (i.e. assign label given by the best feature)
   endif

   endif

Since in our case the output desired is the class label only, the rejection case is counted as misclassification. These results are tabulated under the head 'rejection cases as misclassification.' If rejection is not permitted, the class label given by the best feature is assigned as final class label to improve the recognition rate and the corresponding results are tabulated under the head 'rejection case replaced by the best feature output.' The classifier ensemble results are given in Table 4.

4 Experimentation

The experiments have been conducted on system with Intel Core 2 Duo CPU T5870 @2GHz, 2GB RAM and Windows 7 Ultimate operating system. The standard benchmark ISI database [2, 4] that consists of 22546 grayscale tiff images of isolated handwritten Devanagari numerals has been utilized for experimentation. The details are given in Fig. 2. For preprocessing, the images are binarized using Otsu method and 32×32 size normalization is done for only Gabor feature computation. No other preprocessing technique is used.

Feature Size The feature size for zoning and DDD is 25 (5×5) and 400 ($5 \times 5 \times 16$), respectively. To select the optimal feature size for Zernike and DCT features, a sample set was made taking first 200 characters of ISI database training data of each class for training purpose and next 40 characters from ISI database training data only for testing purpose. Zernike moments were tested with different orders. The maximum recognition obtained for the same was at order $n = 10$, and the best reconstruction achieved was at order $n = 12$. Similarly, various sizes for DCT were tested, and size 6×6 gave the best results. Hence, for both Zernike (order $n = 10$) and DCT (6×6), the feature size opted is 36.

Gabor Feature Computation The computation details of Gabor features along with the criteria of parameter value selection are elaborated by authors in [1]. Accordingly, the following values are used for the purpose: filter size (31×31), frequency scaling factor $= \sqrt{2}$ (half octave spacing), number of frequencies $= 5$, frequency values $= [0.25, 0.1767, 0.125, 0.8838, 0.0625]$, number of orientations $= 8$ in the range $(0, \pi)$, spatial width in x and y direction $\left(\sigma = \sigma_x = \sigma_y\right), (\sigma = \lambda)$, where λ is the wavelength. Four features given by $\left[G_{even\_Mean}, G_{even\_Standard\_Deviation}, G_{Odd\_Mean}, G_{odd\_Standard\_Deviation}\right]$ are calculated for each filter convolved with 32×32 size normalized image. Hence, the feature vector size per numeral image is given by $5 \times 8 \times 4 = 160$ [1].

5 Result and Discussion

The statistical feature extraction techniques of zoning, DDD, Zernike, DCT, and Gabor features are applied individually and in combination with different experimental settings on standard benchmark ISI database for handwritten Devanagari numeral recognition.

5.1 Performance of Individual Features

The accuracy rate obtained for the individual features using the classification techniques of 1-NN and SVM with linear and RBF kernels is given in Table 1.

Observation: The DDD feature is characterized by compact encoding, rich information of both black and white pixel distribution, and direction distance distribution over the whole area of the pattern [11]. The complete information is simply encoded as integer and represents distance in a given direction. The discriminatory strength of the feature is evident from its recognition performance. The DCT feature has quality of excellent energy compaction for highly correlated images, and it has ability to pack input data into few coefficients without compromising coding efficiency [8]. With these characteristics, DDD and DCT with 98.96% and 98.43% recognition accuracies, respectively, have emerged as the best feature extraction techniques as

Table 1 Performance of individual features

Feature	Acronym	1-NN (%)	SVM			
			Linear (%)	RBF (%)	Gamma (γ)	Cost (C)
Zoning	G	95.53	89.74	97.05	0.5	2
DDD	D	95.48	98.48	**98.96**	0.03	32
Zernike	Z	81.34	71.24	86.87	0.5	8
DCT	C	97.68	93.43	98.43	0.5	32
Gabor	B	74.16	93.54	96.99	0.12	32

Table 2 Recognition rate for combining features using 1-NN classifier

Features	Normalized distance metric	1-NN (%)
GC	$\dfrac{G}{100} + \dfrac{C}{42.844}$	97.76
GDC	$\dfrac{G}{100} + \dfrac{D}{203} + \dfrac{C}{42.844}$	97.84
GDZC	$\dfrac{G}{100} + \dfrac{D}{203} + \dfrac{Z}{0.702} + \dfrac{C}{42.844}$	97.84
GDZCB	$\dfrac{G}{100} + \dfrac{D}{203} + \dfrac{Z}{0.702} + \dfrac{C}{42.844} + \dfrac{B}{0.068}$	96.75

observed from Table 1. The recognition ranking order for 1-NN and SVM classifiers with RBF kernel for individual feature extraction techniques is given in (9) and (10).

$$
\begin{array}{llllllllll}
\text{1-NN :} & \text{DCT} & > & \text{Zoning} & > & \text{DDD} & > & \text{Zernike} & > & \text{Gabor} \\
& 97.68\% & & 95.53\% & & 95.48\% & & 81.34\% & & 74.16\%
\end{array}
\tag{9}
$$

$$
\begin{array}{llllllllll}
\text{SVM :} & \text{DDD} & > & \text{DCT} & > & \text{Zoning} & > & \text{Gabor} & > & \text{Zernike} \\
\text{(RBF Kernel)} & 98.96\% & & 98.43\% & & 97.05\% & & 96.99\% & & 86.87\%
\end{array}
\tag{10}
$$

5.2   Performance of Feature Combination Using 1-NN

For the combination using 1-NN, the Euclidean distance metric is first normalized to range (0, 1) using the formula (7) as explained. The features are combined incrementally as per their individual ranking order with 1-NN classifier as given in (9). The result for combining features incrementally using 1-NN along with the formulation used for normalization is given in Table 2.

Table 3 Recognition rate for combining features before classification using SVM

Features	SVM			
	Linear (%)	RBF (%)	Gamma (γ)	Cost (C)
DC	98.51	99.01	0.031	32
DCB	98.67	**99.14**	0.031	8
GDCB	98.64	**99.14**	0.007	128
GDC	98.56	98.99	0.031	8
GDZCB	98.80	99.06	0.031	8

5.3 Performance of Feature Combination Before Using SVM

The experimentation for combining features before classification through SVM is done by concatenating features incrementally as given in (10) after scaling them in range $(-1, 1)$. The result for combining features before classification using SVM is shown in Table 3.

Observation: The maximum recognition obtained with combination is 98.80% and 99.14% for linear and RBF kernel, respectively. The improvement in recognition rate is seen as multiple disjoint properties are captured in the concatenated feature vector. The above results also suggest that combination of complimentary features has better prospects despite of their slightly low individual recognition rate (like DCB recognition better than DCG) and concatenation of sufficient good features (like DCB) is better than combination of all features (like GDZCB) as observed for SVM classifier with RBF kernel.

5.4 Performance of Feature Combination Using SVM Classifier Ensemble

The features selected for combination through the SVM classifier ensemble are as per their ranking order given in (10). Majority voting and winners pooling methods are used, and the result is shown in Table 4.

Observation: The maximum result obtained through majority voting is 99.12% for DCBG while through winners pooling is 99.09% for DCB. Here also, DCB result is better than DCG that reinforces that complimentary feature combination overpowers the total recognition rate.

Comparison of Results: The earlier reported results on ISI database with different feature extraction techniques are Bhattacharya et al. [2] (99.27%), Pal et al. [12] (99.56%), and Jangid et al. [7] (98.98%). The best recognition in our case under individual features is obtained for DDD (98.96%) and DCT (98.43%) through SVM classifier with RBF kernel. The literature survey for feature combination provides

Table 4 Recognition rate for combining features using SVM classifier ensemble

Features	Majority voting		Winners pooling	
	RM^3 (%)	RB^4 (%)	RM^3 (%)	RB^4 (%)
DCB	98.85	99.09	98.85	**99.09**
DCG	98.61	98.72	98.61	98.72
DCBG	98.06	**99.12**	98.96	99.01
DCBGZ	98.59	98.72	98.99	98.99

[3]Rejection cases considered as misclassification
[4]Rejection cases replaced by the best feature output

the following result: Prabhanjan and Dinesh [13] (99.68%), Singh et al. [16] (99.73% accuracy at 6% rejection rate), [15] (99.37%), [17] (98.17%), and Bhattacharya et al. [3] (92.83%). Our results are comparable with peer researchers with techniques used as feature concatenation (99.14%), majority voting (99.12%), and a new proposed technique winners pooling (99.09%).

6 Conclusion

In this paper, we have experimented with statistical feature extraction techniques and proposed feature integration with different experimental settings for the recognition of offline isolated handwritten Devanagari numerals. The DDD feature has been used with extension in its implementation which is not found in the literature as per our study so far, and its results are encouraging compared to other techniques used. The analysis is carried out, and it is observed that the feature integration using SVM has resulted in overall optimal accuracy as 99.14%.

Acknowledgements The authors would like to thank CVPR unit, ISI, Kolkata for providing us the handwritten database of Devnagari Numerals for experimentation purpose. In this work, the first author is funded under the UGC Fellowship given to college teachers for completion of Ph. D. and express gratitude for the same.

References

1. Arya, S., Chhabra, I., Lehal, G.S.: Recognition of Devnagari numerals using Gabor filter. Indian Journal of Science and Technology 8(27) (2015)
2. Bhattacharya, U., Chaudhuri, B.B.: Handwritten numeral databases of Indian scripts and multistage recognition of mixed numerals. IEEE Transactions on Pattern Analysis and Machine Intelligence 31(3), 444–457 March (2009)
3. Bhattacharya, U., Parui, S., Shaw, B., Bhattacharya, K.: Neural combination of ANN and HMM for handwritten Devanagari numeral recognition. In: Tenth International Workshop on Frontiers in Handwriting Recognition. Suvisoft (2006)

4. Bhattacharya, U., Chaudhuri, B.: Databases for research on recognition of handwritten characters of Indian scripts. In: Eighth International Conference on Document Analysis and Recognition (ICDAR'05). pp. 789–793. IEEE (2005)

5. Chang, C.C., Lin, C.J.: Libsvm: a library for support vector machines. ACM Transactions on Intelligent Systems and Technology (TIST) 2(3), 27 (2011)

6. Hsu, C.W., Chang, C.C., Lin, C.J., et al.: A practical guide to support vector classification (2003)

7. Jangid, M., Dhir, R., Rani, R.: Novel approach: Recognition of Devanagari handwritten numerals. International Journal of Electrical, Electronics and Computer Science (IJEECS) 1(2), 41–46 (2011)

8. Khayam, S.A.: The discrete cosine transform (dct): theory and application. Michigan State University 114 (2003)

9. Lam, L., Suen, S.: Application of majority voting to pattern recognition: an analysis of its behavior and performance. IEEE Transactions on Systems, Man, and Cybernetics-Part A: Systems and Humans 27(5), 553–568 (1997)

10. Liu, C., Koga, M., Fujisawa, H.: Gabor feature extraction for character recognition: Comparison with gradient feature. In: Eighth International Conference on Document Analysis and Recognition (ICDAR 2005), 29 August - 1 September 2005, Seoul, Korea. pp. 121–125 (2005), https://doi.org/10.1109/ICDAR.2005.119

11. Oh, I.S., Suen, C.Y.: A Feature for Character Recognition Based on Directional Distance Distributions. In: ICDAR. pp. 288–292. IEEE Computer Society (1997)

12. Pal, U., Sharma, N., Wakabayashi, T., Kimura, F.: Handwritten numeral recognition of six popular Indian scripts. In: Ninth International Conference on Document Analysis and Recognition (ICDAR 2007). vol. 2, pp. 749–753. IEEE (2007)

13. Prabhanjan, S., Dinesh, R.: Handwritten Devanagari numeral recognition by fusion of classifiers. International Journal of Signal Processing, Image Processing and Pattern Recognition 8(7), 41–50 (2015)

14. Singh, C., Upneja, R.: Error analysis in the computation of orthogonal rotation invariant moments 49(1), 251–271 May (2014)

15. Singh, P., Verma, A., Chaudhari, N.: Feature selection based classifier combination approach for handwritten Devanagari numeral recognition. Sadhana 40(6), 1701–1714 (2015)

16. Singh, P., Verma, A., Chaudhari, N.S.: Reliable Devanagri handwritten numeral recognition using multiple classifier and flexible zoning approach. International Journal of Image, Graphics and Signal Processing 6(9), 61 (2014)

17. Singh, P., Verma, A., Chaudhari, N.S.: On the performance improvement of Devanagari handwritten character recognition. Applied Comp. Int. Soft Computing 2015, 193868:1–193868:12 (2015), https://doi.org/10.1155/2015/193868

18. Trier, Ø.D., Jain, A.K., Taxt, T.: Feature extraction methods for character recognition-a survey. Pattern Recognition 29(4), 641–662 (1996), https://doi.org/10.1016/0031-3203(95)00118-2

19. Wang, X., Ding, X., Liu, C.: Gabor filters-based feature extraction for character recognition. Pattern Recogn. 38(3), 369–379 Mar (2005), https://doi.org/10.1016/j.patcog.2004.08.004

Privacy Preserving for Annular Distribution Density Structure Descriptor in CBIR Using Bit-plane Randomization Encryption

Mukul Majhi and Sushila Maheshkar

Abstract With the rapid increase in multimedia services and Internet users over the network, it is crucial to have effective and accurate retrieval while preserving data confidentiality. We propose a simple and effective content-based image retrieval algorithm using annular distribution density structure descriptor (ADDSD) to retrieve the relevant images using encrypted features to preserve the privacy of image content. It exploits the HSV color space of image to generate quantized image. The structure element is obtained using same or similar edge orientation in uniform HSV color space. The structure element is detected using the grid and based on the quantized structure image so formed. Finally, annular histogram is generated from the quantized structure image which is encrypted by bit-plane randomization technique. Experimental analysis illustrates that the proposed method retrieves the relevant images effectively and efficiently without revealing image content information.

Keywords HSV · Feature extraction · Quantization · Edge orientation
Bit-plane randomization · Annular histogram · CBIR · Encryption

1 Introduction

Recently, for human communication, different forms of multimedia resources are available which include image, video, graphics, and audio. These huge amount of information plays a vital role for human to have better understanding of the world. The advancement of imaging techniques has increased the availability of images to the public. Considerable amount of processing and storage is desired for extremely effective and efficient method to retrieve, evaluate, and index visual information

M. Majhi (✉) · S. Maheshkar
Department of Computer Science and Engineering,
Indian Institute of Technology (Indian School of Mines), Dhanbad, Jharkhand, India
e-mail: mukulmajhi@gmail.com

S. Maheshkar
e-mail: sushila_maheshkar@yahoo.com

© Springer Nature Singapore Pte Ltd. 2018
B. B. Chaudhuri et al. (eds.), *Proceedings of 2nd International Conference on Computer Vision & Image Processing*, Advances in Intelligent Systems and Computing 703, https://doi.org/10.1007/978-981-10-7895-8_13

from the image. Moreover, the increasing complication of multimedia contents witnesses an unparalleled evolution in the availability, amount, diversity, complexity, and importance of images in all domains. This demands an acute need for establishing a highly effective image retrieval system to appease the human needs. Thus, image retrieval has emerged as an important research topic for the research community worldwide, and for effective retrieval machine learning, artificial intelligence, and pattern recognition are used in coherence with image processing for effective and efficient retrieval.

Liu et al. [1] have categorized image retrieval system as content-based image retrieval (CBIR), text-based retrieval, and semantic-based retrieval method. In CBIR system, visual analysis of the image content is used as feature to extract relevant images. The features are either low level which include color, texture, and shape of images or high-level features in which semantic of images is considered for retrieval. Smeulders et al. [2] and Datta et al. [3] presented survey which elaborated depicting the evolution of CBIR in the early years and efforts made by authors toward the scope of CBIR, characteristics of domain and sources of knowledge which serve as a preprocessing step during the computation of features. Alzubi et al. [4] presented a detailed survey over the framework of CBIR system and renovation acquired in the fields of preprocessing of image, extraction, and indexing of feature, similarity matching and evaluation of performance.

With the emergence of various applications based on CBIR, privacy preserving plays a vital role for preserving and protecting the private sensitive data from any malicious attacks. Fanti et al. [5] developed one-way privacy model to protect CBIR system. As the database is public, the authorized user wants to keep sensitive information secret. Weng et al. [6] demonstrated SPEED and SRR as two approach for CBIR security. The former rely on cryptographic computation which is homomorphic encryption and multiparty computation, while the later is based on secure index.

In this paper, we propose a secure and effective CBIR system which exploits the annular histogram of color image to generate the quantized structure image, over which encryption is applied. Rest of the paper is organized as follows: In Sect. 2, we discuss the relevant work related to this field, Sect. 3 describes the proposed work with detail explanation, Sect. 4 mainly illustrates the result analysis with discussions, and Sect. 5 concludes the paper.

2 Related Work

Color, texture, and shape provide the most significant feature for CBIR system. Among the various feature extraction algorithm, color histogram has been extensively used in CBIR due to its orientation and scale invariance property.

In order to exploit the spatial information, Manjunath et al. [7] presented an approach on color MPEG-7 descriptor which consisting of color layout descriptor (CLD), dominant color descriptor (DCD), and scalable color descriptor (SCD). Rao et al. [16] proposed annular histogram in which each subset is rotation and translation invariant. Moreover, it can tolerate small movement of camera while capturing the images.

Various feature extraction algorithms based on texture properties of images have been used which includes structure element descriptor, edge histogram descriptor, and gray-level co-occurrence matrix are summarized in Alzubi et al. [4]. Based on edge oriented similarity, Liu et al. [1] proposed microstructure descriptor (MSD) which extracts features by stimulating human visual processing in which whole image information is integrated as color, texture, and shape layout. Alzubi et al. [4] demonstrated moment invariants, Zernike moments, and histogram descriptor as shape feature extraction method.

Combined feature selection algorithms have been adopted for efficient retrieval of image. These algorithms are formed by combining color, texture, and shape features together or by employing different combination of each feature. Yue et al. [8] fused the color histogram feature and texture feature using co-occurrence matrix. Wang and Wang [9] proposed a texture descriptor which effectively describes and represents the local feature of an image using its color and texture property known as structure elements descriptor (SED).

In Zhang et al. [10], secure retrieval for cloud is demonstrated with effective retrieval efficiency. Color, shape, and texture are used as feature vector and locality preserving projection (LPP) is applied for feature dimension reduction. Paillier Homographic encryption provides the privacy to the sensitive data. A concise study to summarize the cryptographic primitives to encrypt the signals and security challenges for retrieval and protection is discussed in Erkin et al. [11]. A hierarchical data structure along with hash indexing scheme in Shashank et al. [12] provides a scalable, accurate, and effective private image retrieval by exploiting image data property. With the involvement of third party such as service provider in transmission and storage of data, Yiu et al. [13] proposed a dataset encoding-based algorithm such that only authorized users have the access of the sensitive content, while third party blindly evaluates the queries. Lu et al. [14] proposed a scheme for feature protection which exhibits comparable performance and good trade-off between computational complexity and security. In order to preserve effective retrieval and protect the images from malicious attack on server, Lu et al. [15] designed secure min hash sketches and secure inverted index scheme which are based on image processing, information retrieval, and cryptography techniques. Both achieve better retrieval and serve as candidate for privacy preserving for retrieval of multimedia.

3 Proposed Work

3.1 HSV Color Space and Color Quantization

Color provides spatial information which is highly reliable for image retrieval and object recognition. The use of RGB color space does not describe human visual perception. On the contrary, HSV color space could mimic human conceptual understanding of colors which provide enough information to distinguish large number of color for image retrieval. It is defined in terms of three basic components: Hue (H), Saturation (S), and Value (V). The H component ranges from 0 to 360° which describe the color type. The S component ranges from 0 to 1 which demonstrate the relative purity. Lower the saturation, more is the grayness presented and more faded will be the color appearance. The V component represents brightness of the color or the amount of black color mixed with H component. It ranges from 0 to 1. In this paper, the HSV color space of image $C(x, y)$ is uniformly quantized into 8, 3, and 3 bins, respectively, which provide a total of $8 \times 3 \times 3 = 72$ color combination. The quantized image obtained is represented as $C'(x, y)$ and $C'(x, y) = q, q\{0, 1, ..., 71\}$.

3.2 Edge Orientation Detection

Edge orientation provides strong influence for processing of low-level image perception. It minimizes the data to be processed. Many existing gradient-based edge detector which include Robert operator, Prewitt operator and Sobel operator are under first-order derivative while Laplacian of Gaussian and Canny edge detection are second-order derivative. If edge detector is applied on each of the three channels separately, some edges which are caused by the spectral variation may not be detected. Similarly, if color image is converted to gray image, and then edge detector is applied, the corresponding chromatic information will be lost. In the proposed work, HSV color space is represented in Cartesian space as (H', S', V'), where $H' = S \cdot cos(H)$, $S' = S \cdot sin(H)$ and $V' = V$.

Edge detection is performed by Sobel operator because as compared to other operator it has less sensitivity to noise. Moreover, it is based on first-order derivative which makes it less expensive as compared to second-order derivative operator such as Canny operator. For two vector, the gradient along x and y direction is denoted as $X(H'_x, S'_x, V'_x)$ and $Y(H'_y, S'_y, V'_y)$, where H_x represents gradient of H' along the horizontal direction and H_y represents gradient of H' along the vertical direction. The dot product and the norm are shown in Eqs. 1–5.

$$|X| = \sqrt{(H'_x)^2 + (S'_x)^2 + (V'_x)^2} \tag{1}$$

$$|Y| = \sqrt{(H'_y)^2 + (S'_y)^2 + (V'_y)^2} \tag{2}$$

$$XY = H'_x H'_y + S'_x S'_y + V'_x V'_y \tag{3}$$

$$\cos(\widehat{X, Y}) = \frac{XY}{|X||Y|} \tag{4}$$

$$\theta = \arccos(\widehat{X, Y}) = \arccos\left[\frac{XY}{|X|.|Y|}\right] \tag{5}$$

The edge orientation $\theta(x, y)$ of size $W \times N$ for pixel is uniformly quantized into m bins, where $m \in \{6, 12, 18, 24, 30, 36\}$. The edge orientation map $\theta(x, y) = \phi$ where $\phi \in \{0, 1, ..., m\}$.

3.3 Structure Map Definition

Color, texture, and shape play an important role in content-based image retrieval. The image is described in terms of descriptors which include color, texture, and shape descriptor. Local structure of same class indicates the presence of common pattern in the images. This property can serve as common base for analysis and comparison of different images. In this context, images can be represented as structure unit and texture decomposition which plays an important role in the final analysis.

In the proposed approach, there are two types of structure map which include color structure map and edge-structure map. The former is based on quantization of HSV color values, while the later is based on quantization of edge-oriented values.

Edge-oriented image $\theta(x, y)$ is used to define the edge-structure map because of its insensitivity to color and illumination. Moreover, it is independent of scaling, translation, and small rotation. The edge-oriented image is quantized into six levels whose value varies from 0 to 5. We move 3×3 block in overlapping manner, starting from the left uppermost pixel and moving it from left-to-right and top-to-bottom to detect edge-structure throughout the image. The step length is of one pixel along both horizontal and vertical directions. In each of the 3×3 block, the eight neighbor pixels having same value as the center pixel are marked active and rest are unchanged. The final edge-structure map is denoted as $M(x, y)$, where $0 \leq x \leq W - 1, 0 \leq y \leq N - 1$. Figure 1 and Fig. 2 show the example to demonstrate the edge-structure map and structure element extraction process, respectively.

The edge-structure $M(x, y)$ is used as a mask to extract the desired underline corresponding colors information from the quantized image $C'(x, y)$. Finally, only those corresponding color pixels of quantized $C'(x, y)$ are preserved which are marked active in edge-extraction map $M(x, y)$. All other color pixels in $C'(x, y)$ are set empty. Finally, quantized image consists of active and empty pixel values. The active color pixel of $C'(x, y)$ is quantized into m bins where $m \in \{6, 12, 18, 24, 30, 36, ...72\}$.

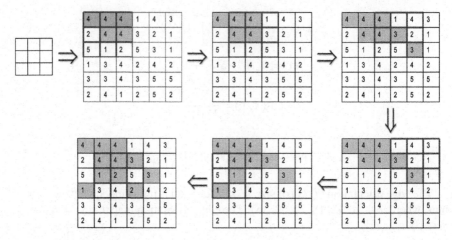

Fig. 1 Edge-structure map extraction process

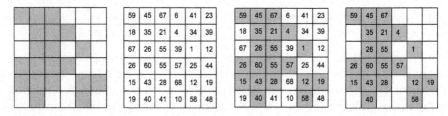

Fig. 2 Structure element generation using edge-structure map

Thus, not only edge features are exploited, but also color features are taken into consideration.

3.4 Annular Distribution Density Structure Descriptor (ADDSD)

The set of preserved pixels of $M(x, y)$ characterizes the distribution of geometric points on a two-dimensional plane. Let $p(i, j)$ represent the color of preserved pixels of $M(x, y)$ where $i \in x$ and $j \in y$. Let m color bins be represented as $B_1, B_2, ..., B_m$.

Suppose, $L_q = (x, y)|(x, y) \in M, p_{xy} \in B_q$ for $1 \leq q \leq m$. Each L_q refers to the set of active pixel whose color is in qth bin termed as histogram subset of bin B_q. For a given L_q of given color pixel bin B_q, suppose $C^n = (x^q, y^q)$ be the centroid for L_q as shown in Eq. 6.

$$x^q = \frac{1}{|L_q|} \sum_{(x,y) \in L_q} x; \, y^q = \frac{1}{|L_q|} \sum_{(x,y) \in L_q} y \qquad (6)$$

Fig. 3 Annular distribution density vector (4, 7, 5, 4) for a bin of structural element image

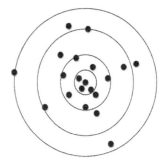

Suppose, d^q be considered as radius of L_q which is shown in Eq. 7.

$$d^q = max_{(x,y) \in L_q} \sqrt{(x - x^q)^2 + (y - y^q)^2} \tag{7}$$

The radius d^q with centroid C^n for each of the L_q can be divided into n concentric circles with kd^q/n radius for each $1 \leq k \leq n$ for generating n annular regions. Figure 3 shows the vector of annular distribution density which calculates the number of points in each and every annular region. Thus, the final histogram formed is of dimension $1 \times k|m|$, where $|m|$ refers to the levels of quantization of $C(x, y)$. The matrix is referred as annular color histogram of image. The centroid and the corresponding annular partition are translation and rotation invariant which can tolerate small camera movement while taking the images. With 72 quantization levels and n (empirically taken as 4) annular region generate 1×288 dimension feature vector termed as annular distribution density structure descriptor (ADDSD).

3.5 Bit-plane Randomization

Image appearance information is best described by most significant bits (MSB) which has been used in scalable encoding for finer granular trade-off between quality and bit-rate. The work is motivated by the fact that feature vectors with smaller distances among their MSB result into similar pattern. Given a feature vector $f = \{f_1, f_2, ..., f_n\} \in \mathbb{R}^n$ and each component f_i is represented in binary form as $\{m_{i1}, ..., m_{il}\}^T$, where m_{i1} and m_{il} denote first MSB and least significant bit (LSB), respectively, where l is total number of bit-planes. Hamming distance between bit-planes is preserved when they are permuted or XORed with same random bit pattern. This property is exploited to encrypt the feature vector from top k bit-planes. Encryption of jth bit-plane of feature vector f is composed of jth MSB with n feature component denoted as $\{m_{ij}, ..., m_{nj}\}$. The randomization of jth bit-plane with $\{r_{ij}, ..., r_{nj}\}$ as binary random vector is shown in Fig. 4.

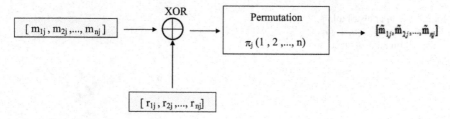

Fig. 4 Encyption of the jth bitplane

The feature vector $\varepsilon(f) = \{f_1', f_2', ..., f_n'\}$ is formed from encrypted bit-plane. Hamming distance is calculated between two encrypted feature vectors $\varepsilon(f)$ and $\varepsilon(g)$ as weighted sum between their individual bit-planes shown in Eq. 8.

$$d_\varepsilon(\varepsilon(f), \varepsilon(g)) = \sum_{i=1}^{n} \sum_{j=1}^{l} |\widetilde{m}_{ij}^{(f)} - \widetilde{m}_{ij}^{(g)}| \times w(j) \tag{8}$$

where $\widetilde{m}_{ij}^{(f)}$ is randomized feature component of input image, $\widetilde{m}_{ij}^{(g)}$ is randomized feature component of output image, and $w(j)$ is the weights assigned to the bit-planes to represent their unequal importance.

$L1$ distance measures the upper bound of the distance metric between randomized features shown in Eq. 9.

$$d_\varepsilon(\varepsilon(f), \varepsilon(g)) = \sum_{i=1}^{n} \sum_{j=1}^{l} |m_{ij}^{(f)} - m_{ij}^{(g)}| \times 2^{-j} \geq \sum_{i=1}^{n} \left| \sum_{j=1}^{l} (m_{ij}^{(f)} - m_{ij}^{(g)}) \times 2^{-j} \right| = ||f - g||_1$$
$$\tag{9}$$

This indicates that there is some disturbance in distance distortion before and after encryption with few of the feature vector whose probability is very low.

3.6 Similarity Measure

For each of the template, 288 dimension feature vector $F = (f_1, f_2, ..., f_i, ..., f_{288})$ is extracted. The query feature vector is represented as $Q = (q_1, q_2,, q_i, ..., q_{288})$. For similarity measure between query and database image, $L1$ norm is used. It simply computes the distance with no square and square root computation involved. It can save a lot of computational cost which is suitable for large-scale retrieval. An overview of proposed method is shown in Fig. 5.

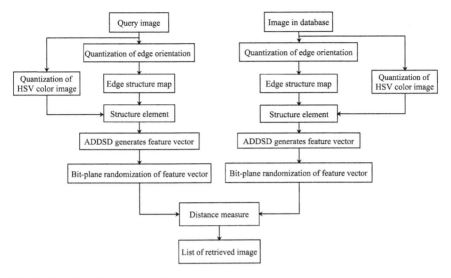

Fig. 5 Overview of proposed method

3.7 *Proposed Method*

Step 1: HSV color space component H, S, and V of color image $C(x, y)$, $0 \leq x \leq W$, $0 \leq y \leq N$ is quantized into $8 \times 3 \times 3$ levels as $C'(x, y) = q, q \in \{0, 1, ..., 71\}$, which generate a total of 72 bins.

Step 2: Sobel operator is applied on $C(x, y)$ on its H, S, and V component to generate horizontal and vertical derivative represented (H_x, S_x, V_x) and (H_y, S_y, V_y) respectively.

Step 3: The components are represented in Cartesian coordinate as (H'_x, S'_x, V'_x) and (H'_x, S'_x, V'_x), where $H = h \cdot \cos(H)$, $S = s \cdot \sin(H)$ and $V = V$.

Step 4: The edge orientation θ obtained as $\theta = \phi$, where $\phi \in \{0, 1, ..., m\}$.

Step 5: The edge orientation map $M(x, y)$ is extracted using 3×3 block masking with step length of one pixel. In each block, the pixels are marked based on the similarity with the central pixel value.

Step 6: The corresponding pixel of $C'(x, y)$ is marked active, and structure element of image is generated.

Step 7: ADDSD for the active pixels in $C'(x, y)$ is obtained having dimension of $1 \times mn$, where n (empirically taken as four) is the number of annular region and m (empirically set as six) is number of quantization levels in $C'(x, y)$.

Step 8: Bit-plane randomization encryption is applied on ADDSD obtained from the query image.

Step 9: L1 norm is used to measure distance between the query and database encrypted feature vectors.

4 Experimental Results and Discussions

Corel database [17] of 1000 images have been used in the experiment for retrieval accuracy and security analysis of CBIR system. The database is in JPEG format and is classified into ten categories, such as African people, buildings, dinosaurs, flowers. Image resolution is either 384×256 or 256×384. MATLAB R2016a with Intel(R)Core(TM) $i3$ processor and 4GB RAM is used to simulate the algorithms.

The proposed work uses HSV color space to evaluate the image retrieval efficiency. As per the evaluated results, the proposed algorithm achieves better performance. Among various standard criterion available to evaluate the accuracy and performance of CBIR systems, precision and recall have been used for evaluating search strategies. Precision refers to the ratio between the number of relevant images retrieved and total relevant images. Recall refers to ratio between the number of relevant images retrieved and total number of relevant images in the corresponding database. Precision and recall measure accuracy and robustness, respectively. For each category, more than 50 test were taken to evaluate precision and recall. Figure 6 displays the comparison between proposed method and Wang and Wang [9].

Query images from different categories of image of coral database with its corresponding first five relevant retrieval are shown in Fig. 7. Thus, we conclude that the image retrieval based on plaintext feature performance of proposed method is better than Wang and Wang [9].

The encryption of feature vector set using the bit-plane randomization and permutation of elements of feature vector securely preserves the privacy. Figure 8 illustrates that both Zhang et al. [10] and proposed algorithm have high search accuracy and close retrieval efficiency. In the experiment, it was analyzed that ciphertext size and encryption time of Zhang et al. [10] are much higher than the proposed encryption algorithm which is shown in Table 1.

Fig. 6 Plaintext retrieval performance on the corel dataset

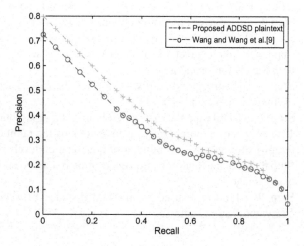

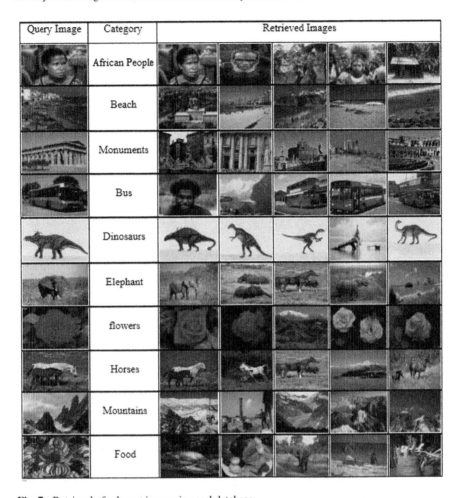

Fig. 7 Retrieval of relevant images in corel database

Table 1 Encryption and ciphertext size analysis

Encryption scheme	Encryption time (s)	Ciphertext size (KB)
Paillier homomorphic	1778.5	32005
Bitplane randomization	0.24	159

Fig. 8 Ciphertext retrieval performance on the corel dataset

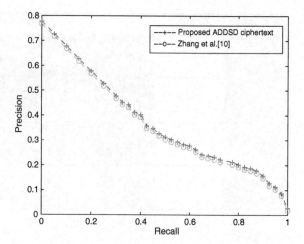

5   Conclusion

The availability of encrypted feature vector brings a new era of secure image retrieval which would protect the privacy and integrity of the user data. In this paper, a secure and effective image retrieval algorithm using ADDSD is proposed. HSV color space is exploited to generate quantized image. Based on same or similar edge orientation in uniform HSV color space, structure element is obtained. Finally, the structure element is detected using the grid and based on that quantized structure of image is formed. Annular histogram is generated from the quantized structure of image which is encrypted by bit-plane randomization technique. L1 norm is used for similarity measure evaluation between the encrypted query feature vector and encrypted database feature vector. Experimental analysis demonstrates that homomorphic-based techniques are secure but possesses high computational complexity and user involvement in various applications, while bit-plane randomization provides high efficiency based on deterministic distance preserving with minimum user involvement. Proposed method retrieved the relevant images effectively and efficiently without revealing image content.

References

1. G.H. Liu, Z.Y. Li, L. Zhang, Y. Xu, Image retrieval based on micro- structure descriptor, Pattern Recog. 44 (9),(2011), 2123–2133.
2. A.W.M. Smeulders, M. Worring, S. Santini, A. Gupta, R. Jain, Content-based image retrieval at the end of the early years, IEEE Trans. Pattern Anal. Mach. Intell., 22 (12), (2000), 1349–1380.
3. R. Datta, J. Li, JZ. Wang, Content-based image retrieval: approaches and trends of the new age, in: Proceedings of the 7th ACM SIGMM International Workshop on Multimedia Information Retrieval, (2005), pp. 253–262.

4. A. Alzubi, A. Amira, N. Ramzan Semantic content-based image retrieval: a comprehensive study J Vis Commun Image Represent, 32, (2015), pp. 20–54.
5. G. Fanti, M. Finiasz, and K. Ramchandran, One-way private media search on public databases: The role of signal processing, IEEE Signal Process. Mag., vol. 30, no. 2, Mar. (2013), pp. 53–61.
6. L. Weng, L. Amsaleg, A. Morton, and S. Marchand-maillet, A Privacy Preserving Framework for Large-Scale Content-Based Information Retrieval, TIFS, vol. 10, no. 1,(2015), pp. 152–167.
7. B.S. Manjunath, J.-R. Ohm, V.V. Vasudevan, and A. Yamada, Color and Texture Descriptors, IEEE Trans. Circuits and Systems for Video Technology, vol. 11, no. 6, (June 2001), pp. 703–715.
8. J. Yue, Z. Li, L. Liu, Z. Fu Content-based image retrieval using color and texture fused features Math. Comput. Modelling, 54,(2011), pp. 1121–1127.
9. X. Wang, Z. Wang, A novel method for image retrieval based on structure elements descriptor, J. Vis. Commun. Image Represent. 24 (1), (2013), 63–74.
10. Zhang, Y., Zhuo, L., Peng, Y., Zhang, J, A secure image retrieval method based on homomorphic encryption for cloud computing. In: 19th International Conference on Digital Signal Processing, IEEE, (2014), pp. 269–274.
11. Z. Erkin, A. Piva, S. Katzenbeisser, R. L. Lagendijk, J. Shokrollahi, G. Neven, et al.,Protection and retrieval of encrypted multimedia content: When cryptography meets signal processing, EURASIP J. Inf. Sec., vol. 7, no. 2, (2007), pp. 1–20.
12. J. Shashank, P. Kowshik, K. Srinathan, and C. Jawahar, Private content based image retrieval, in Proc. IEEE Conf. Comput. Vis. Pattern Recognit., Jun. (2008), pp. 1–8.
13. M.L. Yiu, G. Ghinita, C. S. Jensen, and P. Kalnis, Outsourcing search services on private spatial data,in Proc. IEEE 25th Int. Conf. Data Eng., Apr. (2009), pp. 1140–1143.
14. W. Lu, A. L. Varna, A. Swaminathan, and M. Wu, Secure image retrieval through feature protection, in Proc. IEEE Int. Conf. Acoust. Speech SignalProcessing (ICASSP), Washington, DC, (2009), pp. 1533–1536.
15. W. Lu, A. Swaminathan, A. L. Varna, and M. Wu, Enabling search over encrypted multimedia databases, Proc. SPIE, vol. 7254, Jan. (2009), pp. 7254–7318.
16. Rao, A., Srihari, R. K., and Zhang, Z. Spatial color histograms for content-based image retrieval. Proceedings of the Eleventh IEEE International Conference on Tools with Artificial Intelligence, 1999.
17. The image database used in this paper is available online at: http://wang.ist.psu.edu/docs/related/

Near-Duplicate Video Retrieval Based on Spatiotemporal Pattern Tree

Ajay Kumar Mallick and Sushila Maheshkar

Abstract Recently, due to rapid advancement in multimedia devices and exponential increase in Internet user activities such as video editing, preview, and streaming accumulate enormous amount of near-duplicate videos which cannot be detected or retrieved effectively by conventional video retrieval technique. In this paper, we propose a simple but effective hierarchical spatiotemporal approach for high-quality near-duplicate video retrieval. Pattern generation of encoded key frames using angular distribution density is used which are translation and rotation invariant. Queue pool contributes temporal matching and consistency for the retrieval. Experimental result analysis demonstrates the effectiveness of the proposed method.

Keywords Near-duplicate · Angular density distribution · Encoding · Key frames · Pattern Tree · Queue pool · Video retrieval · CBVR

1 Introduction

With the rapid increase in online video services such as video broadcasting and video sharing attract Internet user to edit, stream, and store videos. It accumulates unpresented amount of multimedia entities in multimedia repositories which demand new innovation in video retrieval techniques. This exponential growth of multimedia content gives rise to the possible existence of near-duplicate videos which occur due to visual or temporal transformation such as scaling, re-encoding frame dropping, slow or fast forwarding. This brings a scenario in which near-duplicate retrieval techniques play an eminent role in video copyright protection, video search, and in many more events. Among the various approaches object oriented data modelling or video

A. K. Mallick (✉) · S. Maheshkar
Department of Computer Science and Engineering, Indian Institute of Technology
(Indian School of Mines), Dhanbad, Jharkhand, India
e-mail: mallickajay6@gmail.com

S. Maheshkar
e-mail: sushila_maheshkar@yahoo.com

© Springer Nature Singapore Pte Ltd. 2018
B. B. Chaudhuri et al. (eds.), *Proceedings of 2nd International Conference on Computer Vision & Image Processing*, Advances in Intelligent Systems and Computing 703, https://doi.org/10.1007/978-981-10-7895-8_14

tagging based method have limitation of manual interpretation for video retrieval is illustrated in Pickering and Ruger [1]. Moreover, exhaustive texture description is needed to tag any video. According to the study conducted by Kim and Vasudev [2], video invariant feature selection is essential, but high-dimensional complex feature vector consecutively lowers indexing efficiency.

Content-based video retrieval (CBVR) and indexing bring a new challenge and opportunity to the research community for effective analysis and evaluation of multimedia in computer vision domain. Bohm et al. [3] demonstrated that the data in the repositories are multimodal as well as multidimensional due to the existence of visual, audio, and temporal information in it which demand better innovative approach. Hartung and Kutter [4] analyzed that watermark and CBVR are two basic methods which can provide the desired retrieval. The former is imperceptible embedding of message in the corresponding media while the later refers to content-based signature extraction scheme of the given media which is compared with the signature of media available in repositories is illustrated in Pickering and Ruger [5]. CBVR technique is more advantageous than watermarking approach.

2 Related Work

Depending upon the video content, there have been various feature extraction approaches that are available in the literature. In the context of CBVR, Liu et al. [6] investigated various existing variants for defining near-duplicate video, state-of-the-art practices, generic framework and finally explored its emerging research trends. There has been lot of work on indexing of high-dimensional data. Bohm et al. [3] provided an insight into principle ideas to overcome the problems related to indexing of high-dimensional space. Color correlation invariance for sequence matching of videos is proposed in Pickering and Ruger [5] which extracted the desired key frames independently. In the study conducted by Rao et al. [7], annular-, angular-, and hybrid-based color histogram are proposed which outperform the traditional histogram. The algorithm satisfy both space and time complexity. Kim and Vasudev [2] demonstrated a computationally effective spatiotemporal matching technique for video copy detection. It computes spatial matching by utilizing ordinal signature for each frame and corresponding temporal matching based on temporal signature. Format conversion such as pillar-box and letter box can be effectively managed. Moreover, its memory requirement for storing or indexing is comparatively low.

Wu et al. [8] demonstrated a content and context fusion approach which consider a joint set of context information and perform varied task with redundancy elimination. Thumbnail and time duration are used to eliminate near-duplicate video which evade the exhaustive pairwise comparison of corresponding key frames of two videos. In a scenario of mismatched alignment among near-duplicate video clips, Wu and Aizawa [9] proposed an approach based on binary frame signature extraction. The extracted signature is encoded into concise matrix through a well-organized self-similarity mining procedure of pattern generated to analyze and investigate the localization and copy capacity of Web videos. In order to handle spatiotemporal variation to detect video copy, Chiu et al. [10] presented a computationally low-cost probabilistic framework that considers video copy detection as partial matching problem and correspondingly transforms it as a shortest path problem. Registration framework of spatiotemporal method in which multimodal features are utilized for frame alignment is illustrated in Rooplakshmi and Reddy [11].

A dynamic programming-based approach using pattern set is demonstrated in Chou et al. [12] which accomplishes relatively high-quality retrieval and localization of near-duplicate videos. Tian et al. [13] presented a transformation-based copy detection for enhanced localization. To overcome the problem related to high-dimensional visual feature, Su et al. [14] proposed a scheme based on pattern indexing and matching. Activity recognition has been a major area of research in content-based video retrieval. Chaudhry et al. [15] proposed a method that model the ongoing activity of any scene which do not require any preprocessing is represented comprehensively as feature of non-Euclidean Histogram Optical flow. In the study conducted by Rosten et al. [16], a simple but effective machine learning approach is proposed which is based on heuristic detector which outperforms the available feature detector significantly. Based on spatiotemporal pattern approach, Chou et al. [17] proposed a hierarchical framework approach for filter and refine near-duplicate video retrieval and corresponding localization. Our work is motivated by the above observation of hierarchical framework to filter and refine near-duplicate videos illustrated in Chou et al. [17]. The proposed method provides better result of the retrieval due to key frame selection and encoding as compared with the method addressed by Chou et al. [17].

The remaining section of paper is categorized as follows. Proposed work is described in Sect. 3. Experimental results and discussion is elaborated in Sect. 4. Finally, Sect. 5 demonstrates the conclusion and future work of the paper.

3 Proposed Work

3.1 Problem Definition

Let V^f be a video of n consecutive frames with resolution of $W \times H$ as shown in Eq. 1.

$$V^f = \langle V_1^f, V_2^f, \ldots, V_l^f \rangle \tag{1}$$

The corresponding ith frame of video V^f with k partition is shown in Eq. 2.

$$V_i^f = \langle V_i^{f1}, V_i^{f2}, \ldots, V_i^{fk} \rangle \tag{2}$$

3.2 Clip Segmentation

In order to reduce the effective computation and enhance retrieval accuracy, video frames are subjected to selection of key frames using its content and subsequently organizing it as clusters of small clips. In the literature, there are basically two approaches for key frame selection which are classified as clip segmentation and shot segmentation. The former technique produces comparatively lesser key frames by just replacing each desired shot by a key frame, while the later technique considers small subsets of frames as single unit produces shots and corresponding key frames which result into effective key frame extraction.

In the proposed work, considered shots may exist in different position instead of consecutive time position in any given video. Shot selection is performed by dividing each frame into $w \times h$ non-overlapping blocks and correspondingly average of each block is computed. The first frame of a clip is termed as pivot frame. Initially, a video is considered as single clip and subsequently many sub-non-overlapping sub-clips are extracted. Starting from the pivot frame of a video, corresponding difference with each frame is evaluated. Frames with value less than a threshold μ is/are considered to be similar to the pivot frame from a part of clip headed by the respective pivot. This process continues until each frame is a part of any clip headed by respective pivot. In our work, threshold is empirically considered as 0.5. Eventually, l clips are generated from any video with n consecutive frames, where $1 \le l \le n$.

3.3 Key Frames Selection

Block-based histogram equalization of frames in a clip is applied as evaluation metric for key frame generation. For each clip ci with k frames, where $1 \le k \le l$, $1 \le i \le l$, starting from the pivot frame, the difference of equalized histogram with other $k - 1$ frames in the clip is calculated and summation of the evaluated value is considered. This process is repeated for each of the frames in the clip until each frame has participated as a key pivot and summation of histogram equalization is computed. This indicates that each frame in a clip has computed its distance with respect to other $k - 1$ frames and the frame with least value is considered as key frame in the corresponding clip. Finally, video with n frames is represented with k frames.

3.4 Key Frame Encoding

Color histogram is considered as an important approach in content- based retrieval. A mathematical model termed as annular distribution density is applied on the extracted key frames mentioned in the above section. It basically characterized set of geometric points distribution state on a two-dimensional plane which provide rotation and translation invariant constraint.

Suppose p_{ij} represent the color of pixel for a key frame of dimension $W \times H$ where $0 \leq i \leq W, 0 \leq j \leq H$. Let $U = (x, y) | 1 \leq x \leq W; 1 \leq y \leq H$. The color space is quantized into B color bins represented as $M_1, M_2, \ldots, M_B$. Suppose $Rq = (x, y) | (x, y) \epsilon U, p_{xy} \epsilon M_q$ for $1 \leq q \leq B$, is the histogram subset of bin M_q, is the set of color pixel in the $q^t h$ bin. For histogram subset R_q with $C^q = (x^q, y^q)$ as centroid of R_q, where x^q an y^q are defined in Eq. 3 and Eq. 4 respectively.

$$x^q = \frac{1}{|R_q|} \sum_{(x,y) \in R_q} x \tag{3}$$

$$y^q = \frac{1}{|R_q|} \sum_{(x,y) \in R_q} y \tag{4}$$

In order to obtain annular partition, each point $(x, y) \epsilon R_q$ direction (principal angle) $\phi(x, y) \epsilon R_q$ in the associated coordinate system is shown in Eq. 5.

$$\phi(x, y) = \arctan\left(\frac{y - y^q}{x - x^q}\right) \pm \pi \tag{5}$$

$$\theta(R_q) = \frac{1}{|R_q|} \sum_{(x,y) \in R_q} \phi(x, y) \tag{6}$$

Similarly, for each of the bin average or principle direction is evaluated by using Eq. 6 and place it in the respective angular partition. Each of the angular regions is encoded by a symbol representation such as A, B, C. For any key frame, the angular region with maximum density is chosen and the frame is represented by the encoded symbol of the angular region. Similarly, whole set of key frames of a video can be represented by sequence of encoded symbols. Key frame encoding of Fig. 1 of key frames is shown in Table 1.

Fig. 1 Key frame encoding

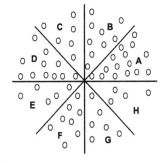

Table 1 Pattern generation of reference and query videos

Video	Encoded sequence	Pattern
v^1	ADBACFHGCAB	ADB, DBA, BAC, ACF, CFH, FHG, HGC, GCA, CAB
v^2	BCFGDABEFA	BCF, CFG, FGD, GDA, DAB, ABE, BEF, EFA
v^3	DCAEBGA	DCA, CAE, AEB, EBG, BGA
v^4	HABDCAFBDAGDC	HAB, ABD, BDC, DCA, CAF, AFB
		FBD, BDA, DAG, AGD, GDC
Q	ACFGDCABE	ACF, CFG, FGD, GDC, DCA, CAB, ABE

3.5   Pattern Tree

In order to facilitate matching of near duplicate videos, a simple but effective data structure in constructed termed as Pattern Tree. It is basically a prefix queue with m levels (empirically m set to 3) generated by overlapping sliding window in which symbol $p_1p_2p_3$ are considered. Similarly, next pattern is generated by sliding the window with unit step length. Similarity among the videos frames does not imply near-duplicate copies; instead, temporal consistency is better cue to detect near-duplicate videos. For any video v^i with $|k|$ encoded symbol $1 \leq k \leq l$, if $p_1p_2p_3$ is the first pattern generated, then 2-tuple $(v^i, 1)$ is inserted into the Pattern Tree based on the value of $p_1p_2p_3$. Encoded sequence for v^1 is *ADBACFHGCAB* as shown in Fig. 2. Patterns generated using m level prefix queue for v^1 are *ADB, BDA, BAC, ACF, CFH, FHG, HGC, GCA*, and *CAB*. The corresponding 2-tuple structures are $(ADB, 1)$, $(BDA, 2)$, $(BAC, 3)$, $(ACF, 4)$, $(CFH, 5)$, $(FHG, 6)$, $(HGC, 7)$, $(GCA, 8)$, and $(CAB, 9)$. In order to consider temporal consistency, $(ADB, 1)$ which belongs to the first pattern of v^1 is queued as $(v^1, 1)$ prefixed by *ADB*. Similarly, $(BDA, 2)$ is queued as $(v^1, 2)$ prefixed by *BDA* as shown in Fig. 3. Finally, the procedure is repeated for every reference video and Pattern Tree is generated. Based on the pattern generated for the reference videos v^1, v^2, v^3, and v^4 in Table 1, a Pattern Tree is generated as shown in Fig. 2.

3.6   Pool Generation

In order to maintain sustainable retrieval and accuracy for near-duplicate videos, queue pool is generated for the query videos using Pattern Tree of reference videos. For each of the N reference videos $v^1, v^2, \ldots, v^N$, queue pool is generated for corresponding temporal matching and consistency among query and known reference videos. For any pattern $p1p2p3$ of a query video, d number of 2-tuple $(v^i, (k-2)d)$, where $0 \leq i \leq N, 1 \leq k \leq l, 0 \leq d \leq N$, is obtained. The obtained value $(v^i, (k-2)d)$

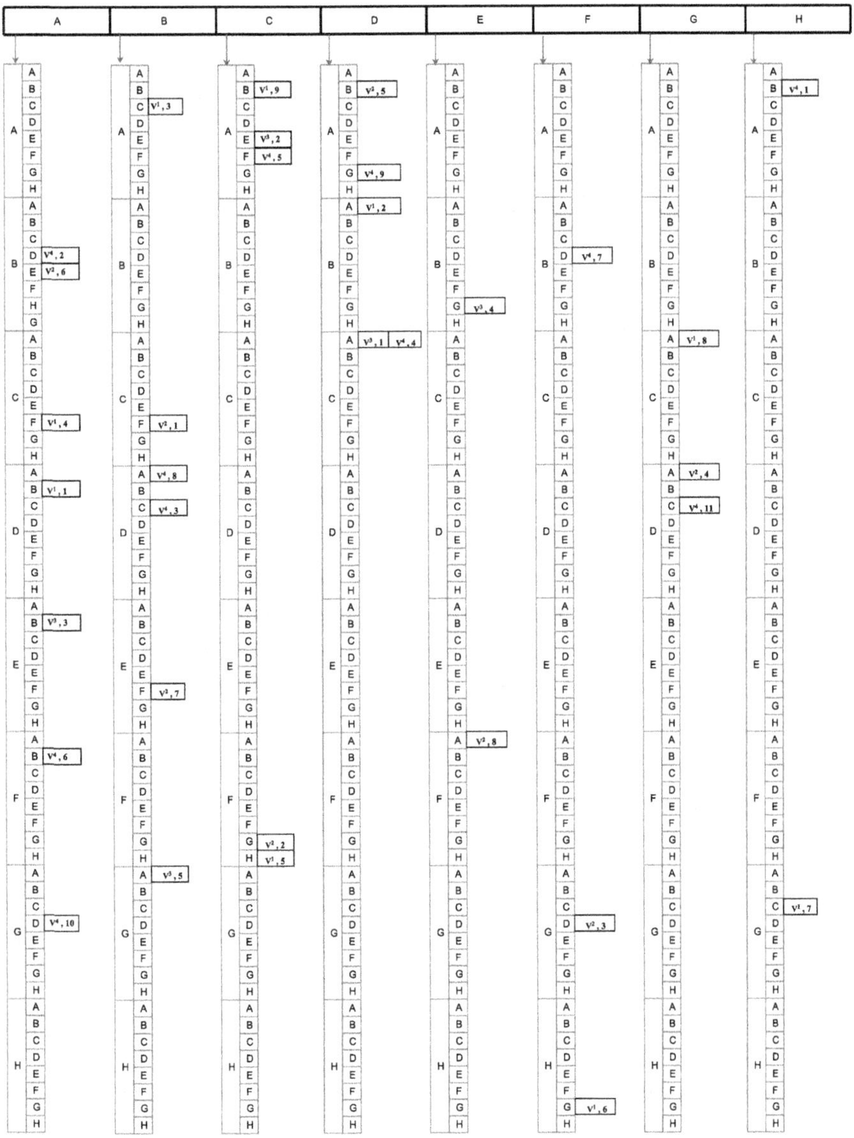

Fig. 2 Pattern Tree generation for reference videos

where v^i is video identity is inserted into the queue q_{ij} of queue pool QPi, where $0 \le i \le N$, $0 \le j \le N$, depending upon the relation between γ and δ. γ is the time stamp difference between two consecutive entries in queue in a particular queue pool, and δ is the threshold empirically taken as 3. If $\gamma \le \delta$, then (v^i, kd) is inserted into the queue q_{ij} of queue pool QPi else the queue is set inactive. Initially, when queue

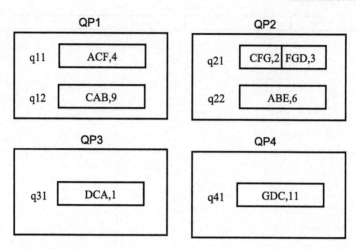

Fig. 3 Queue pool of query video

pool QPi is empty, γ value is set to 0 and unless any pattern of query video matches with any reference pattern in the Pattern Tree all queues are set inactive represented as $\neq$. When any 2-tuple value of query video matches in the Pattern Tree with c^i video pattern, then a new queue pool is constructed with a new queue in it and its status is set active represented by Δ else it is set to be inactive represented as ∇. A counter Cn is used to represent the frequency of query pattern matching with reference pattern matching. If any query pattern $p1p2p3$ is inserted into a queue of queue pool based on its δ and Δ values, the counter Cn value is set to 0 and for all other active queues of the queue pool are incremented by 1. When $Cn \geq \eta$, queue is set inactive and new queue is generated in that particular queue pool which is initially set inactive before any match. Empirically, η is set to 3 in our work. For the given query pattern $ACF, CFG, FGD, GDC, DCA, CAB$, and ABE in Table 1 for reference videos $v^1, v^2, \ldots, v^4$ process of query matching is shown in Table 2.

Initially, no queue and queue pool exist. When the first query pattern ACF finds a match with $(v^1, 4)$, query pool $QP1$ is constructed and $q11$ is generated as active status is obtained represented as Δ and the pattern $(ACF, 4)$ is inserted into queue $q11$ with both counter Cn and γ is set to 0. The second pattern CFG matches with $(v^2, 2)$ in Pattern Tree and Queue Pool $QP2$ is constructed. Queue $q21$ is generated as active in which $(CFG, 2)$ is inserted and counter Cn is set as to 0 while $QP1$ counter is incremented by 1 which signify a pattern miss for $QP1$ for the current queue pattern. Similarly, since γ is $1 \leq \delta$, the matched pattern $(FGD, 3)$ is inserted into the queue $q21$ of $QP1$ with Cn set to 0 and $q11$ queue of queue pool $QP1$ is incremented by 1 making its value as 2. Fourth query pattern is GDC which matches at $(v^4, 11)$. This is to be inserted into $QP4$ queue pool which is created and queue $q41$ is generated and finally $(GDC, 11)$ is inserted in which Cn is set zeros while $QP1$ and $QP2$ Cn are incremented making it as 3 and 1 respectively. Now, Cn of $QP1$ is greater than δ so $q11$ is set inactive denoted by ∇ and new queue $c12$ in $QP1$ is to be generated

Table 2 Queue pool generation for query video using Pattern Tree of reference video

QP1					QP2					QP3					QP4				
Queue	γ	S	Pattern	Cn	Queue	γ	S	Pattern	Cn	Queue	γ	S	Pattern	Cn	Queue	γ	S	Pattern	Cn
q11	0	◁	ACF, 4	0	–	0	≠	–	–	–	0	≠	–	–	–	0	≠	–	–
q11	0	◁	ACF, 4	1	q21	0	◁	CFG, 2	0	–	0	≠	–	–	–	0	≠	–	–
q11	0	◁	ACF, 4	2	q21	1	◁	CFG, 2; FGD, 3	0	–	0	≠	–	–	–	0	≠	–	–
q11	0	▷	ACF, 4	3	q21	1	◁	CFG, 2; FGD, 3	1	–	0	≠	–	–	q41	0	◁	GDC, 11	0
q11	0	▷	ACF, 4	3	q21	1	◁	CFG, 2; FGD, 3	2	q31	0	◁	DCA, 1	0	q41	0	◁	GDC, 11	1
q12	0	◁	CAB, 9	0	q21	1	▷	CFG, 2; FGD, 3	3	q31	0	◁	DCA, 1	1	q41	0	◁	GDC, 11	2
q12	0	◁	CAB, 9	1	q22	0	◁	ABE, 6	0	q31	0	◁	DCA, 1	2	q41	0	▷	GDC, 11	3

for insertion of query pattern $(CAB, 9)$. Similarly, $q21$ becomes inactive and $q22$ is to be generated for inserting $(ABE, 6)$ as shown in Table 2. The pool generation for the query pattern is shown in Fig. 3.

3.7 Distance Measure

In order to measure distinctiveness for the generated pattern p, inverse document frequency (IDF) is computed which is shown in Eq. 7.

$$IDF(p) = \log \left(\frac{|N|}{|PQ(p)|} \right) \tag{7}$$

where $|N|$ refers to the number of reference videos and $|PQ(p)|$ denote number of videos with prefix pattern p. Moreover, temporal relation and its corresponding consistency score (TRC-score) of a queue q_{ij} is represented in Eq. 8.

$$TRC - score(q_{ij}) = \sum_{(q_{ij})in(p_i)} IDF(p_i) \tag{8}$$

where $|q_{ij}|$ is number of pattern inserted into q_{ij} and p_i is the pattern in q_{ij}. When all the query patterns are inserted in the corresponding queue pool QPi for each of the reference videos v^1, v^2, v^3, and v^4, near-duplicate score for each query pool can be generated as shown in Eq. 9.

$$ND - score(v_i) = \sum_{(q_{ij}) \in (QPi)} TRC - score(q_{ij}) \tag{9}$$

3.8 Proposed Algorithm

Step 1:   For a given video V_i^f with l frames, where $1 \le i \le l$, using block-based shot segmentation disjoint set of k clips are generated, where $1 \le k \le l$.

Step 2:   Key frame corresponding to each k clip is generated using summation of equalized histogram differences between pivot and other frames of a clip.

Step 3:   Key frame is encoded using angular distribution density in which symbol assigned to it depends on maximum density angular region.

Step 4:   From the sequence of k symbol generated from k key frame, prefix queue pattern is generated by overlapping sliding window of size m (empirically set to 3).

Step 5:   Based on the generated prefix pattern, Pattern Tree is generated and each pattern which is a 2-tuple $(v^i, (k - 2)d)$, where $1 \le i \le N$, $1 \le d \le N$, is inserted in the node containing corresponding video identity with pattern position, $(k-2)d$ in generated prefix pattern. N refers to the number of videos in the database.

Step 6: Step 1 to Step 5 is repeated for each of the reference videos, and finally, Pattern Tree for whole database is constructed.

Step 7: Initially, all queue pool QPi, $1 \leq i \leq N$ is empty represented as $\neq$.

Step 8: For a query video, $Step1$ to $Step5$ is repeated and for each query pattern, if there is a match at any node in Pattern Tree, then corresponding reference video identity with the pattern position is inserted into the queue q_{ij} of queue pool QPi, where $0 \leq i \leq N$, $0 \leq j \leq N$, if $\gamma < \delta$, where empirically δ is set as 3 and γ is time stamp difference between two consecutive entry in queue in a particular queue pool.

Step 9: Initially, for first insertion in queue of a particular queue pool counter Cn is set 0 and active status Δ.

Step 10: For every insertion in a queue of queue pool, its counter Cn is set 0 and counter of other queue pool is incremented by 1. If any queue pool counter $Cn > \eta$, the queue of the queue pool is set inactive denoted by ∇ and new empty queue is generated in that queue pool.

Step 11: Final generated pool is used to evaluate *NR-score* with respect to every reference video which is calculated for the given query video for detecting near-duplicate reference videos.

4 Experimental Results and Discussions

The proposed method is simulated using MATLAB R2016a on a system with Intel(R) Core(TM) i3 Processor and 4GB RAM. MUSCLE VCD database [18] is used in our work with frame rate of 25 frames per second. Resolution of videos in the corresponding database is 240×240 which comprise of various topics such as educational, ephemeral, documentary. With the perspective of reducing the computational cost for video retrieval, key frame extraction has been considered. Query video key frames with its corresponding retrieved video key frames are shown in Fig. 4 which reflects the effectiveness of the proposed algorithm.

Key frame encoding provided the much needed property for effective retrieval which basically exploits the geometric point distribution of the key frames under constraint of rotation and translation invariance. Generation of queue pool based on the Pattern Tree which not only provide the data structure to match pattern between query and reference video but also temporal consistency is taken into consideration which provide an effective and efficient retrieval. Based on the near-duplicate score of each queue pool, retrieval of near-duplicate videos is computed. Subsequently, precision and recall serve as the fundamental measure to evaluate the accuracy of retrieval. Precision refers to the ratio between numbers of relevant data retrieved and total number of relevant data, while recall refers to the number of relevant data to the total number of data in the corresponding database. In Fig. 5, the retrieval accuracy

Fig. 4 Retrieval of query reference video key frames

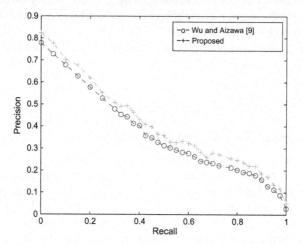

Fig. 5 Retrieval performance analysis

of the proposed algorithm is shown which depicts its effectiveness as compared to Wu and Aizawa [9].

5 Conclusion and Future Work

In this paper, the motivation was to develop a simple but effective hierarchical spatiotemporal approach for effective high-quality near-duplicate video retrieval. Pattern generation of encoded key frames using angular distribution density provided rotation and translation invariance. Queue pool generation using the Pattern Tree generated from reference video encoded pattern contributes temporal matching and consistency. Experimental result analysis on MUSCLE VCD dataset demonstrates promising results of the proposed approach which not only handle spatio and temporal variation, but also contribute cost-effective computation. In future, our focus will be on developing better near-duplicate retrieval system for fast retrieve by using approximate similarity search approach such as Locality Sensitivity Hashing (LSH) and global interest point for better feature representation.

References

1. Pickering M.J, and Ruger S., Evaluation of key frame-based retrieval techniques for video, Computer Vision and Image Understanding, Academic Press Inc Elsevier Science, pp. 217–235, 2003.
2. C. Kim and B. Vasudev, Spatiotemporal sequence matching for efficient video copy detection, IEEE Trans. Circuits Syst. Video Technol., vol. 15, no. 1, pp. 127–132, 2005.
3. C. Bohm, S. Berchitold, and D. A. Keim, Searching in highdimensional spaces: Index structures for improving the performance of multimedia databases, CM Comput. Survey, vol. 33, no. 3, pp. 322–373, 2001.
4. F. Hartung and M. Kutter, Multimedia watermarking techniques, Proc. IEEE, vol. 87, no. 7, pp. 1079–1107, 1999.
5. Yanqiang Lei, WeiqiLuo, Yuangen Wang, Jiwu Huang, Video Sequence Matching Based On The Invariance Of Color Correlation, IEEE Transactions On Circuits And Systems For Video Technology, vol. 22, no. 9, September 2012.
6. J. Liu, Z. Huang, H. Cai, H. T. Shen, C. W. Ngo, and W. Wang, Nearduplicate video retrieval: Current research and future trends, ACM Comput. Surveys, vol. 45, no. 4, pp. 1–23, 2013.
7. Rao, A., Srihari, R. K., AND Zhang, Z. Spatial color histograms for content-based image retrieval. Proceedings of the Eleventh IEEE International Conference on Tools with Artificial Intelligence, 1999.
8. X. Wu, C. W. Ngo, A. Hauptmann, and H. K. Tan, Real-time nearduplicate elimination for web video search with content and context, IEEE Trans. Multimedia, vol. 11, no. 2, pp. 196–207, 2009.
9. Z. Wu and K. Aizawa, Self-similarity-based partial near-duplicate video retrieval and alignment, Int. J. Multimedia Inf. Retrieval, vol. 3, no. 1, pp. 1–14, 2014.
10. C. Y. Chiu, C. S. Chen, and L. F. Chien, A framework for handling spatiotemporal variations in video copy detection, IEEE Trans. Circuits Syst. Video Technol., vol. 18, no. 3, pp. 412–417, 2008.
11. R. Roopalakshmi and G. R. M. Reddy, A novel spatio-temporal registration framework for video copy localization based on multimodal Features, Signal Process., vol. 93, no. 8, pp. 2339–2351, 2013.
12. C. L. Chou, H. T. Chen, Y. C. Chen, C. P. Ho, and S. Y. Lee, Near- duplicate video retrieval and localization using pattern set based dynamic programming, in Proc. 2013 IEEE Int. Conf. Multimedia Expo, pp. 1–6 Jul., 2013.
13. Y. Tian, T. Huang, M. Jiang, and W. Gao, Video copy-detection and localization with a scalable cascading framework, IEEE Multimedia, vol. 20, no. 3, pp. 72–86, Jul. Sep. 2013.
14. J. H. Su, Y. T. Huang, H. H. Yeh, and V. S. Tseng, Effective contentbased video retrieval using pattern indexing and matching techniques, Expert Syst. Appl., vol. 37, no. 7, pp. 5068–5085, 2010.
15. R. Chaudhry, A. Ravichandran, G. Hager, and R. Vidal, Histograms of oriented optical flow and Binet-Cauchy kernels on nonlinear dynamical systems for the recognition of human actions, in Proc. 2009 IEEE Conf. Comput. Vis. Pattern Recog., pp. 1932–1939, Jun. 2009.

16. E. Rosten, R. Porter, and T. Drummond, Faster and better: A machine learning approach to corner detection, IEEE Trans. Pattern Anal. Mach. Intell., vol. 32, no. 1, pp. 105–119, 2010.

17. Chou, C.L., Chen, H.T., Lee, S.Y.: Pattern-based near-duplicate video retrieval and localization on web-scale videos. IEEE Trans. Multimedia 17(3), 382–395, 2015.

18. The Open Video Project. (1998). A Shared Digital Video Collection [Online]. Available: http://www.open-video.org.

Fingerprint Liveness Detection Using Wavelet-Based Completed LBP Descriptor

Jayshree Kundargi and R. G. Karandikar

Abstract Fingerprint-based authentication systems need to be secured against spoof attacks. In this paper, we propose completed local binary pattern (CLBP) texture descriptor with wavelet transform (WT) for fingerprint liveness detection. The fundamental basis of the proposed method is live, and spoof finger images differ in textural characteristics due to gray-level variations. These textural characteristics occur at various scales and orientations. CLBP has high discriminatory power as it takes into account local sign and magnitude difference with average gray level of an image. CLBP extended to 2-D Discrete WT (DWT), and 2-D Real Oriented Dual Tree WT (RODTWT) domain captures texture features at multiple scales and orientations. Each image was decomposed up to four levels, and CLBP features computed at each level are classified using linear and RBF kernel support vector machine (SVM) classifiers. Extensive comparisons are made to evaluate influence of wavelet decomposition level, wavelet type, number of wavelet orientations, and feature normalization method on fingerprint classification performance. CLBP in WT domain has proved to offer effective classification performance with simplicity of computation. While texture features at each scale contribute to performance, higher performance is achieved at lower decomposition levels of high resolution with db2 and db1 wavelets, RBF SVM and mean normalized features.

Keywords Fingerprint liveness detection · Multiscale texture features
Wavelet transform · Completed local binary patterns · Support vector machine

J. Kundargi (✉) · R. G. Karandikar
K. J. Somaiya College of Engineering, Mumbai, India
e-mail: jmkundargi@somaiya.edu

R. G. Karandikar
e-mail: rameshkarandikar@somaiya.edu

© Springer Nature Singapore Pte Ltd. 2018
B. B. Chaudhuri et al. (eds.), *Proceedings of 2nd International Conference on Computer Vision & Image Processing*, Advances in Intelligent Systems and Computing 703, https://doi.org/10.1007/978-981-10-7895-8_15

1 Introduction

Fingerprint-based biometric authentication systems are most commonly deployed for various security-based applications. Fingerprint biometric has the advantages of being unique, reliable, and simplicity of acquisition. It was proved that these systems can be attacked by spoof fingers made of synthetic materials like clay, silicone. [1]. A large number of methods have been proposed in the literature to protect fingerprint authentication systems against spoof attacks. In particular, each of these methods proposed a technique to identify whether an acquired fingerprint image is from a live finger or an artificial finger. Earlier systems are hardware-based which consist of an additional device to acquire some biological trait such as body odor [2]. Since these methods used additional information, they guaranteed reliable protection against spoof attacks. However, they are more costly, rigid, invasive, and inconvenient to the users. Introduction of software-based methods paved the way for economical, flexible, and noninvasive systems. Typically, these methods use image processing techniques to find discriminatory characteristics of a live fingerprint image which would not be present in a spoof fingerprint image. Early software-based systems used change in the dynamic characteristics of a live fingerprint image which would otherwise be constant in a spoof image, by acquiring multiple images of the same finger over a specific time duration [3]. These methods were time consuming. Static software-based methods acquire a single image from the user and are more convenient and user-friendly. Most of the researched methods fall in this category. These methods extract discriminatory features from the images to perform the classification and have proven to be very challenging.

A live finger is characterized by the presence of pores along the ridges through which sweat is released due to perspiration process. Random ridge–valley structure, skin elasticity, and presence of pores and sweat cause significant wide and random gray-level variations in a live finger image. These gray-level variations constitute textures of multiple scales in multiple directions. A spoof finger cannot experience sweat and has a regular ridge–valley structure based on the synthetic material properties resulting in a few gray-level variations in the acquired image. The contribution of the features at each scale will be different in classification performance. In addition, live fingerprint images will exhibit discriminatory features over a large range of scales due to high gray-level variations.

Texture reflects gray-level statistics of an image which includes spatial distribution and structural information of an image [4]. Texture feature plays a very crucial role in many image classification-related works and is widely researched for fingerprint liveness detection. Texture analysis in WT domain has often reported encouraging results. Images are decomposed into multiple levels of resolution from fine to coarse, and features are extracted from low-pass and high-pass subbands of each scale. Subbands at each scale reflect local image characteristics of the original image at respective scale. In addition, high-pass detail subbands reflect directional characteristics in the original image. Therefore, effective texture features computed from all subbands will achieve encouraging results. Many of the proposed WT domain methods for fingerprint liveness detection concatenated texture features at all scales [5–7]. Multiscale features can also be acquired in spatial domain by increasing the radius of feature operator [8, 9]. However, these methods tend to be computationally intensive.

In this paper, we analyze the performance of texture features at each scale. We propose use of CLBP in DWT and RODTWT domain. RODTWT has advantage of better discrimination of features that occur in multiple orientations [10]. It decomposes an image into six detail subbands in six orientations and low-resolution approximation subbands. In our work, we investigated the influence of RODTWT-based features on fingerprint liveness detection performance. Our method computes CLBP texture features from detail and approximation subband coefficients. To analyze the performance of each scale, a feature vector is computed by concatenation of CLBP features of detail and approximation subbands of that level. Local binary pattern (LBP) has proved to be a simple and effective texture descriptor and is used successfully for texture classification [11]. CLBP operator derived as an extension of LBP has higher discriminatory power [12]. We use rotation invariant CLBP operator jointly computed from local sign difference, S, local magnitude difference, M, and average gray level of image, C, computed over a radius of 1 with 8 neighbors designated as $CLBP_{8,1}^{RIU2}\_S\backslash M\backslash C$. We explore Daubechies wavelets db1–db4 to implement DWT. Each of these wavelets has different support area and capture signal features of a particular shape and size. The choice of a wavelet for a particular application is experiment-driven. Extension of CLBP operator to DWT and RODTWT domain results in extraction of powerful discriminatory features at multiple scales and orientations for efficient fingerprint liveness detection. The computed feature vectors have different ranges of values. It is often beneficial to normalize all the features to a common scale. It is essential to choose a suitable normalization method as the SVM performance is sensitive to the method selected. We compare performance of two methods, wherein in the first method (N1) features are scaled to the range 0–1 and in the second method (N2) each feature is normalized by subtracting its mean and dividing by its standard deviation. Supervised classifiers linear SVM and RBF kernel SVM are used in this work as these have proved to offer superior performance in many machine vision-related applications. Being a preliminary study, SVM is operated with default parameters, $C = 1$ and gamma $= 1/($feature dimension$)$, for linear and RBF kernel SVM, respectively. Tuning of SVM parameters to the training dataset is very much likely to improve the performance further. Subsequently, contribution of features at each scale, choice of suitable wavelet, type of WT, normalization method, and SVM classifier type are evaluated based on the classification error. The experimental results are provided on LivDet 2011 and LivDet 2013 datasets that consist of live and spoof images acquired using four sensors [10, 13]. Average classification error (ACE) is used for the performance evaluation.

The rest of the paper consists of four sections. Section 2 contains brief review of existing fingerprint liveness detection methods with emphasis on single-image-based multiscale feature extraction methods. Section 3 describes the proposed method, and Sect. 4 presents experimental results. Conclusions are presented in Sect. 5.

2   Related Work

A live fingerprint image is characterized by abundant and strong textural information due to random ridge–valley structure, specific pore distribution, skin elasticity properties, and sweat on a fingertip surface. Spoof fingers fabricated from synthetic materials like gelatin and latex have constant material and physical properties leading to minimal texture patterns. A large number of texture-based methods have been proposed in the last few years. Among them, multiresolution transform domain methods are widely researched due to their ability to capture multiscale features in multiple orientations. Multiresolution wavelet transform was proposed to analyze fingerprint image liveness by deriving texture features and local ridge frequency features from multiscale pyramid coefficients [14]. It was the first single-image-based effort based on textural features. LBP descriptor, being powerful, has been widely used in many machine vision-related applications [11]. First method to propose use of features derived from wavelet transform detail coefficients of four scales combined with LBP features of image was presented for fingerprint liveness detection in [5]. Each image was decomposed up to four levels using three types of wavelet filters, energy of all detail subbands of four levels for each wavelet filter and of original image constituted wavelet domain feature vector to represent image properties at multiple scales and orientations. Wavelet transform is suitable to highlight point singularities in an image. A fingerprint image consists of curvilinear regions due to ridge–valley structure. Such curvilinear discontinuities can be well captured by curvelet transform. Use of multiscale curvelet transform with texture features derived from three decomposition scales was proposed in [6]. Each image was decomposed up to three scales. Energy of detail subbands of all levels formed one feature vector. Another set of texture features was computed from gray-level co-occurrence matrices (GLCM) derived from each curvelet subband of all levels. The two feature vectors were separately tested. Each of the above two methods used feature selection method due to large feature size. The same authors proposed use of multiscale ridgelet transform to derive texture features [7]. Image was divided into small size subblocks, and ridgelet transform was applied to each subblock. Energy of ridgelet subbands at first decomposition level for each subblock was computed and formed energy feature vector. Co-occurrence features were computed from GLCM derived from each ridgelet subband. Ridgelet transform is suitable to extract line singularities in an image corresponding to ridge lines as compared to wavelet transform. Multiscale transform-based methods rely on features used to extract information from subband coefficients and are computationally simple.

   LBP operator has too small spatial support area as a result of which it cannot handle multiple scale variations in gray levels. Multiresolution form of LBP was presented by using multiple size neighborhoods to capture multiscale texture features in spatial domain [8]. However, the method is computationally intensive. It was pointed out that this approach is sensitive to noise due to sampling at a single pixel position and may cause aliasing [8, 11]. The suggested solutions involved either use of filters to gather information over a region or averaging information over a region [8, 11, 15]. Based on it, Multiscale Block Local Ternary Pattern was proposed for fingerprint liveness detection [9]. Local differences are computed over average values of blocks of pixels to avoid sensitivity to noise, and local differences were encoded into three levels which make it more discriminatory compared to LBP. These methods are computationally intensive and parameter-driven as the exact size of block is a free parameter and size suitable for efficient feature extraction for a given application has to be determined experimentally.

There are other carefully researched state-of-the-art methods for fingerprint liveness detection. A Local Phase Quantization descriptor obtained by binarizing phase information corresponding to four low-frequency components in a local window was proposed in [16]. Known for its insensitivity to blurriness, the performance on LivDet 2011 database is moderate. A learning-based Binarized Statistical Image Feature descriptor hat use filter designed from a large number of natural images was proposed in [17]. Experiments reported significantly low error rate on LivDet 2011 dataset at the cost of large feature size. Local Contrast Phase Descriptor feature consisting of local phase information and local amplitude contrast was proposed in [18]. The best results were reported on LivDet 2011 dataset. Spoof images in LivDet 2011 dataset are obtained using cooperative method and hence are more challenging than LivDet 2013, wherein spoof images are obtained using noncooperative method. In this paper, we have carried out work on LivDet 2011 dataset.

3 Proposed Method

The procedure for the proposed method for fingerprint liveness detection is shown in Fig. 1. It involves basic steps of preprocessing, decomposition using DWT/RODTWT up to four levels, $CLBP_{8,1}^{RIU2}\_S\backslash M\backslash C$ feature extraction from wavelet subbands, feature normalization, training of SVM classifiers with training data, and finally classification of test dataset using trained SVM classifier. In the next five subsections, we describe the method in brief.

3.1 Image Preprocessing

Image preprocessing involves conversion of fingerprint images from RGB to gray format and image resizing by symmetric extension of image boundaries to enable wavelet decomposition up to four levels. We compare our method with other methods in which no other preprocessing operations such as segmentation and filtering are performed on database images. In feature-based image classification task, features are derived from the entire image. The background area and fingerprint image area are different for fingerprint images acquired using different sensors

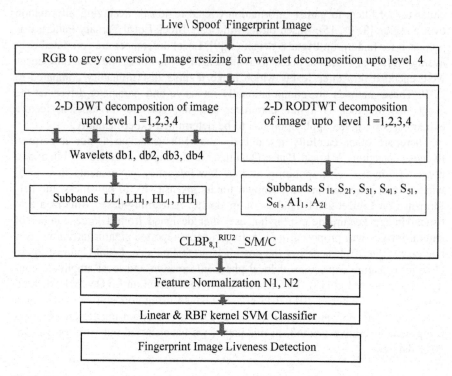

Fig. 1 Block diagram of proposed investigation of CLBP descriptor with multiscale wavelet representation for fingerprint liveness detection

and are likely to influence the classification performance. Hence, we evaluate our proposed method directly on the images without any operations.

3.2 2-D Discrete Wavelet Transform (DWT)

DWT is suitable for multiscale analysis as it can represent an image at multiple levels of resolution. DWT has been efficiently employed in many image processing applications such as image de-noising, image analysis, image segmentation, image compression, texture classification. Another important characteristic of DWT is its good space-frequency localization property which can preserve intrinsic spatial and textural information for efficient fingerprint liveness detection. DWT consists of three wavelet functions to produce high-frequency detail subbands to measure gray-level variations corresponding to textural information in horizontal, vertical, and diagonal directions and one scaling function which produces low-frequency approximation subband at the next coarse level of resolution. Wavelets by

I. Daubechies are orthogonal, exhibit time invariance, produce real coefficients, and have sharp transition bands to minimize boundary effects [19]. Each Daubechies wavelet has different support area that determines the nature of the signals that can be encoded. We use db1, db2, db3, and db4 to compare their performance for liveness detection. The approximation and detail subbands at each level 1 are denoted by LL_1, LH_1, HL_1, HH_1, respectively. Each image is decomposed up to four levels, and $CLBP_{8,1}^{RIU2}$_S\M\C features extracted from all bands at each level are denoted by LL_1_CLBP, LH_1_CLBP, HL_1_CLBP, and HH_1_CLBP where 1 varies from 1 to 4 for each decomposition level and CLBP stands for $CLBP_{8,1}^{RIU2}$_S\M\C. The feature size of $CLBP_{8,1}^{RIU2}$_S\M\C operator is 200.

3.3 2-D Real Oriented Dual Tree Wavelet Transform (RODTWT)

RODTWT is implemented using two separate DWTs in parallel. Six oriented detail subbands and two approximation subbands are formed by taking the sum and difference of pairs of DWT subbands. For mathematical details, refer to [10]. In our work, we used RODTWT to determine its effectiveness over DWT. The detail subbands can capture gray-level variations in the original image in six directions, $\{+15°, -15°, +45°, -45°, +75°, -75°\}$ and are designated as S_{1l}, S_{2l}, S_{3l}, S_{4l}, S_{5l}, S_{6l}, respectively, at each decomposition level l. The two approximation subbands are denoted as A_{1l} and A_{2l}, respectively. Each image is decomposed up to four levels, and $CLBP_{8,1}^{RIU2}$_S\M\C features extracted from all subbands are denoted by S_{1l}_CLBP, S_{2l}_CLBP, S_{3l}_CLBP, S_{4l}_CLBP, S_{5l}_CLBP, S_{6l}_CLBP, A_{1l}_CLBP, and A_{2l}_CLBP where l varies from 1 to 4 for each decomposition level and CLBP stands for $CLBP_{8,1}^{RIU2}$_S\M\C.

3.4 Completed Local Binary Pattern (CLBP)

CLBP was proposed to enhance discrimination capability of original LBP for effective feature extraction. CLBP consists of three operators CLBP_S, CLBP_M, and CLBP_C obtained by encoding local difference sign (S), local difference magnitude (M), and average gray level of image (C), respectively. M and S are obtained by using local difference sign magnitude transform (LDSMT) as specified in [12]:

$$d_p = g_p - g_c = s_p * m_p, \begin{cases} s_p = sign\left(d_p\right) \\ m_p = \left|d_p\right| \end{cases} \tag{1}$$

$m_p, s_p = \begin{cases} 1, d_p \geq 0 \\ -1, d_p < 0 \end{cases}$ are magnitude and sign of d_p

The three operators are jointly combined to produce CLBP feature histogram. In our work, we use uniform rotation invariant CLBP computed with radius of one over eight neighbors represented as $CLBP_{8,1}^{RIU2}\_S\backslash M\backslash C$.

3.5 Support Vector Machine (SVM)

The SVM is a state-of-the-art supervised classifier introduced in [20] and is successfully applied for many image classification tasks due to its high accuracy and ability to handle high-dimensional data. Originally proposed as a binary classifier, it classifies the data with a hyperplane into two regions. The boundary between the two regions is called decision boundary. Linear SVM with linear decision boundary is capable of classifying linearly separable data. However data which is not linearly separable, a nonlinear SVM with nonlinear decision boundary need to be used. A computationally simple way to convert linear decision boundary into nonlinear is use of kernel function. Commonly used kernel functions are Radial Basis Function (Gaussian), Polynomial and Sigmoid. In image classification tasks, it is not possible to know nature of data as features are of high dimensions. Choice of SVM is decided based on experimentation. In our work, we used linear SVM and RBF kernel nonlinear SVM, with default parameters, from LIBSVM package [21].

4 Experimental Results

In this section, after describing dataset used to evaluate performance of our proposed fingerprint liveness detection method, we present the investigated results. We conclude with comparison of achieved results with related state-of-the-art techniques and results of LivDet 2011 Fingerprint Liveness Detection competition.

4.1 Dataset

We conducted experiments on the challenging database made available for LivDet 2011 competition jointly organized by University of Cagliari and Clarkson University [22]. More details of LivDet 2011 dataset are reported in Table 1. For each scanner, the training and test datasets consist of 1000 live and 1000 spoof images with no overlap of images in train and test sets.

Table 1 LivDet 2011 dataset characteristics

Dataset	1	2	3	4
Scanner	Biometrika	Italdata	Digital persona	Sagem
Model no.	FX2000	ET10	400B	MSO300
Resolution (dpi)	500	500	500	500
Image size	312 * 372	640 * 480	355 * 391	352 * 384
Live samples	2000	2000	2000	2000
Spoof samples	2000	2000	2000	2000
Material used for spoof samples	Ecoflex, gelatine latex, silicone wood glue	Ecoflex, gelatine latex, silicone wood glue	Gelatine, latex play doh, silicone wood glue	Gelatine, latex play doh, silicone wood glue
Live fingers	200	200	200	112
Spoof fingers	81	81	100	100
Live subjects	200	200	52	20
Spoof subjects	34	34	42	68

5 Results

The experimental results are presented in terms of average classification error (ACE) as per the requirement of [22] and computed as:

$$ACE = \frac{FLR + FFR}{2} \tag{2}$$

False Living Rate (FLR) and False Fake Rate (FFR) are the percentages of spoof images classified as live and of live images classified as spoof, computed with liveness score threshold of 50 available from SVM classifier. In Table 2, we present comparison of ACE obtained, when $CLBP_{8,1}^{RIU2}\_S\backslash M\backslash C$ operator was applied to LivDet 2011 dataset in spatial domain with normalization methods N1 and N2 and linear and RBF SVM, with LBP results from the literature. Encouraged by these results, we extended CLBP operator to DWT domain. In Table 3, we present the results obtained in DWT domain for db1–db4 wavelets, normalization methods N1 −N2, decomposition levels l = 1–4 and linear and RBF SVM. Table 4 presents ACE over all the four scanners, used in [22] for overall performance comparison with CLBP-DWT, to determine the best combination of the involved factors. In Table 5, we present the results obtained using CLBP operator in RODTWT domain. Table 6 presents ACE over all the four scanners, used in [22] for overall performance comparison with CLBP-RODTWT.

Table 2 Comparison of ACE with original LBP in spatial domain

Method in spatial domain	Biometrika				Italdata				Best Ave. ACE
CLBP$_{8,1}^{RIU2}$_S\M\C	Linear SVM		RBF SVM		Linear SVM		RBF SVM		9.6
	N1	N2	N1	N2	N1	N2	N1	N2	
	11.5	10.3	19.5	**9.3**	22.3	24.5	29.1	**19.6**	
Original LBP [11]	13.0				24.1				
Method spatial domain	*Digital persona*				*Sagem*				
CLBP$_{8,1}^{RIU2}$_S\M\C	Linear SVM		RBF SVM		Linear SVM		RBF SVM		
	N1	N2	N1	N2	N1	N2	N1	N2	
	5.0	5.3	9.7	**4.8**	5.9	4.2	11.5	**4.7**	
Original LBP [11]	10.8				11.5				14.9

In Table 7, we present FLR and FFR obtained for four scanners using CLBP-DWT at first decomposition level using db2, RBF SVM, and normalization method N2. Table 8 presents comparison of the liveness detection result obtained by the proposed method with the related existing solutions in terms of average ACE. Figure 2 presents receiver operating characteristics for the four scanners obtained with CLBP-DWT, db2, RBF SVM, and N2 normalization method. From Tables 3 and 4, we see that in DWT domain the best liveness detection result was obtained with db2 wavelet at first decomposition level with RBF kernel SVM and mean normalized features, N2, for Biometrika and Italdata scanners. For Digital Persona wavelet db3 at level 1 with RBF kernel SVM produced the highest result with mean normalized features. For Sagem, we achieved the highest result at level 2 with db1 wavelet, RBF SVM with mean normalized features. The overall average classification error rate over four scanners was highest at level 1 with db2, RBF SVM, and mean normalized features. Tables 5 and 6 indicate that there is no improvement in classification performance with ORSTWT over DWT as was expected due to its multi-orientation analysis capability. Highest result was obtained for Biometrika and Italdata at first level with RBF SVM and mean normalized features. The best result for Digital Persona was achieved with linear SVM and N1 normalization method at first decomposition level. For Sagem, the highest result was obtained at third decomposition level with RBF SVM and mean normalized features.

For both DWT and RODTWT, not much difference was observed with the two normalization methods, i.e., N1 and N2 with linear SVM. However, the performance was RBF kernel SVM differed by significant margin for N1 and N2 indicating that RBF kernel SVM is sensitive to feature normalization methods. Superior results are obtained by the proposed method in spite of being a simple and general method.

Table 3 ACE comparison for all scanners to investigate influence of DWT wavelet type, decomposition level, normalization method, and SVM classifier, best results highlighted in bold with underline

CLBP-DWT domain wavelet	db1				db2				db3				db4			
	Linear SVM		RBF SVM		Linear SVM		RBF SVM		Linear SVM		RBF SVM		Linear SVM		RBF SVM	
Scanner, decomp. level no.	N1	N2	N1	N2	N1	N2	N1	N2	N1	N2	N1	N2	N1	N2	N1	N2
Biometrika																
Level 1	8.4	8.0	19.9	8.8	5.8	6.2	15.3	**4.5**	8.0	8.7	15.5	6.6	8.3	9.1	17.0	5.9
Level 2	16.7	18.2	26	15.3	16.0	18.1	21.9	12.4	17.4	18.4	20.4	12.0	18.1	18.8	23.1	14.1
Level 3	23.0	22.8	21.7	23.0	19.9	20.5	25.3	17.2	24.8	25.5	26.8	19.0	24.4	24.5	28.6	19.9
Level 4	28.6	29.4	21.7	23.0	30.8	31.7	21.9	23.6	32.8	37.0	28.7	25.6	30.5	30.5	28.6	23.2
Italdata																
Level 1	18.9	21.8	23.7	17.2	18.6	19.3	24.4	**16.7**	21.7	25.2	27.1	18.7	23.6	24.3	25.3	17.9
Level 2	40.5	40.8	37.1	30.7	27.0	28.9	30.0	24.5	27.3	27.5	29.2	21.1	29.9	27.3	33.7	23.9
Level 3	35.9	35.0	40.4	29.2	46.0	46.7	43.5	40.4	38.8	37.7	38.4	31.8	36.6	36.7	37.6	30.0
Level 4	38.6	39.0	44.0	33.7	39.1	42.6	45.2	40.2	46.9	47.6	45.2	42.3	43.0	43.2	43.4	36.9
Digital per.																
Level 1	4.9	4.9	12.2	5.0	4.7	5.1	11.6	5.0	4.7	5.0	10.9	**4.1**	5.7	5.3	12.4	4.5
Level 2	7.6	8.0	12.7	7.9	7.1	7.4	13.5	6.5	8.2	9.5	18.1	7.2	9.2	9.9	14.5	7.5
Level 3	13.2	15.2	17.5	9.8	12.6	13.7	17.7	10.0	14.1	18.4	19.6	10.7	15.9	16.9	19.6	12.3
Level 4	13.5	15.0	19.7	8.5	11.7	12.6	16.1	6.6	12.6	12.9	16.0	7.5	13.8	15.3	15.7	9.8
Sagem																
Level 1	8.0	8.6	14.9	6.1	8.0	8.8	17.1	6.9	8.7	9.1	18.8	7.05	10.0	10.5	21.5	7.6
Level 2	10.0	10.8	13.8	**4.5**	9.9	10.7	21.4	8.8	11.1	10.6	24.1	8.7	12.1	13.5	25.1	9.3
Level 3	13.1	13.8	15.9	7.3	13.4	14.5	15.6	8.2	12.9	12.9	12.9	8.1	15.0	15.9	14.7	8.2
Level 4	14.2	14.5	22.1	9.0	16.7	17.2	23.9	10.5	16.6	18.3	23.0	10.8	16.4	18.3	21.0	10.7

Table 4 Avg. ACE computed over all four scanners, best result indicated in bold with underline

CLBP-DWT domain wavelet	db1				db2				db3				db4			
	Linear SVM		RBF SVM		Linear SVM		RBF SVM		Linear SVM		RBF SVM		Linear SVM		RBF SVM	
Decomposition level no.	N1	N2	N1	N2	N1	N2	N1	N2	N1	N2	N1	N2	N1	N2	N1	N2
Level 1	10.0	10.8	17.7	9.2	9.3	9.8	17.1	**8.3**	10.8	12.2	19.0	9.0	11.9	12.3	19.0	9.0
Level 2	18.7	19.4	22.4	15.4	15.0	16.2	21.7	13.0	16.0	16.5	22.9	12.2	17.3	17.3	24.1	13.7
Level 3	21.3	21.7	24.3	16.0	23.0	23.9	25.5	19.1	22.6	23.6	24.4	17.4	23.0	23.3	25.1	17.6
Level 4	23.7	24.4	26.9	18.5	24.4	26.0	28.7	20.2	27.2	28.9	28.2	21.5	25.9	26.8	27.2	20.1

Table 5 ACE comparison for all scanners to investigate influence of RODTWT, decomposition level, normalization method, and SVM classifier, best results highlighted in bold with underline

CLBP-RODTWT domain		Biometrika				Italdata				Digital persona				Sagem			
		Decomposition level				Decomposition level				Decomposition level				Decomposition level			
		1	2	3	4	1	2	3	4	1	2	3	4	1	2	3	4
Linear SVM	N1	6.9	12.4	21.1	24.0	22.5	30.3	31.3	45.5	**3.9**	6.9	9.2	9.6	8.2	12.1	11.7	11.9
	N2	7.2	12.7	20.6	23.3	23.7	29.5	33.3	44.7	4.1	7.8	11.6	9.7	8.9	11.6	11.7	12.7
RBF SVM	N1	16.0	21.0	25.6	26.7	25.9	31.7	36.4	45.5	12.2	13.0	15.9	12.5	18.0	23.7	10.7	20.6
	N2	**5.7**	9.4	18.0	20.4	**17.7**	25.0	29.2	44.7	5.5	5.9	7.7	6.2	7.2	9.2	**6.1**	8.4

Table 6 Avg. ACE computed over all four scanners, best result indicated in bold with underline, LivDet 2011

CLBP-RODTWT domain		ACE over four scanners			
		Decomposition level			
		1	2	3	4
Linear SVM	N1	10.4	15.4	18.3	22.6
	N2	10.9	15.4	19.2	22.6
RBF SVM	N1	18.0	22.3	22.1	26.3
	N2	**9.0**	12.3	15.2	19.9

Table 7 Comparison of FLR and FFR for LivDet 2011

Scanner	FLR	FFR
Biometrika	5.3	3.6
Italdata	16.8	16.6
Digital persona	6.4	3.6
Sagem	4.3	9.5

Table 8 Comparison of ACE with existing methods and LivDet 2011 competition results

Method	Biometrika	Italdata	Digital persona	Sagem	Avg. ACE
Proposed. method CLBP_DWT	4.5	16.7	5.0	6.9	**8.3**
MBLTP [9]	10.0	16.3	6.9	5.9	9.77
MLBP [8]	10.8	16.6	7.1	6.4	10.22
LCPD [18]	4.9	11.0	4.2	2.7	5.7
BSIF [17]	6.8	13.6	3.5	4.9	7.2
LPQ [16]	12.8	15.6	9.7	8.4	12.3
LBP [11]	13.0	24.1	10.8	11.5	14.85
Curvelet [6]	45.2	47.9	21.9	28.5	35.87
C. GLCM [6]	22.9	30.7	18.3	28.0	24.97
Wavelet [5]	50.2	46.8	14.0	22.0	33.25
Darma.g [22]	22.0	21.8	36.11	13.8	22.92
Federico [22]	40.0	40.0	8.9	13.4	25.57
CASIA [22]	33.9	26.7	25.4	22.8	27.2

Fig. 2 ROC for four
scanners of LivDet 2011
dataset

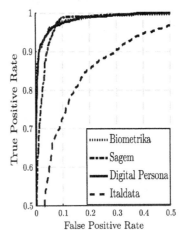

6 Conclusion

In this paper, a completed local binary pattern with multiscale wavelet representation is proposed. It has demonstrated a superior discriminative power with promising results for fingerprint liveness detection. It has achieved a significant performance in spite of being a general purpose texture descriptor. The method demonstrates the influence of texture features at different scales on fingerprint liveness detection. Daubechies wavelets db2 and db1 have achieved superior classification performance indicating their suitability. The proposed method is computationally simple and efficient than multiscale spatial domain LBP operator. Fingerprint images from each scanner have unique properties. Careful selection of features, feature normalization method, and classifier are essential to achieve better results. Features at each scale offer different classification performance. Careful study of contribution of features at each scale is likely to improve the fingerprint liveness detection performance.

References

1. T. Matsumoto, H. Matsumoto, K. Yamada, and S. Hoshino: Impact of Artificial Gummy Fingers on Fingerprint Systems. In: Proc. of SPIE, vol. 4677, pp. 275–289, (2002).
2. D. Baldissera, A. Franco, D. Maio, and D. Maltoni: Fake Fingerprint Detection by Odor Analysis. In: Proc. of International Conference on Biometric Authentication, (2006).
3. S.T.V. Parthasaradhi, R. Derakhshani, L.A. Hornak, and S. A C Schuckers: Time-series detection of perspiration as a liveness test in fingerprint devices. In: IEEE Transactions on

Systems, Man, and Cybernetics, Part C, Applications and Reviews, vol. 35, no. 3, pp. 335–343, (2005).

4. Haralick R M, Shanmugam K, Dinstein I H.: Textural features for image classification. In: IEEE Transactions on Systems, Man and Cybernetics, vol. SMC-3,no. 6, pp. 610–621, (1973).

5. S. Nikam and S. Agarwal.: Texture and Wavelet-Based Spoof Fingerprint Detection for Fingerprint Biometric Systems. In: First International Conference on Emerging Trends in Engineering and Technology, pp. 675–680, (2008).

6. S. Nikam and S. Agarwal.: Fingerprint Liveness Detection using Curvelet Energy and Co-occurrence Signatures. In: IEEE Fifth International Conference on Computer Graphics, Imaging and Visualisation (CGIV), pp. 217–222, (2008).

7. S. Nikam and S. Agarwal.: Ridgelet-Based Fake Fingerprint Detection. In: Neurocomputing, 72, 2491–2506, (2009).

8. T. Ojala, M. Pietikainen, and T. Maenpaa.: Multiresolution gray-scale and rotation invariant texture classification with local binary patterns. In: IEEE Transactions on Pattern Analysis and Machine Intelligence, vol. 24, no. 7, pp. 971–987, (2002).

9. X. Jia, X. Yang, Y. Zang, N. Zhang, R. Dai, J. Tian, and J. Zhao.: Multi-scale Block Local Ternary Patterns for Fingerprints Vitality Detection. In: International Conference on Biometrics (ICB), pages 1–6, (2013).

10. I. Selesnick, R. Baraniuk, and N. Kingsbury.: The dual-tree complex wavelet transform. In: IEEE Signal Process. Mag., vol. 22, no. 6, pp. 123–151, (2005).

11. Timo Ojala, Matti Pietikinen, and David Harwood.: A comparative study of texture measures with classification based on featured distributions. In: Pattern Recognition, vol. 29, no. 1, pp. 51–59, (1996).

12. Z. Guo, D. Zhang.: A completed modeling of local binary pattern operator for texture classification. In: IEEE Transaction on Image Processing, vol. 19, pp. 1657–1663, (2010).

13. L. Ghiani, D. Yambay, V. Mura, S. Tocco, G.L. Marcialis, F. Roli, and S. Schuckers.: LivDet 2013 Fingerprint liveness detection competition 2013. In: Proc. Int. Conf. Biometrics, pp. 1–6, (2013).

14. A. Abhyankar and S. Schuckers.: Fingerprint liveness detection using local ridge frequencies and multiresolution texture analysis techniques. In: Proc. IEEE International Conference on Image Processing, pp. 321–324, (2006).

15. S. Liao, X. Zhu, Z. Lei, L. Zhang, S. Z. Li.: Learning multi-scale block local binary patterns for face recognition. In: Proceedings of the ICB, (2007).

16. L. Ghiani, G. Marcialis, F. Roli.: Fingerprint Liveness Detection By Local Phase Quantization. In: International Conference on Pattern Recognition, pp. 537–540, (2012).

17. L. Ghiani, A. Hadid, G. Marcialis, F. Roli.: Fingerprint Liveness Detection Using Binarized Statistical Image Features. In: IEEE International Conference on Biometrics: Theory, Applications and Systems-BTAS, 2013, pp. 1–6, (2013).

18. D. Gragnaniello, G. Poggi, C. Sansone, and L. Verdoliva.: Local Contrast Phase Descriptor for Fingerprint Liveness Detection. In: Pattern Recognit., vol. 48, no. 4, pp. 1050–1058, (2015).

19. I. Daubechies.: Ten Lectures on Wavelets.: SIAM, (1992).

20. Boser, B.E., Guyon, I.M., and Vapnik, V.N.: A training algorithm for optimal margin classifiers. In: 5th Annual ACM Workshop on COLT, pp. 144–152, (1992).

21. Chih-Chung Chang and Chih-Jen Lin.: LIBSVM: a library for support vector machines. In: ACM Transactions on Intelligent Systems and Technology, 2:27:1—27:27, (2011).

22. Yambay, L. Ghiani, P. Denti, G. L. Marcialis, F. Roli, and S. Schuckers.: LivDet 2011 Fingerprint Liveness Detection Competition 2011. In: Proc. 5th IAPR/IEEE Int. Conf. Biometrics, pp. 208–215, (2012).

Silhouette-Based Real-Time Object Detection and Tracking

Bhaumik Vaidya, Harendra Panchal and Chirag Paunwala

Abstract Object detection and tracking in the video sequence is a challenging task and time-consuming process. Intrinsic factors like pose, appearance, variation in scale and extrinsic factors like variation in illumination, occlusion, and clutter are major challenges in object detection and tracking. The main objective of the tracking algorithm is accuracy and speed in each frame. We propose the best combination of detection and tracking algorithm which performs very efficiently in real time. In proposed algorithm, object detection task is performed from given sketch using Fast Directional Chamfer Matching (FDCM) which is capable of handling some amount of deformation in edges. To deal with the articulation condition, part decomposition algorithm is used in the proposed algorithm. Combination of these two parts is capable enough to handle deformation in shape automatically. Amount of time taken to perform this algorithm depends on the size and edge segment in the input frame. For object tracking, Speeded up Robust Features (SURF) algorithm is used because of its rotation invariant and fast performance features. The proposed algorithm works in all situations without the prior knowledge about number of frames.

Keywords Convexity defects · Fast directional chamfer matching
Part decomposition · Speeded up robust features

1 Introduction

The ongoing research topic in computer technology that makes efforts to detect, recognize, and track objects through a series of picture frames is known as object detection and tracking. Along with that, it makes an attempt to determine and

B. Vaidya (✉)
GTU, Ahmedabad, India
e-mail: vaidya.bhaumik@gmail.com

H. Panchal · C. Paunwala
EC Department, SCET, Surat, India

© Springer Nature Singapore Pte Ltd. 2018
B. B. Chaudhuri et al. (eds.), *Proceedings of 2nd International Conference on Computer Vision & Image Processing*, Advances in Intelligent Systems and Computing 703, https://doi.org/10.1007/978-981-10-7895-8_16

narrate attributes of an object, thereby superseding the obsolete traditional method of supervising cameras manually. Object detection and tracking is consequential and challenging task in large number of vision applications such as vehicle navigation system, self-govern robot navigation and surveillance, traffic control.

Accurate object detection is an important step in object tracking. It needs an object detection step either in every frame or the frame in which the object appears first. While going for object tracking, basic operation is to make separation of interested object called 'foreground' from 'background' [1]. There are three important steps in video analysis—detection of objects, tracking of these objects from every frame to frame, and analysis of their tracks to determine their attributes. Some work in literature focused on developing algorithms for automatic detection and tracking which reduces the need of human surveillance. In some environments, the background is not available and can always be changed under critical situations like illumination changes, objects being introduced or removed from the scene. So, the background representation model must be more robust and adaptive [1].

Representation of shapes for matching is well-studied problem, and there are existing two types of categories: (1) Appearance-related cues of objects and (2) Contour or silhouettes or voxel sets [2]. The proposed algorithm is based on second type of category as it can deal with deformation in object shape easily [2].

The remaining paper is arranged as follows: Sect. 2 summarizes the related work done in object detection and tracking; Sect. 3 describes problem formulation; Sect. 4 describes FDCM algorithm in detail; Sect. 5 describes SURF algorithm in detail; Sect. 6 summarizes experiments and results of proposed algorithm.

2 Related Work

For detection of object, background modeling is very important. In basic model, background is modeled using an average, a median, or a histogram analysis over time [3]. In this case, once the model is computed, the pixels of the current image are categorized as foreground by thresholding the difference between the background image and the current frame. The statistical models are more robust to illumination changes and for dynamic backgrounds [4]. They can be categorized as Gaussian models, support vector models, and subspace learning models. Gaussian is the easiest way to represent a background [4]. It expects the history over the time of pixel's intensity values. But a single model cannot deal with the dynamic background, for example, waving trees, rippling water. To overcome this problem, the Mixture of Gaussians (MoG) or Gaussian Mixture Model (GMM) is used [4].

The second category utilizes more sophisticated statistical models, for example, Support Vector Machine (SVM) [5], Support Vector Regression (SVR) [6]. In estimation models, the filter is used to estimate the background. The filter may be a Wiener filter [7] or a Kalman filter [8] or Chebychev filter [9]. The Wiener filter works well for periodically changing pixels and it produces a larger value of the threshold for random changes that are utilized in the foreground detection. The main advantage of the Wiener filter is that it lessens the uncertainty of a pixel value by representing how it varies with time. A drawback occurs when a moving object corrupts the history values [7]. The Chebychev filter slowly updates the background for changes in lighting and scenery while making the use of a small memory footprint with low computational complexity [9]. For cluster modeling K-means models, codebook models, neural network models can be used [10, 11].

It can be seen that most of the algorithms detect the moving object with different accuracy and conditions. The background initialization and updating of background after specific time are necessary in all of them. Some of the other techniques are based on contour-based approach where background initialization is not needed. Contour-based approach gives best result in most of the conditions like illumination change, bootstrapping, dynamic backgrounds [12]. Shape band algorithm [12] which is a contour-based model detects an object within a bandwidth of its sketch/contour. However, it not only detects/matches the identical template and input image but also captures the reasonable variation and deformation of object in the same class. Sreyasee and Anurag [2] presented an algorithm for object detection using part-based deformable template. They are using part-wise hierarchical structure decomposition for matching the template with the parts of object. For this, Gopalan's algorithm [13] is used for estimating the parts of the shape through approximate convex decomposition by measure of convexity and decomposed convex shape with junction of concave region.

The proposed algorithm uses convexity defect for finding junction points from where object convex parts are decomposed. Then parts are matched using FDCM and then SURF is used for tracking.

3 Proposed Algorithm

Figure 1 shows the flowchart of proposed algorithm for object detection and tracking. From the given input image or sketch, shape segmentation and part decomposition are done automatically and stored in the database on the server side. On client side from a test video, first frame is extracted and parts are matched using FDCM. Then using SURF, key points are extracted and matched. Individual steps are explained in detail, and results for individual steps are shown in next section.

3.1 Part-wise Decomposition

Figure 2 shows the flowchart for part-wise decomposition.

A. Input Sketch Model

The object sketches of human or animal are common articulation sketches used. It has large deformation if the viewpoint changes or part of object is in action and

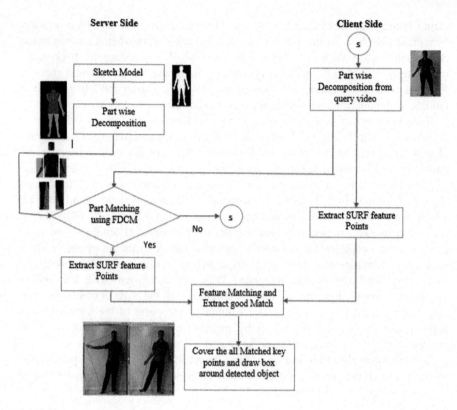

Fig. 1 Flowchart of proposed algorithm

Fig. 2 Flowchart for
part-wise decomposition

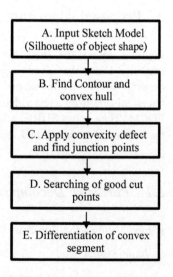

Fig. 3 **a** Human silhouette
b green line denotes contour
and blue line convex hull

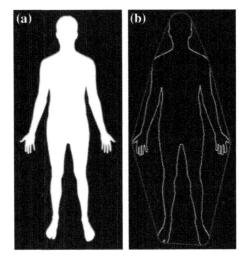

movement. To deal with this situation, the proposed algorithm uses convexity defect to identify the joint from where the articulating shape can undergo movement.

B. Contour extraction and Convex Hull Creation

Contours can be explained simply as a curve joining all the continuous points (along the boundary), having same color or intensity. The contours are useful tool for shape analysis, object detection, and recognition. The Suzuki's algorithm [14] for retrieving contours from the binary image is used in proposed algorithm which gives only the outermost borders.

The convex hull [15] of a planar shape P is defined by finding the smallest convex region enclosing shape P. For example, to fit shape P into a polygon, it is necessary and sufficient that the convex hull CH(P) of shape P fits. It is usually the case that CH(P) has less points than shape P.

C. Convexity defects

The convexity defect [16] is actually the space between the convex hull and actual object. As shown in Fig. 5, human body silhouette shown in Fig. 3 can be described by five defect triangles (A, B, C, D, and E). Each triangle represents three coordinate points: (i) defect starting point (x_{ds}, y_{ds}), (ii) defect position point (x_{dp}, y_{dp}), and (iii) defect end point (x_{de}, y_{de}). Identification of defect points is a crucial and important task. If triangle A is considered, the defect is defined by considering each points of object contour in triangle A and making perpendicular online passes from points (x_{ds}, y_{ds}) and (x_{de}, y_{de}) and that is called as depth of that point. So the defect points of the triangle A can be described as the following vector:

Fig. 4 Non-convex polygon
P and its convex hull CH(P)
(dashed line) [20]

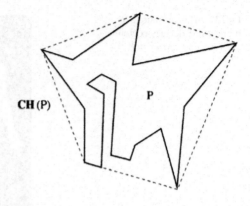

$$V_d^A = <A(x_{ds}, y_{ds})A(x_{dp}, y_{dp})A(x_{de}, y_{de})>$$

In real world, we have to deal with different shapes with uneven surface and also more complicated shapes. So finding only convexity defect position is not sufficient. It needs some modifications to deal with those artifacts. In proposed algorithm, for removing the unnecessary defect points from computation, all defect points which have less depth than depth threshold are discarded and remaining points are taken into account for further computation (Fig. 4).

D. Cut Points search

This step is used to find a set of points from the defect points that can be used to segment different parts from the sketch. The algorithm to find this set of cut point is given below:

Fig. 5 Convexity defects
human body silhouette

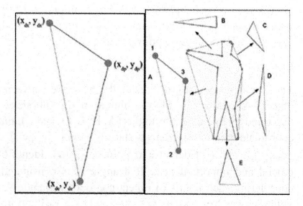

```
Program  Cut-Points (Defect Points)

d1= defect point1
ROI =  region of interest (ROI) around the defect point
M = Appropriate Mask Size
Repeat till No_of_contors > 1 in ROI :
   if No_of_contour < 1:
       Find d2= defect point 2 on Same Contour
   elseif No_of_defect_point == 2
       Joint (d1,d2) with line
   Else:
       M = 0.1*M
d2= Make first guess of defect point on another contour
d3= starting or end point of convexity defect in ROI
f1=  minimum Euclidean distance defect point
   if d2 is present :
       Join(d1,d2) with line
   elseif d3 is present:
       join(d1,d3) with line
   else :
       join(d1,f1) with line
   Move to Next Defect Point
end
```
Follow same procedure for each defect point and differentiate all segment from each other

E. *Differentiation of convex segment*

This step is used to segment different parts from the sketch by using defect positions and cut points found in previous steps. The algorithm for part segmentation is explained below, and results are shown in Fig. 6a and b:

Fig. 6 a Defect position with cut point **b** differentiation of convex segment

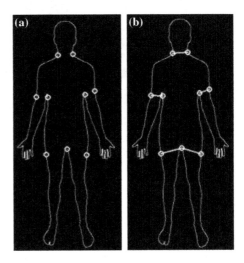

```
Program Part Segmentation (Defect Points, Cut-Points)
for all cut points c1:
   Find nearest cut point c2
   Joint (c1, c2) with a line
   Repeat for all part segments:
   D1 = starting defect point
   D2 = Iterating defect point
   Contour points = [D1]
   While D2 != D1:
      Contour Points += D2
      D2= Next Defect Point
end
```
Contour points array represent close contour part. At the end fill different colors to represent different segment parts.

All experiment results for part identification and segmentation are shown in Fig. 7. The performance of proposed system is evaluated on ETHZ dataset [17] which is suitable for investigating performance of this algorithm. The dataset contains sketches of various objects like animals, mug, apple, and bottle. Main focus of work is to decompose the whole structure into the part-wise structure as the human can visualize.

Table 1 shows the experimental setup for ETHZ shape class dataset which contains sketches and images for five classes of objects: bottle, apple, mug, giraffe,

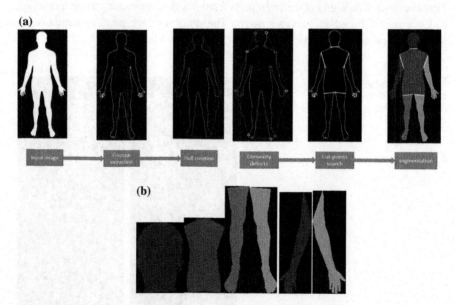

Fig. 7 a Step-wise analysis for human of proposed algorithm **b** segmentation of object in part-wise structure

Table 1 Experimental setup data

Sketch	Image size	Depth threshold	Mask size	Sketch
Human	294 × 549	20	150 × 150	Human
Apple	93 × 114	10	40 × 40	Apple
Bottle	200 × 300	7	60 × 60	Bottle
Giraffe	250 × 250	20	130 × 130	Giraffe
Mug	200 × 200	15	40 × 40	Mug

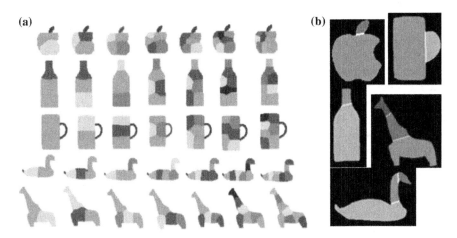

Fig. 8 **a** Result of part decomposition using the method by Gopalan [13], each row represents the results for specific shape as the user estimate for the number of part varies from 2 to 8. **b** Results using the proposed part decomposition method without user input

and swan [17]. As shown in the table, depth threshold and mask size values are decided based on the structure and convex hull area. First input data is silhouette of human body which is not from ETHZ dataset.

Figure 8 shows comparison between the results of part decomposition using Gopalan's algorithm [13] and the proposed algorithm. It can be observed that proposed algorithm outperforms Gopalan's algorithm because it gives different parts of object as human can visualize automatically without user explicitly specifying the number of parts as it was required in case of Gopalan's algorithm [13]. So the proposed algorithm will work well with objects of different shapes without worrying about the number of parts for decomposition.

4 Part Matching Using Fast Directional Chamfer Matching (FDCM)

After segmenting sketch into several parts, there is a need to match the parts obtained from the client-side sketch to the parts that are already stored in server database. For real-time operation, this function should be as fast as possible. There are many shape-matching algorithms available, but the chamfer matching is the most preferred one based on speed and robustness compared to other algorithms [18]. There are different types of variations in Chamfer Matching algorithm which are described in literature like (i) Chamfer matching (ii) Directional Chamfer Matching (iii) Fast Directional Chamfer Matching. Chamfer matching provides a fairly smooth measure of fitness and can tolerate small rotations, misalignments, occlusions, and deformations. Chamfer matching becomes less reliable in the presence of background clutter [18]. To improve robustness, Directional Chamfer matching has been introduced by incorporating edge orientation information into the matching cost. In DCM, match cost is more in terms of time and memory and to overcome that FDCM was introduced. In that directional matching cost is optimized in three stages by linear representation of the template edges, describing a three-dimensional distance transform representation and presenting a directional integral image representation over distance transforms. Using these intermediate structures, exact computation of the matching score can be performed in sublinear time in the large number of edge points. In the presence of many shape templates, the memory requirement also reduces drastically. In addition, smooth cost function allows binding the cost distribution of large neighborhoods and skipping the bad hypotheses [18]. Experimental results for fast direction chamfer matching are shown in Fig. 9.

Hand-drawn sketch shown in white box and green box indicates detected object using given sketch. After detecting parts using FDCM, SURF is used for continuous detection and tracking extracted feature point in image sequences.

5 Speeded Up Robust Features (SURF)

One of the most important tasks is to select the 'interest points' at distinct location in an image. An output of FDCM is considered as region of interest (ROI). SURF detector and descriptor is used for finding minimum feature points in ROI [19].

SURF uses Hessian matrix approximation for interest point detection which performs well in terms of accuracy [19]. SURF detects blob-like structure at locations with the maximum value of determinant. For reducing the computational complexity, integral images are used which uses only four memory locations and three addition operations to calculate the intensities inside a rectangular region of any size. Hence, the calculation time of this is independent from the size of an image [19]. For making SURF descriptor invariant to image rotation, Haar wavelet

Fig. 9 Output of FDCM algorithm on ETHZ dataset

response is calculated in x- and y-direction with the radius of 6 s around the interest point, where 's' is the scale at which the interest point is detected. Then wavelet response is computed and Gaussian weight is given centered at interested point. The sum of all responses within the window $\pi/3$ is calculated, and with use of this, the dominant orientation is estimated. The horizontal and vertical responses are summed, and longest vector of the window defines the orientation of the interest point. Figure 10 shows the experimental results using SURF algorithm for different conditions like scaling and rotation.

Fig. 10 Output of surf, left-side image is input image, and right-side image shows detected object image

Fig. 11 Results of human tracking in two different frames with different shapes

6 Experiments and Results

Experiment results for object tracking for different objects like human, book, and bottle are shown in Figs. 11, 12, and 13. The results are taken in different environments to prove the effectiveness of the proposed algorithm. Figure 11 shows human tracking in two different frames, where shape of the body is different and there is deformation in parts of body. In Fig. 12, the object of interest is book so the algorithm tracks book in different frames even when bottle is introduced or moved in the frame. In Fig. 13, bottle is tracked but book is not tracked. Both results are shown with start frame, two intermediate frames, and end frames. The results indicate that proposed algorithm performs well in challenging conditions.

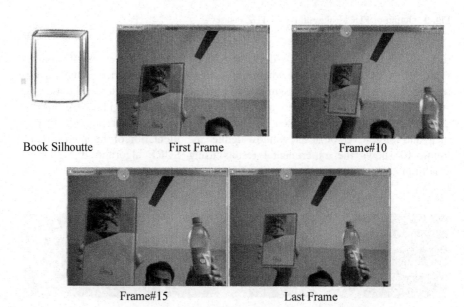

Fig. 12 Results of book tracking in the presence of bottle movements in different frames

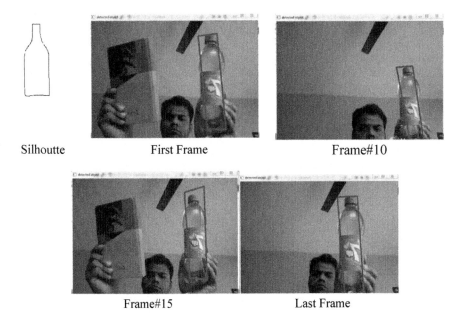

| Silhoutte | First Frame | Frame#10 |

| Frame#15 | Last Frame |

Fig. 13 Results of bottle tracking in the presence of book movement in different frames

7 Conclusion

In this paper, novel algorithm has been proposed for object detection and tracking using part decomposition and SURF. A method to decompose non-planar structure in part-wise planar structure using silhouette of object is also proposed. The method works very efficiently when input silhouette is without distortion. The computation complexity is also less than state-of-the-art algorithms. As in Gopalan's [13] algorithm and other state-of-the-art algorithms required prior knowledge about number of parts in given structure or user has to decide number of parts to be decomposed, but this method works in all situation without the prior knowledge about number of parts. As shown in experiment results, proposed algorithm gives better results in terms of part decomposition for all input sketches of ETHZ dataset. SURF can fulfill the requirement of real time though part decomposition and matching take more time, so in future work time taken for part decomposition can be modified.

Informed consent Additional informed consent was obtained from all individual participants for whom identifying information is included in this article.

References

1. Bouwmans, T.: Traditional and recent approaches in background modeling for foreground detection: An overview. Computer Science Review, 11, pp. 31–66 (2014).
2. Bhattacharjee, S. D., Mittal, A.: Part-based deformable object detection with a single sketch. Computer Vision and Image Understanding, vol. 139, pp. 73–87 (2015).
3. Bouwmans, T.: Recent advanced statistical background modeling for foreground detection-a systematic survey. Recent Patents on Computer Science, vol. 4 no. 3, pp. 147–176 (2011).
4. Stauffer, C., Grimson, W. E. L.: Adaptive background mixture models for real-time tracking. In: IEEE Computer Society Conference on. Computer Vision and Pattern Recognition, Vol. 2, pp. 246–252, IEEE (1999).
5. Lin, H. H., Liu, T. L., & Chuang, J. H.: A probabilistic SVM approach for background scene initialization. In: International Conference on Image Processing Vol. 3, pp. 893–896. IEEE (2002).
6. Wang, J., Bebis, G., Miller, R.: Robust video-based surveillance by integrating target detection with tracking. In: Conference on Computer Vision and Pattern Recognition Workshop, CVPRW'06, IEEE (2006).
7. Toyama, K., Krumm, J., Brumitt, B., Meyers, B.: Wallflower: Principles and practice of background maintenance. In: The Proceedings of the Seventh IEEE International Conference on Computer Vision, Vol. 1, pp. 255–261, IEEE (1999).
8. Ridder, C., Munkelt, O., Kirchner, H.: Adaptive background estimation and foreground detection using kalman-filtering. In: Proceedings of International Conference on recent Advances in Mechatronics, pp. 193–199 (1995).
9. Chang, R., Gandhi, T., & Trivedi, M. M.: Vision modules for a multi-sensory bridge monitoring approach. In: Proceedings of 7th International IEEE Conference on Intelligent Transportation Systems, pp. 971–976, IEEE (2004).
10. Butler, D. E., Bove, V. M., & Sridharan, S.: Real-time adaptive foreground/background segmentation. In: EURASIP Journal on Advances in Signal Processing (2005).
11. Kim, K., Chalidabhongse, T. H., Harwood, D., Davis, L.: Background modeling and subtraction by codebook construction. In: International Conference on Image Processing (ICIP'04), Vol. 5, pp. 3061–3064, IEEE., 2004.
12. Bai, X., Li, Q., Latecki, L. J., Liu, W., Tu, Z.: Shape band: A deformable object detection approach. In: IEEE Conference on Computer Vision and Pattern Recognition, CVPR 2009, pp. 1335–1342, IEEE (2009).
13. Gopalan, R., Turaga, P., Chellappa, R.: Articulation-invariant representation of non-planar shapes. In: Computer Vision–ECCV 2010, 286–299 (2010).
14. Suzuki, S.: Topological structural analysis of digitized binary images by border following. In: Computer vision, graphics, and image processing, vol 30 no. 1, pp. 32–46 (1985).
15. Graham, R. L., Yao, F. F.: Finding the convex hull of a simple polygon. In: Journal of Algorithms, vol 4 no. 4, pp. 324–331 (1983).
16. Youssef, M. M., Asari, V. K.: Human action recognition using hull convexity defect features with multi-modality setups. In: Pattern Recognition Letters, vol. 34 no. 15, pp. 1971–1979 (2013).
17. ETHZ datasets, http://www.vision.ee.ethz.ch/datasets/.
18. Liu, M. Y., Tuzel, O., Veeraraghavan, A., Chellappa, R.: Fast directional chamfer matching. In: IEEE Conference on Computer Vision and Pattern Recognition (CVPR), pp. 1696–1703. IEEE (2010).
19. Bay, H., Ess, A., Tuytelaars, T., Van Gool, L. Speeded-up robust features (SURF). In: Computer vision and image understanding, vol. 110 no. 3, pp. 346–359 (2008).
20. Zunic, J., Rosin, P. L.: A new convexity measure for polygons. In: IEEE Transactions on Pattern Analysis and Machine Intelligence, vol. 26 no. 7, pp. 923–934 (2004).

Visual Object Detection for an Autonomous Indoor Robotic System

Anima M. Sharma, Imran A. Syed, Bishwajit Sharma, Arshad Jamal and Dipti Deodhare

Abstract This paper discusses an indoor robotic system that integrates a state-of-the-art object detection algorithm trained with data augmented for an indoor scenario and enabled with mechanisms to localize and position objects in 3D and display them interactively to a user. Size, weight, and power constraints in a mobile robot constrain the type of computing hardware that can be integrated with the robotic platform. However, on the other hand, the robot's mobility if leveraged properly can provide enough opportunity to detect objects from different distances and viewpoints as the robot approaches them giving more robust results. This work adapts a CNN-based algorithm, YOLO, to run on a GPU-enabled board, the Jetson TX1. An innovative method to calculate the object position in the 3D environment map is discussed along with the problems therein, such as that of duplicate detections that need to be suppressed. Since multiple objects of different or same class may be detected, the user is overloaded with information and management of the visualization through human–machine interaction gains an important role. A scheme for informative display of objects is implemented which lets the user interactively view object images as well as their position in the scene. The complete robotic system including the interactive visualization tool can be put to various uses such as search and rescue, indoor assistance, patrolling and surveillance.

A. M. Sharma (✉) · I. A. Syed · B. Sharma · A. Jamal · D. Deodhare
Centre for Artificial Intelligence and Robotics, Bangalore 560093, India
e-mail: anima@cair.drdo.in

I. A. Syed
e-mail: imran@cair.drdo.in

B. Sharma
e-mail: b_sharma@cair.drdo.in

A. Jamal
e-mail: arshad@cair.drdo.in

D. Deodhare
e-mail: dipti@cair.drdo.in

© Springer Nature Singapore Pte Ltd. 2018
B. B. Chaudhuri et al. (eds.), *Proceedings of 2nd International Conference on Computer Vision & Image Processing*, Advances in Intelligent Systems and Computing 703, https://doi.org/10.1007/978-981-10-7895-8_17

Keywords Object detection · CNN · Duplicate suppression
3D localization · Mapping · Visualization

1 Introduction

There are important challenges that need to be addressed in developing an autonomous robotic system which can look for objects in a 3D environment. While mobility of the platform restricts the computational capability of the system, applications demand higher accuracy of algorithms. In the present context, an autonomous robotic system is required to perform object detection and locate them in a 3D point cloud. One crucial challenge is the selection of compute capability for a robotic system because it affects the mobility of the robot and the time durations for which the robot can endure. Another challenge is the selection of object detection algorithms that run in real time on the restricted computer onboard the robot with high detection rate and low false positives.

Computer vision techniques for object detection have remained a topic of research for almost four decades. In a robotic application such as autonomous driving [1], search and rescue, or object manipulation [2, 3], various sensors that are already available on the platform can be reused for object detection. Object detection can be performed in images, point cloud or range data, or a combination thereof. 2D and 3D LiDAR can be used to obtain point clouds which can be used to perform object detection in 3D. While camera images can be used for image-based object detection [4], stereo cameras and RGBD cameras are capable of finding depth in the scene as well which can be used for multimodal object detection [5–9]. Recently, convolutional neural network (CNN)-based methods [7, 10, 11] have boosted the classification accuracies in images to acceptable limits while exploiting GPUs for faster execution, making them suitable for practical applications. These detection algorithms are usually benchmarked on outdoor data such as KITTI [1, 12, 13] for autonomous driving application, but to the best of our knowledge, no benchmark dataset exists for indoor scenarios for search or human assistance applications.

This work uses a CNN classifier [14] running on a portable CUDA-enabled board trained for indoor scenarios, capturing the detected objects in images from an RGBD camera and their position in the scene using range information present in the depth images. Our literature survey indicates that such an integrated indoor robotic system featuring a reliable object detection algorithm has not been reported yet. The novel contributions of this paper are as follows:

1. Evaluation and selection of an object detection algorithm suitable for running onboard a mobile robotic platform,
2. Re-training the detection algorithm using indoor images for improved performance,
3. Finding 3D position of detected objects and its integration in the map,

4. Suppression of duplicate detection when the same object is detected multiple times, and
5. Interactive visualization of objects relating them to their position in the map.

In the following sections, our robotic platform setup is described, followed by our approaches for adapting object detection for onboard and indoor scenarios, 3D localization of detected objects while suppressing duplicates, and interactive visualization of the detection data in a 3D point cloud. The results presented indicate the usefulness of this autonomous system for the targeted scenario.

2 Setup

Our hardware setup consists of a base platform, sensors, compute, and power supply. A Pioneer 3DX is selected as the base platform, mounted with a SICK 2D scanning LiDAR and an xTion Pro Live RGBD camera, as shown in Fig. 1. CNN-based algorithms require CUDA-enabled GPU cards. Though most object detection works report tests done on high-end CUDA-enabled graphics cards, these cards cannot be used onboard on a robotic platform. As of today, only NVIDIA produces a CPU–GPU hybrid board, targeted for robotic applications. The version used here is the Jetson TX1 with 256 GPU cores. To run CNN-based algorithms on a robotic platform, we are left with only this choice. Hence, two computing platforms are mounted on the platform. While one is a dual-core CPU-based box-pc, the other is a 256 core GPU–CPU hybrid board, the Jetson TX1.

The software setup uses the ROS framework, with the sensors and computer hardware connected on a network, publishing and subscribing data and commands through nodes running on both the computing hardware. The exploration, path planning, and Simultaneous Localization And Mapping (SLAM) algorithms run on

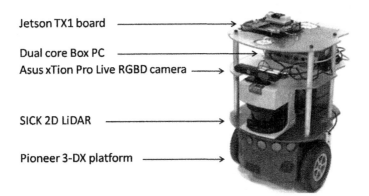

Jetson TX1 board

Dual core Box PC
Asus xTion Pro Live RGBD camera

SICK 2D LiDAR

Pioneer 3-DX platform

Fig. 1 Pioneer 3DX platform with SICK LiDAR and xTion Pro RGBD camera, and the Jetson TX1

the box-pc using the LiDAR and RGBD data, details of which are beyond the scope of this paper. In the purview of object detection, the Jetson board is employed to run a CNN-based algorithm on the images obtained from the RGBD camera.

3 Approach

This section details the visual object detection, 3D localization of objects, and the interactive visualization approaches developed in this work. First, various object detection algorithms suitable for a robotic platform are evaluated and the requirement of adapting it to the indoor scenario is then brought out. Our novel method of localization of objects in 3D point cloud is then described, followed by details of design of an interactive visualization scheme for displaying the detected objects in a 3D map of the scene.

3.1 Visual Object Detection

Our robotic platform performs localization and mapping as it moves around exploring an indoor scene. While the generated map provides information about obstacles and traversable regions, a user would also be interested in locating objects of interest. To fulfill this requirement, an object detection algorithm is required to run simultaneously in near real time.

Driven by the need for accuracy and execution speed, various landmark algorithms available in the literature, such as HOG [15], DPM [16], and YOLO [14], were evaluated. While DPM and HOG use handcrafted features, YOLO is a CNN-based classifier. The HOG was originally proposed for human detection, using a histogram of local intensity gradients for each cellular window in an image. A part-based model consisting of a star structure, comprising a root and parts, and its deformation model was later proposed. This framework could use features like HOG to detect root and part positions. Since information of each part is corroborated with the corresponding deformation model, the detections are more robust. A real-time version of DPM reports detections at 30 fps, which is desirable for this work. However, CNN-based detection algorithms have outperformed DPM by a considerable margin, thereby attracting our attention.

YOLO, being a CNN-based classifier, learns the features on its own for the classification problem. This is in contrast to the handcrafted features and frameworks developed erstwhile. While early CNNs provided only image classification, more recent works endeavor to provide the bounding boxes in order to localize the object in the image [17]. YOLO [14] and R-CNN [18–20] are two of the most prominent CNN algorithms reporting much higher accuracies compared to traditional handcrafted feature-based methods, and providing object localization as well. However, both these algorithms have different approaches for localizing or segmenting the object in the image. While R-CNN uses region proposals to generate bounding boxes, YOLO divides the image into grids and predicts bounding boxes for each grid.

While various datasets that are published for benchmarking object detection algorithms try to include diverse classes, all such objects may not appear in any single application. For example, the VOC 2012 test includes aeroplane with furniture and animals. In the present context, a single camera mounted on an indoor robot might not be exposed to all these classes. The indoor test scenario used in this work consists of an office-like environment with walls, galleries, doors, stairs, fire extinguishers, plants, etc. In our initial tests, blank walls were often misclassified as 'car' or 'bird.' YOLO was slightly better than R-CNN in averting classification errors particularly with respect to indoor scenes. However, re-training was inevitable to include indoor object classes fit for this application. Note that in the above example, misclassifications are primarily due to the absence of the 'wall' class. This work augments the PASCAL VOC database [21] with few more classes and uses them for re-training. The additional classes are 'wall,' 'fire extinguisher,' 'door,' 'window,' 'switchboard,' 'stairs,' 'cupboard,' etc., with around 2000 additional images. Figure 2 shows sample images of some augmenting classes. Figure 3 compares results on indoor images as detected by YOLO trained using the original PASCAL VOC, and after augmentation with additional classes. It is clear from Fig. 3 that standard datasets do not suffice for indoor scenarios and re-training is a necessity.

It is reported that the performance of the YOLO algorithm is relatively poor for small objects [14], and the localization error is higher for smaller detection boxes. Our application paradigm minimizes the impact of these problems by leveraging the mobility of the platform. The mobile platform is capable of moving closer to objects so that larger images of objects are obtained. Further, since this work also includes depth information to find 3D object location, faraway detections which are beyond the range of the depth sensor are discarded. Finding the 3D object location is detailed in a later section.

In addition to a slightly better accuracy, YOLO was adaptable to lower number of GPUs. This version of YOLO, called Fast YOLO, retains acceptable detection speed and accuracy, making it suitable for the application at hand. The full version of YOLO has a trained model file size exceeding 1 GB and hence could not be

Fig. 2 Sample images for some of the additional indoor object classes used in training. These classes augment the PASCAL VOC dataset [21] to meet the requirements of the presented work. The images are best viewed in color

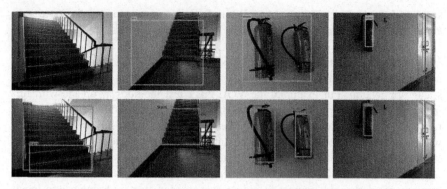

Fig. 3 Top row shows 'car,' 'bird,' 'bicycle,' and 'bottle' classes detected from PASCAL VOC training, compared to detections after re-training with additional indoor object classes. The bottom row shows detection of 'stair' in orange boxes, and 'extinguisher' in blue boxes. The images are best viewed in color

Table 1 Comparison table for object detection methods with respect to speed, hardware requirement and accuracy. The detection speeds were tested on our setup. mAP values are as reported in [14] for Pascal VOC 2007 dataset

Method	Speed (FPS)	Test hardware	Mean average precision (mAP)
HOG	7	2.6 GHz CPU	–
DPM	0.2	2.6 GHz CPU	26.1
YOLO	25	2.6 GHz Intel CPU + K20 GPU	63.4
Faster R-CNN	18	2.6 GHz Intel CPU + K20 GPU	62.1
Fast YOLO	11	Jetson TX1	52.7
Faster R-CNN	0.8	Jetson TX1	62.1

loaded on the Jetson TX1 which has limited RAM. However, Fast YOLO was successfully run on this hardware. Table 1 compares the above methods based on required hardware, achievable speed, and reported accuracies. In our application, the Fast YOLO network was trained offline for 135 epochs and a batch size of 64 on a Tesla K20 GPU card. The trained model file is under 200 MB and can be easily loaded on the Jetson TX1.

3.2 3D Localization of Objects

The detected objects are enclosed in bounding boxes in the camera image. However, this is not sufficient to facilitate operator decision. The actual location of the object in 3D space has to be computed, and the information is used to merge the object in the 3D point cloud generated by the mapping module. Note that the mapping module operates on point clouds obtained by the LiDAR sensor and the SLAM is performed using this input. As the robot moves around the object or sees it again from a different viewpoint, multiple detections are possible. These need to be suppressed. This section details the work required to meet the above requirements.

The mapping algorithm onboard the platform uses the computed pose for each RGBD frame to find the position where the current point cloud should be inserted. After detection, though the 2D bounding box is available in the RGB image, the 3D bounding box still needs to be found. The depth image does provide range information for each pixel in the bounding box, but these include background pixels as well as foreground pixels from an occlusion. Assuming that firstly the object is convex and has no holes, and secondly, the central region on the bounding box clearly sees the object without any occlusions, this work selects a small 10×10 pixels window in the center of the bounding box. Selecting the central region gives a reasonable assurance that all the background pixels are discarded. It is also important to note that the object may have oblong dimensions, in which case the computed object position is of the portion of the surface visible in the depth image. Since the exact dimensions of objects are not recorded and their orientation is also unavailable, a unit cube (1 m^3) is placed around the object to mark its bounding box in 3D. Finding the exact bounding box is possible if for each class a standard set of dimensions is recorded, and post-object detection and its orientation are also computed by matching it with a set of views [5]. Alternatively, a 3D object detection [3, 13, 22] algorithm may also provide the required dimensions for the bounding box. However, this has not been attempted here.

After detection, the 3D position and class label of the object are recorded. To suppress duplicate detections, the presence of another object of the same class label is verified within a threshold distance. The localization algorithm may itself have slight errors in addition to the 3D object localization. To accommodate these errors, the threshold for duplicate detection is set at 50 cm. Thus, a duplicate detection within a sphere of 1 m diameter is suppressed. Figure 4 provides a flowchart for performing the tasks of object detection, 3D position finding, and suppression of duplicates. Figure 5 shows detected objects and their locations bounded in unit cubes in a scene. This method is good enough for static objects in the scene. However, moving objects such as humans may get detected multiple times as they move around and appear multiple times in the view.

3.3 Interactive Visualization

The interface for displaying the point cloud allows zooming, panning, and rotation for interactive 3D viewpoint manipulation. The object positions are required to be shown as unit cubes enclosing the detected object. However, displaying all the object classes detected by the algorithm in the generated point cloud overcrowds the

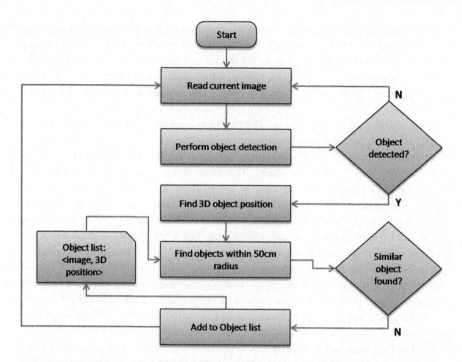

Fig. 4 Flowchart for the object detection, 3D position finding, and suppression of duplicates

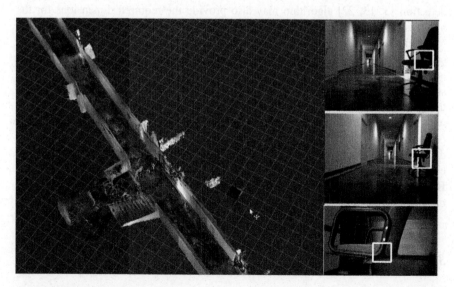

Fig. 5 Unit cubes painted on each chair object detected in the images shown at the right. Multiple detections appeared in adjacent frames, but were suppressed. Note that it is difficult to find which chair was located where

Fig. 6 Interactive visualization showing the clicked image shown in red frame is disambiguated in the map by changing the color of its bounding box

map. Another issue with point cloud is that its resolution is not as good as the image itself, which makes it difficult for a user to identify the class. The user interface has been designed with features that cater to above issues for better-managed visualization of the detected objects in the scene.

The interface consists of an object–class selection interface that lists all the classes detectable using the classifier. The display panel consists of an image list and a point cloud display. The image list is a scrolling pane which shows images in which only the selected object class was detected. The image list is cleared and updated whenever another class is selected.

Another pane displays the point cloud of the environment traversed by the robot, with unit cubes drawn at the computed object locations. Though the image list and point cloud are shown juxtaposed, correlating an image with multiple unit cubes strewn across the point cloud is difficult. Hence, whenever an image is clicked upon, the corresponding cube is shown in a different color, as shown in Fig. 6. This enables users to identify the actual location of the currently selected object.

4 Results and Discussion

Using the augmented classes for training, the performance of the system improves with fewer misclassifications. The mean average precision (mAP) is 0.77 computed over the additional classes alone for easy understanding. It was noticed that stairs

Table 2 Average precision for augmented classes computed over corresponding number of testing images. The mAP was computed over augmented classes alone

Augmenting classes	Training images	Testing images	Average precision
Stairs	320	80	0.92
Fire extinguisher	224	56	0.78
Switchboard	184	46	0.89
Door	152	38	0.5
Window	216	54	0.833
Wall	400	100	0.637
Text	133	33	0.9
Cupboard	72	18	0.667
	Total = 1701	Total = 425	mAP = 0.7659

and window classes had fewer misclassification than door, wall, and cupboard classes. Also, doors and walls were often confused with each other. This is partially due to the inclusion of contextual information along with the object of interest while training using YOLO. Our door images almost always have walls present in the same image. Further, our door examples consisted of plain surfaces usually having same color as the wall. YOLO could learn stairs, text on nameplates, switchboards, and fire extinguishers quite well, as can be seen from Table 2. It is worthwhile to note that the wall class was included primarily to reduce misclassifications.

The 3D positions of objects were without any perceptible error, unless the object contains holes or is concave. Figure 4 shows the correct location of chair in a unit

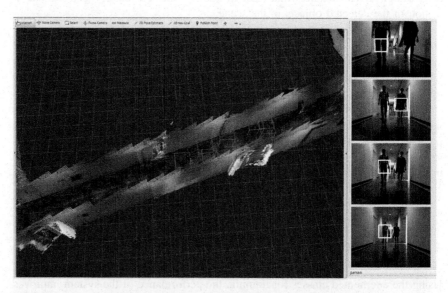

Fig. 7 Same person is detected multiple times but at different positions. All detections are shown in the list on the right. The corresponding cubes appear as a trail in the 3D map

cube bounding box painted in the point cloud. Note that the same object was detected multiple times as the platform crossed the gallery. The duplicate suppression algorithm filtered later detections and retained only one. However, with dynamic objects changing position, duplicate suppression will require tracking them and suppressing detection on the track. Figure 7 shows a trail of boxes painted by a group of moving people in the scene. Suppressing these will be attempted in a future work.

The object detection algorithm integrated with 3D mapping was useful in visualizing a unified display for the area explored by the robotic platform. However, a list of images with detected object was also displayed as images have better resolution than the point cloud. The robotic system works as an assistant to a user, and human–machine interface is an important aspect for the usability of the system. The visualization developed here assures information is not overloaded, and the displayed images of detected objects and their position in the map are visually linked. This ensures clarity and ease of operation for the user.

5   Conclusion

The developed indoor robotic system successfully integrates object detections with 3D mapping and displays them interactively to a user. The mobility of the robot constrains the type of compute hardware, but on the other hand, it provides enough opportunity to detect objects as the robot approaches them. In our experiments, detection of at least 10 fps was sufficient to capture most objects present in the scene. However, higher accuracies are required to avoid misclassifications and must be chosen according to the computer hardware available. In this work, YOLO was chosen since a CUDA-enabled portable hardware was available. When only a box-pc is available, the DPM would be a better choice though the detection accuracies would suffer. This work also suggested an innovative method to calculate the object position in the 3D environment map and discussed the problems therein. It brings out the necessity of also recording object dimensions and orientation detection in order to present a tighter bounding box in comparison to the currently used unit cubes. An attempt is made to suppress duplicate detections, considering the errors encountered in detection in the presence of localization and mapping errors. Finally, a scheme for informative display of objects is implemented which lets the user interactively view object images as well as their position in the scene.

The developed system may be put to various uses such as robotic assistance, search and rescue, or autonomous remote exploration. With inclusion of a robotic arm, miniaturization of computing hardware, and improved detection algorithms, future work could incorporate object manipulation and enable robots to autonomously search for and interact with selected objects present in its environment.

References

1. A. Geiger, P. Lenz, and R. Urtasun. Are we ready for autonomous driving? the KITTI vision benchmark suite. In CVPR, 2012.
2. Menglong Zhu, Konstantinos G. Derpanis, Yinfei Yang, Samarth Brahmbhatt, Mabel Zhang, Cody Phillips, Matthieu Lecce and Kostas Daniilidis, Single Image 3D Object Detection and Pose Estimation for Grasping, ICRA, 2014.
3. Ian Lenz, Honglak Lee and Ashutosh Saxena, Deep Learning for Detecting Robotic Grasps, arXiv 2014.
4. Ling Cai, Lei He, Yiren Xu, Yuming Zhao, Xin Yang, Multi-object detection and tracking by stereo vision, Pattern Recognition, 2010.
5. Arjun Singh, James Sha, Karthik S. Narayan, Tudor Achim, Pieter Abbeel, BigBIRD: A Large-Scale 3D Database of Object Instances, ICRA, 2014.
6. Omid Hosseini Jafari, Dennis Mitzel, Bastian Leibe, Real-Time RGB-D based People Detection and Tracking for Mobile Robots and Head-Worn Cameras, ICRA, 2014.
7. Pierre Sermanet, David Eigen, Xiang Zhang, Michael Mathieu, Rob Fergus, Yann Le Cun, OverFeat: Integrated Recognition, Localization and Detection using Convolutional Networks, arXiv, 2014.
8. Saurabh Gupta, Ross Girshick, Pablo Arbelaez, and Jitendra Malik, Learning Rich Features from RGB-D Images for Object Detection and Segmentation, arXiv, 2014.
9. Yulan Guo, Mohammed Bennamoun, Ferdous Sohel, Min Lu, and Jianwei Wan, 3D Object Recognition in Cluttered Scenes with Local Surface Features: A Survey, IEEE Transactions on Pattern Analysis and Machine Intelligence, vol. 36, no. 11, November 2014.
10. Christian Szegedy, Alexander Toshev, Dumitru Erhan, Deep Neural Networks for Object Detection, NIPS, 2013.
11. Dumitru Erhan, Christian Szegedy, Alexander Toshev, and Dragomir Anguelov, Scalable Object Detection using Deep Neural Networks, CVPR, 2014.
12. Yu Xiang, Roozbeh Mottaghi, Silvio Savarese, Beyond PASCAL: A Benchmark for 3D Object Detection in the Wild, WACV, 2014.
13. Xiaozhi Chen, Kaustav Kundu, Yukun Zhu, Andrew Berneshawi, Huimin Ma, SanjaFidler, Raquel Urtasun, 3D Object Proposals for Accurate Object Class Detection, NIPS, 2015.
14. Joseph Redmon, Santosh Divvala, Ross Girshick, Ali Farhadi, You Only Look Once: Unified, Real-Time Object Detection, CVPR, 2016.
15. Navneet Dalal and Bill Triggs, Histograms of Oriented Gradients for Human Detection, CVPR 2005.
16. Pedro F. Felzenszwalb, Ross B. Girshick, David McAllester and Deva Ramanan, Object Detection with Discriminatively Trained Part Based Models, PAMI 2010.
17. J. Dong, Q. Chen, S. Yan, and A. Yuille. Towards unified object detection and semantic segmentation. In Computer Vision–ECCV 2014, pages 299–314. Springer, 2014.
18. Ross Girshick, Jeff Donahue, Trevor Darrell, Jitendra Malik, Rich feature hierarchies for accurate object detection and semantic segmentation, CVPR, 2014.
19. Shaoqing Ren, Kaiming He, Ross Girshick, Jian Sun, Faster R-CNN: Towards Real-Time Object Detection with Region Proposal Networks, NIPS, 2015.
20. Ross Girshick, Jeff Donahue, Trevor Darrell, Jitendra Malik, Region-Based Convolutional Networks for Accurate Object Detection and Segmentation, IEEE Transactions on Pattern Analysis and Machine Intelligence, vol. 38, no. 1, January 2016.
21. M. Everingham, S. M. A. Eslami, L. Van Gool, C. K. I. Williams, J. Winn, and A. Zisserman. The pascal visual object classes challenge: A retrospective. International Journal of Computer Vision, 111(1):98–136, Jan. 2015.
22. Khaled Alhamzi, Mohammed Elmogy, Sherif Barakat, 3D Object Recognition Based on Local and Global Features Using Point Cloud Library, IJACT, 2015.

Engineering the Perception of Recognition Through Interactive Raw Primal Sketch by HNFGS and CNN-MRF

Apurba Das and Nitin Ajithkumar

Abstract The impression of a scene on human brain, specifically the primary visual cortex, is still a far-reached goal by the computer vision research community. This work is a proposal of a novel system to engineer the human perception of recognizing a subject of interest. This end-to-end solution implements all the stages from entropy-based unbiased cognitive interview to the final reconstruction of human perception in terms of machine sketch in the framework of forensic sketch of suspects. The lower mid-level vision as designed behaviorally in primary visual cortex honoring the scale-space concept of object identification has been modeled by hierarchical 2D filters, namely hierarchical neuro-visually inspired figure-ground segregation (HNFGS) for interactive sketch rendering. The aforementioned human–machine interaction is twofold: in gross structural design layer and finer/granular modification of the pre-realized digital perception. Pre-realized sketches are formed learning the characteristics of human artists while sketching an object through integrated framework of deep convolutional neural network (D-CNN) and Markov Random field (MRF). After few iterations of interactive fine-tuning of the sketch, a psychovisual experiment has been designed and performed to evaluate the feasibility and effectiveness of the proposed algorithm.

1 Introduction

The process of recognition is perhaps the most important aspect of visual perception [1]. Translating the perception of human to a machine through human–machine interaction is even more difficult. As vision is the most important sense of perception, memory of a perceived vision is the most important source of data for modeling

A. Das (✉)
Embedded Innovation Lab., Tata Consultancy Services, Bengaluru, India
e-mail: das.apurba@tcs.com

N. Ajithkumar
Amrita School of Engineering, Amrita Vishwavidyapeetham, Kollam, Kerala, India
e-mail: nitinakumar@gmail.com

© Springer Nature Singapore Pte Ltd. 2018
B. B. Chaudhuri et al. (eds.), *Proceedings of 2nd International Conference on Computer Vision & Image Processing*, Advances in Intelligent Systems and Computing 703, https://doi.org/10.1007/978-981-10-7895-8_18

the scene digitally. The proposed system intends to engineer the individual's (e.g., witness in forensic sketching procedure) perception of recognition, and convert this information into a form very close to the subject's (e.g., suspect in forensic sketch) identity.

Countless criminal cases have taken a turn with the help of forensic sketches. The project intends to automate the forensic face sketching process, i.e., the process of converting the witness's visual perception of recognizing suspect into a raw primal sketch which is identifiable, shareable, and modifiable by possible ornamentations like beard, bald digitally. The available forensic sketching softwares FACE [2], EvoFIT [3] use composite techniques to render face sketches from template options. Both Laughery et al. [4] and Willis et al. [5] have proved the effectiveness of the aforementioned software to be unacceptable in practice due to face composition and psychological bias while using image templates in cognitive interview, respectively.

The general sequence of forensic face sketching process consists of three parts, namely cognitive interview, raw primal sketching, and sketch refinement and correction through iterative interactions. The sketching process is often done by the professional sketch artist in such a way that the witness does not see the sketch while in progress in order to prevent memory bias. Hence, showing possible set of template faces to the witness and coming up with composite face is not an option for practical machine sketching. We have formulated the optimized question bank based on the mathematical model of Entropy maximization as described in the Sect. 2. Based on the cognitive interview, a gross face template with essential fiducial regions is created automatically. Honoring the scale-space concept of recognition by visual cortex, a hierarchical neuro-visually inspired figure-ground segregation (HNFGS) forms the smoothened tentative raw primal sketch representing shades by quantized gray intensities as discussed in Sect. 3. Next, the face sketch is generated by integrated deep convolutional neural network (D-CNN) and Markov Random Field (MRF) [6] imitating the process of sketching by professional human artists as discussed in Sect. 4. At this stage, for the first time the output sketch is shown to the user (i.e., witness) for confirmation. As discussed in Sect. 5, through a number of iterations of interactions between user and machine, the draft sketch is fine-tuned by thin plate spline warping [7] method, next. Section 6 has formulated and executed a psycho-visual experiment to prove the statistical significance of the proposed algorithm using ANOVA. Finally, in Sect. 7 we have concluded our findings with a direction to the future research.

2 Cognitive Interview: Face Wire-Frame Synthesis

The cognitive interview is conducted by selecting and asking questions from a question bank and populating the next layer of questions triggered by the previous answers. A question bank is first created by listing out the possible questions that are required to create a perceptually accurate sketch of the face (e.g., shape of eye, structure of face contour, eyebrow curve). The goal of the cognitive interview is to ask minimum number of relevant questions to get the maximum possible information. We have generated a layered question bank honoring the concept of Entropy maximization as discussed in the subsequent subsections.

2.1 Entropy Optimization

Entropy is a measure that represents the average information per message. In the proposed cognitive interview, Entropy optimization [8] is dependent on the path of control across the layers in the question bank. The layers in the question bank have been illustrated in Fig. 1. The first layer contains sections that are divided on the basis of the facial feature of priorities for recognition by Das et al. [9], e.g., the first section pertains to eyebrow, the second to eye, the third to lips. Each section contains questions that pertain to that feature. Considering one section, the second layer consists of sections that are relevant to different aspects of the facial feature. Considering the probabilities of the sections in the current layer to be $p_1, p_2, \ldots p_n$ (where n denotes the number of sections in the current layer all probabilities are uniformly distributed), the information that can be obtained from the current layer is given by Eq. 1. In this equation, n and m are the number of sections in the current layer and number of possible outcomes, i.e., number of sections in the next layer, respectively. The probability of choosing section i in the current layer p_i. It is noticed that the base of the logarithmic term is m, as the possible choices from the current layer is given by the number of sections in the next layer. Hence the calculated Entropy for ith layer H_i is defined as follows.

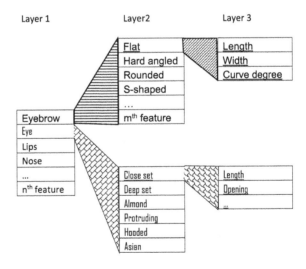

Fig. 1 Layered question bank for entropy maximization

$$H_i = -\sum_{i=1}^{n} p_i log_m p_i \tag{1}$$

Equation 1 is applicable solely for the interaction between 1st and 2nd layer, whereas the concept of mutual information and information gain would pitch in for subsequent layers. Considering section i of the first layer, the probabilities of the second layer are given by $p_{i1}, p_{i2}, \ldots, p_{im}$ (where m denotes the number of sections in the second layer). However, these values will be joint probabilities as it depends on the selection of section i in layer 1. Thus information that is obtained in the second layer will also contain information from the first layer. To address information redundancy, we consider information gain. The mutual information is obtained by considering the conditional probabilities of the question section for each layer. By extending this pattern of calculation to each layer, upto the question level, we can say that *Information from each question, H = (Information for Layer 1) + (Information gain in Layer 2) + ⋯ (Information gain in layer k)*, where k denotes the number of layers. The path of control from one layer to another can be considered as a channel whose capacity has been incorporated into the calculation of Entropy H to model the cognitive interview has been depicted in Eq. 2.

$$H = max \left(\sum_{K=1}^{A} H_k \right)$$

$$= max \sum_{k=1}^{A} \left(-\sum_{i=1}^{n} p_i log_m p_i - \sum_{j=1}^{m} p_j log_\xi p_j - \cdots + \sum_{j=1}^{m} p_{ij} log_\xi p_{ij} \cdots \right) \tag{2}$$

where, A denotes the number of questions in the question bank, n denotes the number of sections in layer 1, m denotes the number of questions in layer 2, ξ denotes the number of questions in layer 3, p_i is the probability of selecting a section in a particular layer, and p_{ij} is the conditional probability of a particular layer upon the previous layer. Thus, the overall design is in the form of a computer adaptive test which maximizes the information that can be gained from the witness with minimum number of questions.

2.2 Gross Face Template Synthesis on Golden Ratio

This system is used to build a gross template face based on the input of the cognitive interview. First a database was created where the gross templates of eyes, eyebrows, lips, nose, and their corresponding labels are stored. The system takes the input from the cognitive interview through voice-enabled interaction engine and maps the input to the images in the database by cross-referencing the labels intentionally hiding the database feature templates to remove psychological bias [10]. The corresponding

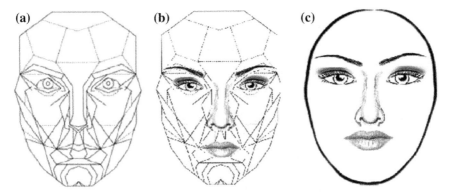

Fig. 2 **a** Female Face Golden Ratio (GR), **b** Gross synthesis of face from cognitive interview onto GR, **c** Synthesized face for next stage of operations

feature templates are then transferred onto a base template face formed by gender golden ratio as depicted in Fig. 2.

3 Machine Sketching by Hierarchical Neuro-Visually Inspired Figure-Ground Segregation (HNFGS)

Ever since Marr [11] proposed his famous theory, a lot of work has been done by psychologists and computer vision engineers to mathematically model the human vision system. By creating such models that closely resemble the way human vision operates, they believed that they could make machines perform complex tasks like pattern recognition and understanding. An import part of this step in the process involves searching for zero-crossings in the image and presenting the image using two distinct levels. Alternatively, it can be said that a perception-based binarization has been done on the image using Laplacian of Gaussian (LoG) operator [12]. This can be considered as equivalent to centre-surround receptive field model upon which lateral inhibition is based, viz. the Difference of Gaussians (DoG) [13].

3.1 NFGS Filter for Binarization

It has been observed by physiologists for a fairly long time (almost forty years [14]) that the cell's behavior is greatly affected by a number of cells lying outside the Classical Receptive Field (CRF). This modulation is called extended/extra-classical receptive field (ECRF) and is known to evoke nonlinear responses. It is based in the concept of ECRF that Ghosh et al. proposed a model for neuro-visual figure-ground

(a)   **(b)**

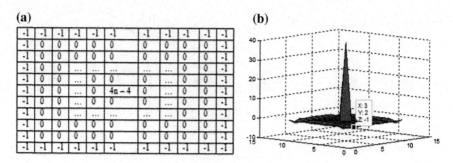

Fig. 3 **a** Tabular and **b** Graphical representation of ECRF kernel (size 11 × 11)

segregation which has been modified into the form of a digital mask [15, 17] as shown in Fig. 3.

Here 'n' denotes the filter size ($n = 2m + 1 > 1, m \in Z+$). The number of zeros from the centre to the periphery of the digital mask is always $m - 1$ for both row and column. The finite difference approximation for the classical Laplacian mask is reduced by the filter to it lowest value at m= 1. The convolved image is then binarized by thresholding each of the elements at zero. Das et al. have shown the efficiency of face recognition [17] systems using NFGS. Another work by the him [18] has described the importance of filter size in recognition through cellular neural network. The following subsection would address the choice of relative size of NFGS filter kernels adaptively.

3.2   Scale-Space Hierarchy: HNFGS

Inspired from the scale-space concept [16] of recognizing an object and figure-ground segregation by human brain [15], it has been established that extraction of information is complete when addressed both in wholistic and local/granular fashion. The proposed hierarchical NFGS (HNFGS) algorithm has honored the same. The smaller the size of the NFGS kernel, the more details can be captured. However, we do not want to capture all details as it will make the image distorted and noisy as shown in Fig. 4b, i. The gross synthesized template image (*CogIntGRimage*) through cognitive interview (*CogInt*) and golden ratio (*GR*) overlay has been filtered through NFGS kernels of different size in hierarchy. The first NFGS kernel is of size $N = 2X + 1$ where $X = ImageSize/4$. The NFGS filtered image, *ImgN* is considered as a mask to extract the region of interest (RoI) from *CogIntGRimage*. All the pixel values in the extracted RoI image have been assigned to an intensity quantized as $256/NumberOfLevels$. In our experiment, we have used four levels of quantization for generating raw primal sketch. The same procedure is followed until the smallest mask of size 3 × 3 is obtained by iterative reduction of mask size as depicted in the

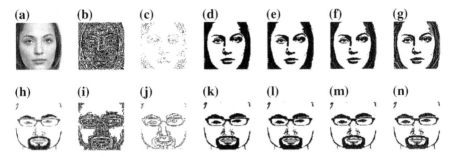

Fig. 4 Efficiency of HNFGS: **a–c, h–j** Comparisons between classical figure-ground segregation and edge detectors, **d–g, k–n** Stages of HNFGS filtering

Pseudocode 3.1. This ensures that detailing happens only in a manner that is required to produce a sketch-like image.

Algorithm 3.1. MACHINESKETCHHNFGS($CogInt, GR$)

---

comment: Output = Machine Sketch

$ImgN = NFGS(CogIntGRimage, SizeN = 2X + 1)$
$ImgNew := ImgN$
while $N >= 3$

$$\text{do} \begin{cases} ConvrtImgNewtoMask \\ RoI := ApplyMask2CogIntGRimage \\ NFGS(CogIntGRimageANDRoI, Size = N) \\ ImgNew := ReplaceAllPixValInRoIbyQVal \\ N := N/NumberOfLevels \\ \textbf{if } EVEN(N) \\ \quad \textbf{then } N := N + 1 \\ ImgN := ImgN + ImgNew \end{cases}$$

Figure 4 shows the efficiency of HNFGS over conventional edge detectors which produce images that fail to catch all the features of importance let alone the detailing. One-shot binarization technique (e.g., NFGS) also failed to address both wholistic and granular feature extraction. However, the HNFGS image produces a smooth image with all the important aspects captured and the detailing done hierarchically. NFGS filter of kernel size 3 is directly applied to the grayscale face image Fig. 4a resulting Fig. 4b. It is observed that although it captures all the details, the output is noisy. The Canny edge detector failed to extract the face feature details eliminating noise as shown in Fig. 4c. The same is illustrated using a sample template face Fig. 4h. The application of NFGS filter with window size 3 captures more details than necessary as can be seen in Fig. 4i. Also, the Canny image gives poor results as can be seen in Fig. 4j. The figure also depicts the different layers of HNFGS filtering response. The image Fig. 4d, k shows the output for the corresponding input images when the NFGS filter of window size $N = 2X + 1$ is applied in the first stage ($X = ImageSize/4$), and the corresponding regions are colored with the quantized value following the pseudocode 3.1. Similarly, Fig. 4e, l is the output from the second iteration with odd kernel size $N/4$, Fig. 4f, m from the third iteration for filter size $N/16$ and finally Fig. 4g, h from the fourth iteration for filter size 3 which is the final output.

4 Imitating Artist's Style of Sketching by CNN-MRF

In order to convert the HNFGS generated raw primal sketches to professional forensic artist's style of sketching, we have employed the cascaded CNN-MRF system as proposed by Zhang et al. [6]. The combination of generative Markov Random Field (MRF) models [19] and trained Deep Convolutional Neural Nets (D-CNN) learns the style of a forensic sketch artist and transforms HNFGS filtered face into a forensic sketch. It does this by dividing the training photographs and their corresponding sketches into patches and learning how each patch is translated into its corresponding sketch version. The gross-level mapping and placement of the patches in the test image are done using the MRF and is optimized for local feature details using the trained CNN. The system was trained by giving the HNGFS filtered images and its corresponding forensic sketches (Fig. 5).

5 Refining the Facial Sketch by Iterative Human–Machine Interaction

The first iteration of the forensic sketch rendering may have errors due to lack of accuracy in the feature description. The current section deals with the method of iterative human–machine interaction to fine-tune the face features. At this stage, the

Fig. 5 Refinement of eye feature by Thin Plate Spline warping: **a** Sketch at iteration 1, **b** Sketch at iteration 3, **c** Sketch at iteration 5 for more elongated eye

Fig. 6 Flowchart of the perception driven raw primal machine-sketching system

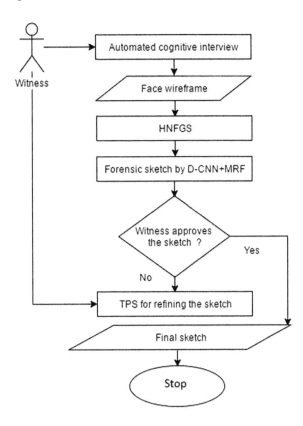

output sketch is shown to the witness for the first time. Primarily, the witness is asked to provide feedback regarding the general appearance of the face and its similarity to the face in memory. The specific details regarding the features are asked and the responses are used to control changes that are to be made on the face. For example, if the witness says the eyes are wider and it needs to be changed, or that the nose is little bigger, etc., the necessary changes are made in the next stage. The detailed process is described in terms of a flowchart as shown in Fig. 6. We have employed thin plate spline (TPS) [7] warping for the fine-tuning the sketch.

5.1 Fine-Tuning the Facial Features by Thin Plate Spline (TPS)

The error correction system is based on the idea of TPS warping [7]. The control points for each template points are decided beforehand, and the system stores the location of these points. When a change in feature proportion is required, the system automatically selects the necessary control points and moves it in the required direction. This causes the warping of the feature template and perceptually gives the desired effect; e.g., if the change to made is to make the eye longer, the system moves the corresponding control points and it changes the shape of the eye as illustrated in Fig. 5. In this manner, minor modifications and adjustments can be made to increase the similarity between the suspect and the rendered facial sketch.

6 Psycho-Visual Experiment and Results

To validate the effectiveness of the proposed algorithm, a psycho-visual experimentation has been designed and performed in terms of a survey. The outcome of the survey has been finally analyzed through statistical significance measure. Before the survey was performed, five faces were machine sketched following the proposed method (Fig. 7 shows three sample sketches). In the survey phase, 100 new subjects were identified randomly in two age groups (fifty subjects from age group <30 and fifty subjects from age group ≥30). The goal of the survey was to determine if the sketch produced by the system was recognizable. A GUI was developed in which the five sketches and ten photographs were shown in a matrix form (Sketches in the rows and photographs in the columns). The five sketches were shuffled each time the survey was done. Five of the photographs, which corresponded to the sketch photographs, were also displayed in a random order and the rest of the five faces were chosen randomly from the database of 50 photographs. Few sketches were camouflaged intentionally by ornamentation like adding or removing spectacles and/or

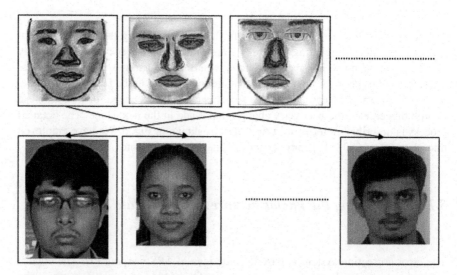

Fig. 7 Machine sketch correspondence used for psycho-visual survey (3 sample matches are shown)

Table 1 Psycho-visual survey report for 3 machine sketch recognition

	χ_{11}	χ_{12}	χ_{13}	χ_{1R}	χ_{21}	χ_{22}	χ_{23}	χ_{2R}	χ_{31}	χ_{32}	χ_{33}	χ_{3R}
Age <30 years	43	0	0	7	0	45	0	5	0	0	39	11
Age ≥30 years	45	0	0	5	0	48	0	2	0	0	44	6

facial hair. The subjects were allowed to add their confidence level also to each of the recognition between 0 to 100%. In the Table 1, we have weighted the confidence level to number of recognition and finally rounded off wherever required. Recognizing jth face from ith sketch is represented as χ_{ij} in Table 1. In the table, only three instances have been shown (randomly chosen) and $j = R$ when some other person is recognized out of the given set. With regard to survey, it needs to noted that it was conducted with local groups in and around the Author's circles, in a controlled setup, with the full consent of the subject. The photographs of the subjects that have been displayed in Fig. 7 have been done so with the written consent of the subjects.

6.1 ANOVA: Validation Through Analysis of Psycho-Visual Survey

Analysis of Variance (ANOVA) is a collection of statistical models that are used to find the sources of variances that are felected in a variable as observed variance. It provides a way of knowing, statistically, if the means of several groups are equal or not [17]. From the Table 1, it is clear that most of the subjects could recognize persons from machine sketches correctly. The box plot (Fig. 9) approves the fact.

In this experiment, a one-way ANOVA is performed. It is done so by comparing the means of 12 columns of data in the 2-by-12 matrix X, as presented in Table 1. In this table, each column represents an independent sample containing two mutually independent observations for the two predefined age groups. The output of the test is a p-value for the null hypothesis that all samples in X are drawn from the same population.

The p-value determines if the null hypothesis is true or not, i.e., whether atleast one sample is significantly different from other sample means. Generally, this conclusion is arrived at if the p-value is less 0.05. The columns in the ANOVA table (Fig. 8) show the following: column 1 shows the source of the variability, column 2 shows the Sum of Squares (SS) due to each source, column 3 shows the degrees of freedom (df) associated with each source, column 4 shows the Mean Squares (MS) for each source, which is the ratio SS/df, column 5 shows the F statistic, which is

ANOVA Table					
Source	SS	df	MS	F	Prob>F
Columns	8132	11	739.273	233.45	3.13979e-12
Error	38	12	3.167		
Total	8170	23			

Fig. 8 ANOVA for the psycho-visual survey presented in Table 1

Fig. 9 Box plot for the ANOVA table presented in Fig. 8 for survey of Table 1

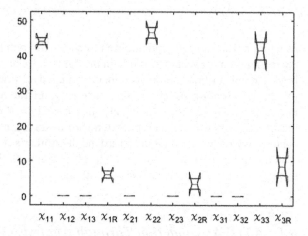

the ratio of the MS's and column 6 shows the p-value, which is derived from the cdf of F. As F increases, the p-value decreases. Figures 8 and 9 confirm the significance of correct recognition ($\chi_{11}, \chi_{22}, \chi_{33}$) of human from machine sketches synthesized by proposed algorithm.

7 Conclusion

Engineering the perception of recognition is an important aspect of computer vision to ensure the machine-based object recognition to be even more accurate. The current work has proposed a novel algorithm of cognitive interview-based machine sketching using HNFGS and analyzed the effectiveness of the sketches synthesized through psycho-visual experiments. To the best of our knowledge, this is the first attempt to enable a machine to sketch a human face through iterative interaction honoring the concept of Entropy maximization. In future work, the impact of ornamentation (facial hair, spectacles etc.) on recognition should be examined. Researchers may also look for modeling the machine sketches in terms of linear combination of multiple faces available in database of law enforcement agencies. The measure of confidence of the witness and weightage of emphasis on feature should also be mapped in possible future work to make the system even more adaptive and agile.

References

1. Ullman, S.: High Level Vision *MIT Press, Cambridge, Massachussets*, 1996.
2. Paterson, A., Squad, C. I., Police, V.: Computerised facial construction and reconstruction. *Proceedings of the Asia Pacific Police Technology Conference*, 135–144 (1991)
3. Frowd, C. D., Hancock, P. J., Carson, D.: EvoFIT: A holistic, evolutionary facial imaging technique for creating composites. *ACM Transactions on applied perception (TAP)*1, no. 1, 19–39 (2004)
4. Laughery, K. R., Fowler, R. H.: Sketch artist and Identi-kit procedures for recalling faces. *Journal of Applied Psychology* 65, no. 3, 307 (1980)
5. Willis, G. B.: Cognitive interviewing and questionnaire design: a training manual. *US Department of Health and Human Services*, Centers for Disease Control and Prevention, National Center for Health Statistics(1994)
6. Zhang, L., Lin, L., Wu, X., Ding, S., Zhang, L. : End-to-end photo-sketch generation via fully convolutional representation learning. *5th ACM on International Conference on Multimedia Retrieval*, 627–634 (2015)
7. Whitbeck, M.,Guo, H.: Multiple Landmark Warping Using Thin-plate Splines. *IPCV*, 6, 256–263 (2006)
8. Das, A.: Digital Communication: Principles and system modelling. *Springer Science and Business Media*, 169–172 (2010)
9. Parua, S; Das, A; Mazumdar D.; Mitra S.: Determination of Feature Hierarchy from Gabor and SIFT Features for Face Recognition. *Second International Conference on Emerging Applications of Information Technology*, 2011, pp. 257–260
10. Wise, R. A., Fishman, C. S., Safer, M. A.: *How to analyze the accuracy of eyewitness testimony in a criminal case.*, Conn. L. Rev., 42, 435 (2009)
11. Marr, D.: Vision: A computational investigation into the human representation and processing of visual information. *MIT press*, 2010.
12. D. Marr and E. Hildreth. Theory of edge detection. In *Proceedings of the Royal Society of London*, 1980, 207, 187217.
13. R. W. Rodieck and J. Stone. Analysis of receptive fields of cat retinal ganglion cells. *Journal of Neurophysiology*, 1965, 28:833849.
14. Ikeda, H. and Wright, J. H.: Functional organization of the periphery effect in retinal ganglion cells *Vision Research*, 1972, 12, 1857–1879
15. Ghosh, K., Roy, A.: Neuro-visually inspired figure-ground segregation. *International Conference on Image Information Processing (ICIIP)*, 1–6 (2011)
16. Lindeberg: Scale-space theory: A basic tool for analyzing structures at different scales. *Journal of Applied Statistics*, 21(2):224270, 1994.
17. Das, A. and Ghosh, K.: Enhancing face matching in a suitable binary environment. *International Conference on Image Information Processing (ICIIP)*, 1–6 (2011)
18. Das, A., Roy, A., and Ghosh, K.: Proposing a CNN Based Architecture of Mid-Level Vision for Feeding the WHERE and WHAT Pathways in the Brain, *FANCCO, LNCS* 7076/2011, pp. 559–568.
19. Wu, Z., Lin, D., Tang, X.: Deep Markov Random Field for Image Modeling. *14th European Conference on Computer Vision* (2016)

An Efficient Algorithm for Medical Image Fusion Using Nonsubsampled Shearlet Transform

Amit Vishwakarma, M. K. Bhuyan and Yuji Iwahori

Abstract Multimodal medical image fusion techniques are utilized to fuse two images obtained from dissimilar sensors for obtaining additional information. These methods are used to fuse computed tomography (CT) images with magnetic resonance images (MRI), MR-T1 images with MR-T2 images, and MR images with single photon emission computed tomography (SPECT) images. In proposed method, nonsubsampled shearlet transform (NSST) is used for decomposition of source images to attain the low-frequency and high-frequency bands. The low-frequency bands are fused using weighted saliency-based fusion criteria, and high-frequency bands are fused with the help of phase stretch transform (PST) features. Applying inverse NSST operation, fused image is obtained. The results show the proposed method produces better results compared to state-of-the-art methods.

Keywords Medical image fusion · Nonsubsampled shearlet transform (NSST) Phase stretch transform (PST)

1 Introduction

Three primary requirements defined for the excellent image fusion [2]. First, to transfer the prominent attributes of input images into the fused version of the image. Second, any relevant details of input images should preserve after the fusion process. Third, fused image should be free from any undesirable artifacts. The CT images show the hard tissue details such as bones, while MRI images show the soft tissues

A. Vishwakarma (✉) · M. K. Bhuyan
Indian Institute of Technology Guwahati, Guwahati 781039, India
e-mail: amitvishwakarma2625@gmail.com

M. K. Bhuyan
e-mail: mkb@iitg.ac.in

Y. Iwahori
Department of Computer Science, Chubu University, Kasugai, Aichi 487-8501, Japan
e-mail: iwahori@cs.chubu.ac.jp

© Springer Nature Singapore Pte Ltd. 2018
B. B. Chaudhuri et al. (eds.), *Proceedings of 2nd International Conference on Computer Vision & Image Processing*, Advances in Intelligent Systems and Computing 703, https://doi.org/10.1007/978-981-10-7895-8_19

particulars of the body. Image fusion approaches are implemented to fuse CT and MRI images to perceive the hard and soft tissue details simultaneously in a single fused image. In general, radiologist also used to observe a fused form of SPECT and MR images for the better diagnosis because SPECT images show the biological activity of tissues such as blood flow. Hence, fused image shows both biological activities and soft tissue information [11, 18].

Image fusion algorithms are categorized into three categories: pixel-, feature-, and decision-level. The pixel-level fusion techniques directly applied to the pixels. These techniques can further classify into two types: spatial domain- and transform domain-based approaches. In the spatial domain, fusion is performed using local spatial features like pixel intensities and local energy. While transform-based fusion approaches, first decomposed the source images into multiple subimages, then apply appropriate fusion rules. Then, fused image is attained using inverse transform. The multiscale transforms (MSTs)-based fusion approaches are gradient pyramid (GP) [16], discrete wavelet transform (DWT) [13], etc. Moreover, MST-based methods are shift variant due to the use of subsampling operations. If the source images are misregistered, then these sampling operations can cause serious artifacts such as ringing effect near edges in the fused image [22].

Recently, to enrich the fused image with high-frequency information, multigeometrical analysis (MGA) methods, for example, curvelet transform [5], nonsubsampled contourlet transform (NSCT) [22], and NSST [12] are applied for image fusion. Due to direct operations on pixels, pixel-level fused image suffers from contrast and spatial distortion. While, feature-level fusion associated with the features attained from the source images [14, 15, 21]. At last, decision-level fusion includes attributes of forecast, fuzzy logic, and voting [17] to perform the fusion. Feature- and decision-level fusion methods suffer from loss of spatial and spectral information during the feature extraction process, which degrades the quality of fused image. The proposed method uses NSST for fusion over curvelet and NSCT due to lower implementation complexity and more directional bands. Furthermore, NSST has good localization in spatial and frequency domain, optimally sparse and directional sensitivity with parabolic scaling [7–10].

2   Preliminaries

Nonsubsampled Shearlet Transform (NSST): NSST [7, 9, 10] is applied to obtain low- and high-frequency bands of the input images. In dimension $n = 2$, continuous shearlet transform of the signal $f(x)$ is given as a mapping:

$$SH_\psi f(a, s, t) = \langle f, \psi_{ast} \rangle \tag{1}$$

where a is scale, s is orientation, t is location, $\psi_{ast}(x) = |\det \mathbf{M}_{as}|^{-1/2}\psi(\mathbf{M}_{as}^{-1}$ $(x-t))$, $\mathbf{M}_{as} = \begin{pmatrix} a & -\sqrt{as} \\ 0 & \sqrt{a} \end{pmatrix}$ for $a > 0$, $s \in R$ and $t \in R^2$. $\psi_{ast}(x)$ are called shearlets. Matrix $\mathbf{M}_{as} = \mathbf{B}_s\mathbf{A}_a$ is associated with the two different matrices: anisotropic dilation done by matrix $\mathbf{A}_a = \begin{pmatrix} a & 0 \\ 0 & \sqrt{a} \end{pmatrix}$ and shearing done by matrix $\mathbf{B}_s = \begin{pmatrix} 1 & -s \\ 0 & 1 \end{pmatrix}$. The NSST is the nonsubsampled version of shearlet transform (ST). The NSST is not using sampling operations, which make NSST shift invariant. Due to that ringing artifacts reduces in the fused image [7, 9, 10].

Phase Stretch Transform (PST): The PST [1] is a multistep process for detecting the high-frequency features of the images. First, noise reduction is done by Gaussian smoothing, then PST is applied on the smooth image. PST having a nonlinear frequency dependent transfer function. In PST, 2-D phase function is applied to the smooth image in frequency domain. The extent of phase applied to image is depends upon frequency. Lower phase implemented to the low-frequency contents such as texture features and fine details. However, higher phase applied to high-frequency contents such as edge features. Then a threshold is applied to the PST output to highlight the prominent high-frequency features. Then using some morphological operations, reliable edge features obtained by removing noise artifacts. Let any image $\mathbf{I}(m, n)$, where m, n are the 2-D space variables. PST applied to $\mathbf{I}(m, n)$ is given as follows [1]:

$$\mathbf{PST}^l(m, n) = \angle \left\langle \text{IFFT2} \left\{ \tilde{\mathbf{K}}(p, q)\tilde{\mathbf{L}}(p, q)\text{FFT2} \left\{ \mathbf{I}(m, n) \right\} \right\} \right\rangle \qquad (2)$$

where, $\mathbf{PST}^l(m, n)$ is output phase image, $\angle \langle . \rangle$ is angle operation, FFT2 is 2-D fast Fourier transform, IFFT2 is 2-D inverse FFT, p and q are frequency variables. $\tilde{\mathbf{L}}(p, q)$ is frequency response of localize smoothing kernel and $\tilde{\mathbf{K}}(p, q)$ is warped phase kernel (in frequency domain), which is given as follows [1]:

$$\tilde{\mathbf{K}}(p, q) = e^{j\varphi(p,q)} \qquad (3)$$

where,

$$\varphi(p, q) = \varphi_{\text{polar}}(r, \theta) = \varphi_{\text{polar}}(r) = \frac{SWr\tan^{-1}(Wr)-(1/2)\ln(1+(Wr)^2)}{Wr_{\max}\tan^{-1}(Wr_{\max})-(1/2)\ln(1+(Wr_{\max})^2)} \qquad (4)$$

where, $r = \sqrt{p^2 + q^2}$, $\theta = \tan^{-1}(q/p)$, $\ln(.)$ is the natural logarithm and r_{max} is the maximum frequency r. S and W are the real-valued numbers associated with strength (S) and warp (W) of phase profile function [1].

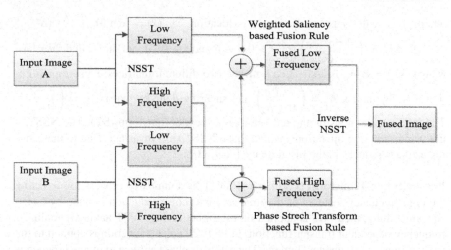

Fig. 1 Schematic block diagram of proposed image fusion technique [3]

3   The Proposed Method

Figure 1 shows the block diagram of proposed image fusion method. In proposed method, NSST decomposed the input images **A** and **B** into low-frequency bands: $\mathbf{L}_A(m, n)$ and $\mathbf{L}_B(m, n)$, in addition into high-frequency bands $\mathbf{H}_{k,l}^A(m, n)$ and $\mathbf{H}_{k,l}^B$ (m, n). Here, k and l shows the decomposed high-frequency band in kth direction at lth decomposition level, respectively. NSST domain variables are denoted by m and n. Then, applying the proposed fusion rules and inverse NSST, fused image **F** is acquired. In the 2-level of NSST decomposition of an image produces one low-frequency coefficients band, with four high-frequency coefficients bands.

Low-Frequency Band Fusion Rule: Low-frequency bands acquired after NSST decomposition are fused using weighted saliency-based method [4]. Weighted saliency fusion rule fused the coefficients of NSST in such a way that fused image efficiently captures the texture, fine, and edge information. This is done by calculating energy and correlation between the pixels of low-frequency bands. This fusion rule is implemented in the two steps: first implementing the salience measure then implementing the match measure. The salience measure measured at $\mathbf{L}(m, n)$ as local energy, where **L** is low-frequency band of image **I** and m, n is coordinate positions in **L**. The salience measure $\mathbf{S}_I(m, n)$ within a neighborhood **p** of size $\hat{m} \times \hat{n}$ is obtained as [4]:

$$\mathbf{S}_I(m, n) = \sum_{\hat{m}, \hat{n}} \mathbf{p}(\hat{m}, \hat{n}) \mathbf{L}_I(m + \hat{m}, n + \hat{n})^2 \tag{5}$$

If $\mathbf{S}_I(m, n)$ of corresponding pixels of both the low-frequency bands $\mathbf{L}_A$ and $\mathbf{L}_B$ are equal, then their average and if saliency measure is different, then pixel with maximum saliency measure is considered. Furthermore, match measure is used to find, which combined method to use (if saliency measures are different) either selection or averaging. However, an alternative way to use a correlation between the low-frequency bands. At sample (m, n) within a neighborhood $\mathbf{p}$, match as local normalized correlation is [4] given as follows:

$$\mathbf{M}_{AB}(m, n) = \frac{\sum_{\hat{m}, \hat{n}} \mathbf{p}(\hat{m}, \hat{n})\mathbf{L}_A(m + \hat{m}, n + \hat{n})\mathbf{L}_B(m + \hat{m}, n + \hat{n})}{0.5(\mathbf{S}_A(m, n) + \mathbf{S}_B(m, n))} \quad (6)$$

M_{AB} has value 1 for the same structures and less than 1 for remaining. If the match measured at the respective sample of $\mathbf{L}_A$ and $\mathbf{L}_B$ is high, then the average of the decomposed coefficient is taken else coefficient with the high saliency measure is considered. This rule represents the weighted averaging. Summarizing, in selection mode, if match measure $M_{AB}(m, n)$ at position m, n is below the threshold $\alpha < 0.75$ then weights $W_A(m, n)$ and $W_B(m, n)$ in (8) are either 1 and 0, depending upon the saliency measure $S(m, n)$ obtained using (5). Now, assign $\mathbf{W}_A(m, n) = 1$ and $\mathbf{W}_B(m, n) = 0$ in (8), if $\mathbf{S}_{L_A}(m, n) > \mathbf{S}_{L_B}(m, n)$. But, assign $\mathbf{W}_A(m, n) = 0$ and $\mathbf{W}_B(m, n) = 1$ in (8), if $\mathbf{S}_{L_A}(m, n) < \mathbf{S}_{L_B}(m, n)$.

Now if $\mathbf{M}_{AB}(m, n) > 0.75$ (which is basically averaging mode), weights $W_A(m, n)$ and $W_B(m, n)$ in (8) are evaluated using (7). First apply (7) and find the minimum and maximum weights $\mathbf{W}_{\min}(m, n)$ and $\mathbf{W}_{\max}(m, n)$. Now, assign $\mathbf{W}_A(m, n) = \mathbf{W}_{\min}(m, n)$ and $\mathbf{W}_B(m, n) = \mathbf{W}_{\max}(m, n)$ in (8), if $\mathbf{S}_{L_A}(m, n) < \mathbf{S}_{L_B}(m, n)$. But, assign $\mathbf{W}_A(m, n) = \mathbf{W}_{\min}(m, n)$ and $\mathbf{W}_B(m, n) = \mathbf{W}_{\max}(m, n)$ in (8), if $\mathbf{S}_{L_A}(m, n) > \mathbf{S}_{L_B}(m, n)$. The $\mathbf{W}_{\min}(m, n)$ and $\mathbf{W}_{\max}(m, n)$ are given as [4]:

$$\mathbf{W}_{\min}(m, n) = \frac{1}{2} - \frac{1}{2}\left(\frac{1 - \mathbf{M}_{AB}(m, n)}{1 - \alpha}\right) \quad \text{and} \quad \mathbf{W}_{\max}(m, n) = 1 - \mathbf{W}_{\min}(m, n). \quad (7)$$

The fused low-frequency band $\mathbf{L}_F(m, n)$ is obtained as:

$$\mathbf{L}_F(m, n) = \mathbf{W}_A(m, n)\mathbf{L}_A(m, n) + \mathbf{W}_B(m, n)\mathbf{L}_B(m, n) \quad (8)$$

Novel Fusion Rule for High-Frequency Bands: New fusion criteria based on PST features [1] is employed to attain fuse high-frequency coefficients. PST features-based fusion rule enhances the high-frequency attributes such as line, edges, corners, and fine details. Applying PST on any image gives a feature descriptor. This feature descriptor consists of information regarding edges, structural and fine information present in the applied image. Based on this information, coefficients of high-frequency band are selected because human visual system is very sensitive to this information compared to other information present on the images. Let $\mathbf{H}_{k,l}^A(m, n)$, $\mathbf{H}_{k,l}^B(m, n)$, and $\mathbf{H}_{k,l}^F(m, n)$ are high-frequency bands in lth directional band at the kth decomposition level at (m, n) position, of input images $\mathbf{A}$, $\mathbf{B}$, and fused image $\mathbf{F}$.

Algorithm:

The outline of proposed algorithm is as follows:

(a) Decomposed the preregistered input images using NSST.

(b) Fuse the low-frequency band using weighted saliency [4] fusion rule given in (8), and fuse the high-frequency bands using PST features [1] given in (9). Apply PST on each high-frequency bands and obtain the features of high-frequency bands $\mathbf{PST}^A_{k,l}(m, n)$ and $\mathbf{PST}^B_{k,l}(m, n)$ using (2). Then, using (9) obtain fused high-frequency bands as follows:

$$\mathbf{H}^F_{k,l}(m, n) = \begin{cases} \mathbf{H}^A_{k,l}(m, n), & \mathbf{PST}^A_{k,l}(m, n) \geq \mathbf{PST}^B_{k,l}(m, n) \\ \mathbf{H}^B_{k,l}(m, n), & Otherwise. \end{cases} \tag{9}$$

(c) Fused image $F(i, j)$ is attained by applying inverse NSST transform operation. Proposed method also applicable for the fusion of PET/SPECT images with MRI images. The fusion has done by first, calculating the luminance components (Y) of the source images by RGB to YIQ transformation [19]. When fusion of Y components are done, then reconstruct the fused color image by applying YIQ to RGB transformation.

4 Results and Comparison

To carry out the experiments with two datasets of medical images are considered. The source images downloaded from the following Web site [20]. The proposed technique is compared to following state-of-the-art techniques such as GP [16], DWT [13], NSCT Fuzzy [25], NSCT phase congruency (NSCT PC) [3], SWT NSCT [2] previously proposed for medical image fusion. Analysis of image fusion techniques carried out subjective and objective manner.

The proposed method is tested using three metrics such as mutual information (MI) [6], structure similarity-based index (Q_S) [24], and edge-based similarity measure ($Q^{AB/F}$) [23]. Proposed technique is tested in two datasets of medical images. The dataset-1 for brain images of MR-T1 and MR-T2 modalities, dataset-2 for brain images of MR-T2 and SPECT modalities. The different parameter settings are adjusted to attain good quality fused image. Only 2-level of NSST decomposition is used to minimize the effect of misregistration and noise. The weighted saliency [4] rule is implemented to fuse low-frequency band, and PST features-based fusion rule is implemented to fuse high-frequency bands. Apply inverse NSST to get the fused image. Considering image dataset-1 of MR-T2 and MR-T1 images Fig. 2a1–a3 and Fig. 2b1–b3, respectively. Some of the regions in the brain are specifically selected. These areas are highlighted with the arrows to make the easier subjective comparison of different fusion methods. Compared to previous methods, proposed approach

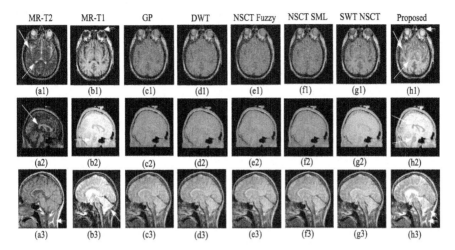

Fig. 2 Subjective comparison of the proposed technique for dataset-1. Images **a1–a3** MR-T2, **b1–b3** MR-T1, **c1–c3** GP [16], **d1–d3** DWT [13], **e1–e3** NSCT fuzzy [25], **f1–f3** NSCT PC [3], **g1–g3** SWT NSCT [2], **h1–h3** proposed

efficiently captures the most of the salient details of the input images. The claim is justified subjectively as shown in Fig. 2h1–h3. Objectively, this can be justified from results given in Table 1. From Table 1, Q_S is higher in SWT NSCT method for a couple of images. However, overall our proposed method gives better performance.

In Fig. 3 analyzing the structures of dataset-2, arrows are shown for the simple subjective comparative study of the proposed method with previous techniques. Figure 3 shows the proposed method able to preserve the prominent details of input images such as lesions and infraction in fused image, which is desirable for diagnosis. From the observation of Table 1, this claim justified. The main drawback of GP and DWT methods is the effect of ringing artifacts in the fused images and inefficient to represent the edge singularities of input source images. The NSCT Fuzzy approach suffers from the low contrast due to applied fusion rules, which can create a problem for diagnosis. The NSCT PC method is unable to capture prominent information of the source images such as texture because PC mainly enhances the high-frequency information, not the low-frequency information.

The NSST is less computationally complex than NSCT. While SWT NSCT method suffers from the low contrast fused image due to the principal component analysis (PCA)-based fusion rule because this rule distorts the spectral information of fused image. Moreover, SWT NSCT method suffers from noise artifacts due to maximum selection fusion rule. However, SWT NSCT method able to capture more structural information more efficiently due to SWT and NSCT basis. The SWT can detect texture information more effectively compare to NSCT. The reasons for outperforming our methods are proposed fusion rules and NSST decomposition and

Table 1 Objective evaluation for fused medical images of dataset-1 and dataset-2

Images	Indices	GP	DWT	NSCT fuzzy	NSCT PC	SWT NSCT	Proposed
Fig. 2a1–b1	MI	3.738	3.714	3.673	3.722	4.015	**4.439**
	$Q^{AB/F}$	0.404	0.372	0.340	0.378	0.461	**0.557**
	Q_S	0.762	0.755	0.745	0.757	**0.885**	0.829
Fig. 2a2–b2	MI	3.386	3.341	3.431	3.430	3.894	**4.342**
	$Q^{AB/F}$	0.449	0.429	0.444	0.453	0.547	**0.645**
	Q_S	0.711	0.719	0.733	0.723	0.940	**0.981**
Fig. 2a3–b3	MI	3.544	3.509	3.635	3.590	3.720	**4.132**
	$Q^{AB/F}$	0.458	0.426	0.399	0.444	0.455	**0.496**
	Q_S	0.719	0.71	0.690	0.725	**0.844**	0.708
Fig. 3a1–b1	MI	2.219	2.193	2.315	2.224	2.700	**2.711**
	$Q^{AB/F}$	0.505	0.440	0.400	0.468	**0.700**	0.656
	Q_S	0.838	0.831	0.804	0.832	**0.967**	0.903
Fig. 3a2–b2	MI	3.171	3.127	3.256	3.171	3.282	**3.704**
	$Q^{AB/F}$	0.471	0.402	0.333	0.439	0.436	**0.479**
	Q_S	0.834	0.815	0.784	0.819	**0.838**	0.812
Fig. 3a3–b3	MI	3.091	3.026	3.115	3.097	3.48	**3.693**
	$Q^{AB/F}$	0.528	0.452	0.448	0.486	0.642	**0.643**
	Q_S	0.822	0.808	0.876	0.810	0.91	**0.920**

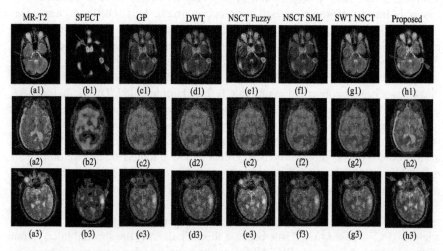

MR-T2	SPECT	GP	DWT	NSCT Fuzzy	NSCT SML	SWT NSCT	Proposed
(a1)	(b1)	(c1)	(d1)	(e1)	(f1)	(g1)	(h1)
(a2)	(b2)	(c2)	(d2)	(e2)	(f2)	(g2)	(h2)
(a3)	(b3)	(c3)	(d3)	(e3)	(f3)	(g3)	(h3)

Fig. 3 Subjective comparison of the proposed technique for dataset-2. Images **a1–a3** MR-T2, **b1–b3** SPECT, **c1–c3** GP [16], **d1–d3** DWT [13], **e1–e3** NSCT fuzzy [25], **f1–f3** NSCT PC [3], **g1–g3** SWT NSCT [2], **h1–h3** proposed

reconstruction (role of directional basis functions). Also, due to window based fusion rules for the low-frequency bands, proposed method becomes slight robust to fine misregistration error and noise, as most of the image information lies in the low-frequency bands. Because image pixels intensities are highly correlated to neighboring pixels. Hence, proposed method performs the fusion by considering local window instead of considering pixel intensity of single pixels.

5   Conclusion

The medical imaging is taking a progressively critical part in health care. The proposed method uses novel NSST based image fusion approach for multimodal medical images. The proposed approach utilizes new fusion rule based on the PST features to select the fused coefficients of high-frequency bands. Low-frequency bands are fused by employing weighted saliency fusion criteria. The proposed method gives reliable and consistent results for the medical images.

References

1. Asghari, M.H., Jalali, B.: Edge detection in digital images using dispersive phase stretch transform. Journal of Biomedical Imaging 2015, 6 (2015)
2. Bhateja, V., Patel, H., Krishn, A., Sahu, A., Lay-Ekuakille, A.: Multimodal medical image sensor fusion framework using cascade of wavelet and contourlet transform domains. IEEE Sensors Journal 15(12), 6783–6790 (2015)
3. Bhatnagar, G., Wu, Q.J., Liu, Z.: Directive contrast based multimodal medical image fusion in NSCT domain. IEEE transactions on multimedia 15(5), 1014–1024 (2013)
4. Burt, P.J., Kolczynski, R.J.: Enhanced image capture through fusion. In: Computer Vision, 1993. Proceedings., Fourth International Conference on. pp. 173–182. IEEE (1993)
5. Choi, M., Kim, R.Y., Kim, M.G.: The curvelet transform for image fusion. International Society for Photogrammetry and Remote Sensing, ISPRS 2004 35, 59–64 (2004)
6. Collignon, A., Maes, F., Delaere, D., Vandermeulen, D., Suetens, P., Marchal, G.: Automated multi-modality image registration based on information theory. In: Information processing in medical imaging. vol. 3, pp. 263–274 (1995)
7. Easley, G., Labate, D., Lim, W.Q.: Sparse directional image representations using the discrete shearlet transform. Applied and Computational Harmonic Analysis 25(1), 25–46 (2008)
8. Grohs, P., Keiper, S., Kutyniok, G., Schaefer, M.: Alpha molecules: curvelets, shearlets, ridgelets, and beyond. In: SPIE Optical Engineering + Applications. pp. 885804–885804. International Society for Optics and Photonics (2013)
9. Guo, K., Labate, D.: Optimally sparse multidimensional representation using shearlets. SIAM journal on mathematical analysis 39(1), 298–318 (2007)
10. Guorong, G., Luping, X., Dongzhu, F.: Multi-focus image fusion based on non-subsampled shearlet transform. IET Image Processing 7(6), 633–639 (2013)
11. Kaplan, I., Oldenburg, N.E., Meskell, P., Blake, M., Church, P., Holupka, E.J.: Real time MRI-ultrasound image guided stereotactic prostate biopsy. Magnetic resonance imaging 20(3), 295–299 (2002)
12. Kong, W., Liu, J.: Technique for image fusion based on nonsubsampled shearlet transform and improved pulse-coupled neural network. Optical Engineering 52(1), 017001–017001 (2013)

13. Li, H., Manjunath, B., Mitra, S.K.: Multisensor image fusion using the wavelet transform. Graphical models and image processing 57(3), 235–245 (1995)
14. Liu, Y., Liu, S., Wang, Z.: Multi-focus image fusion with dense sift. Information Fusion 23, 139–155 (2015)
15. Mitianoudis, N., Stathaki, T.: Optimal contrast correction for ICA-based fusion of multimodal images. IEEE sensors journal 8(12), 2016–2026 (2008)
16. Petrovic, V.S., Xydeas, C.S.: Gradient-based multiresolution image fusion. IEEE Transactions on Image processing 13(2), 228–237 (2004)
17. Prabhakar, S., Jain, A.K.: Decision-level fusion in fingerprint verification. Pattern Recognition 35(4), 861–874 (2002)
18. Schoder, H., Yeung, H.W., Gonen, M., Kraus, D., Larson, S.M.: Head and neck cancer: Clinical usefulness and accuracy of pet/ct image fusion 1. Radiology 231(1), 65–72 (2004)
19. Shih, P., Liu, C.: Comparative assessment of content-based face image retrieval in different color spaces. International Journal of Pattern Recognition and Artificial Intelligence 19(07), 873–893 (2005)
20. Summers, D.: Harvard whole brain atlas: http://www.med.harvard.edu/aanlib/home.html. Journal of neurology, neurosurgery, and psychiatry 74(3), 288 (2003)
21. Wang, T., Zhu, Z., Blasch, E.: Bio-inspired adaptive hyperspectral imaging for real-time target tracking. IEEE Sensors Journal 10(3), 647–654 (2010)
22. Xiao-Bo, Q., Jing-Wen, Y., Hong-Zhi, X., Zi-Qian, Z.: Image fusion algorithm based on spatial frequency-motivated pulse coupled neural networks in nonsubsampled contourlet transform domain. Acta Automatica Sinica 34(12), 1508–1514 (2008)
23. Xydeas, C., Petrovic, V.: Objective image fusion performance measure. Electronics letters 36(4), 308–309 (2000)
24. Yang, C., Zhang, J.Q., Wang, X.R., Liu, X.: A novel similarity based quality metric for image fusion. Information Fusion 9(2), 156–160 (2008)
25. Yang, Y., Que, Y., Huang, S., Lin, P.: Multimodal sensor medical image fusion based on type-2 fuzzy logic in nsct domain. IEEE Sensors Journal 16(10), 3735–3745 (2016)

A Novel Text Localization Scheme for Camera Captured Document Images

Tauseef Khan and Ayatullah Faruk Mollah

Abstract In this paper, a hybrid model for detecting text regions from scene images as well as document image is presented. At first, background is suppressed to isolate foreground regions. Then, morphological operations are applied on isolated foreground regions to ensure appropriate region boundary of such objects. Statistical features are extracted from these objects to classify them as text or non-text using a multi-layer perceptron. Classified text components are localized, and non-text ones are ignored. Experimenting on a data set of 227 camera captured images, it is found that the object isolation accuracy is 0.8638 and text non-text classification accuracy is 0.9648. It may be stated that for images with near homogenous background, the present method yields reasonably satisfactory accuracy for practical applications.

Keywords Text detection · Feature map · Background suppression
Textness features · Text non-text classification · MLP

1 Introduction

Camera captured scene or document images containing text is one of the most expressive means of effective communication. So, detecting and localizing text components from natural images have been most pioneer work in recent research trend, where many researchers have implemented several algorithms for efficient extraction of text blocks. The automated localization, extraction and recognition of scene text in unconstrained environments are still open research problems. The core of the problem lies in the extensive variability of scene text in terms of its location, physical appearance and design. Although, standard optical character recognition (OCR) is a solved problem for document images acquired with flatbed scanners,

T. Khan (✉) · A. F. Mollah
Department of Computer Science and Engineering,
Aliah University, Kolkata 700156, India
e-mail: tauseef.hit2013@gmail.com

© Springer Nature Singapore Pte Ltd. 2018
B. B. Chaudhuri et al. (eds.), *Proceedings of 2nd International Conference on Computer Vision & Image Processing*, Advances in Intelligent Systems and Computing 703, https://doi.org/10.1007/978-981-10-7895-8_20

detection and recognition from camera captured scene images are an active research problem. Complex background, variation of text layouts, orientation of handwritten text, various font size, uneven illumination and multilingual scripts lead to severe challenges compared to well-formatted scanned document images.

There are several approaches for extraction of text regions that have been implemented. Zhang et al. [1] proposed a fully convolutional network (FCN)-based multi-oriented text line detection from natural scene images and Chen et al. [2] used a strong AdaBoost classifier for recognizing text regions from natural images. Text extraction using connected component analysis [3–6] is another approach where candidate text components are identified and then they are classified as text or non-text based on various statistical features. Epshtein et al. [7] proposed a stroke width transformation function (SWT) to distinguish text components from non-text regions in scene images. SWT is used to detect colour independent text from video frames [8]. There are also several classifier-based approaches such as support vector machine (SVM), neural network-based machine learning algorithms, clustering based algorithms [9–11] for efficient discrimination of text lines from non-text objects. A multi-scale kernel based approach [12] where multi directional filter is used on the image to determine foreground intensity and generates a scale map. From this scale map, different shape, colour and texture-based features have been extracted and fed into different classifiers to distinguish text components from background and other regions. Also, HOG-based approaches [13] have been used for localization of text components.

Dalal et al. [14] proposed a histogram based on different orientation of gradient (HOG) for human detection. Minetto et al. [15] proposed an advanced texture-based HOG (T-HOG) descriptor for detecting single line text. In recent times, HOG has been implemented using co-occurrence matrix (CO-HOG) for multilingual character recognition [16]. Though this approach is not fully accurate for images where texts are hard to segment also time complexity of this algorithm is a constraint. Local binary pattern (LBP) is another efficient texture descriptor proposed by Ojala et al. [17]. Later, Multi-Scale LBP (MS-LBP) is used which is an extended version of LBP [18]. Wavelet transform, discrete cosine transform (DCT), Fourier transform descriptor have been extensively used for texture analysis for component classification [19, 20].

Feature extraction is one of the most important phases of text information extraction (TIE). Choosing discriminating and scale invariant features leads to better result for classifying text and non-text components. CC-based method uses bottom-up approach by merging all small foreground components to get larger components until all possible regions of the image get identified. In CC-based method after component extraction, non-text components need to prune out by applying discriminating, scale invariant, rotation invariant features computed from foreground components [21, 22].

Song et al. [23] designed a text extraction model based on K-means clustering algorithm where first multi-scale technique is applied to localize the text components and then, colour-based K-means clustering algorithm has been applied for character segmentation step. Colour image segmentation is much more suitable compared to conventional greyscale image segmentation for extracting text regions from natural scene images. Chen et al. [24] have developed an approach for extracting text from natural scene images based on support vector regression (SVR). They have extracted features from edge map, and then, support vector regression model has been applied for final classification. Hao et al. [25] have used some morphology-based features for detection of license plate from complex background.

Though several approaches have been made till now, efficient and accurate text detection from complex background is an unsolved problem. There are very few works for real-time application for hand-held devices such as mobile phones and other portable hand-held digital devices. In this paper, a novel text detection and extraction scheme is presented for camera captured text images. The technique is discussed in detail in Sect. 2, and experimental results are presented in Sect. 3. Finally, conclusion is made in Sect. 4.

2 Present Work

At first, input images are converted into greyscale images using a weighted average of red, green and blue channels as discussed in [26]. Then, reversed images are converted into normal images and foreground components of such normal images are segmented from their backgrounds. After that, various statistical features are extracted from foreground objects and fed into a binary classifier for further classification as text components or non-text components.

2.1 Correction of Reverse Image

In normal cases, scene images containing text components are contrasting with background, where text strokes are in darker side with light background, which clearly separate text from background. But, in some images it has been observed that background is relatively dark and text lines are light. These images are reverse images. Such reverse images need to be converted into normal images. It has been observed that for normal images global mean intensity is less than global median, but in reverse image mean is greater than median of the image. By analysing this, property reverse images are detected and corrected using Eq. 1.

$$f(x, y) = \begin{cases} 255 - f(x, y), & \text{if } \mu_f > m_f \\ f(x, y), & \text{otherwise} \end{cases} \tag{1}$$

where μ_f denotes mean grey level intensity of input image denoted as $[f(x, y)]_{M \times N}$ and m_f denotes median value. Here if mean is greater than median, we just inverse the image to get the normal image; otherwise, the normal image is retained as usual for further processing.

2.2 Foreground/Background Separation

Let $p[i]$ denote the probability density of grey level i where $i \in [0, 255]$ and $p[i] \in [0, 1]$, as $p[i] = n_i/(M \times N)$, where n_i is the number of pixels having grey level i. Then, from this distribution, background mean intensity μ_{BG} is calculated using Eq. 2.

$$\sum_{i=0}^{t_{otsu}} p[i] \times i \qquad (2)$$

where t_{otsu} denotes the Otsu threshold [27].

Then, adaptive parameter τ is generated using the expression given in Eq. 3. It may be noted that $\tau \in [t_{otsu}, \mu_{BG}]$ and the value of τ depends upon k. A pictorial representation of mean background intensity (μ_{BG}) and mean foreground intensity (μ_{FG}), Otsu threshold (t_{otsu}) and τ is shown in Fig. 1.

$$\tau = \mu_{BG} - k(\mu_{BG} - t_{otsu}) \qquad (3)$$

where $0 \leq k \leq 1$. Although, τ, if considered a threshold may coarsely divide the image into background and foreground, it is avoided as it may distort the foreground object appearance in the foreground image. Instead, a seed $s \in [\tau, 255]$ is chosen from around the boundary regions of the greyscale image and background expansion is started from the seed using 8-connected neighbours. Two pixels p and q are said to be neighbours if both $p, q \in [\tau, 255]$ and finally, a single background connected component is formed and foreground components become isolated.

This approach has a potential advantage in protecting foreground objects from being over segmented. Firstly pixels within a foreground region, having intensities $f(x, y) \in [\tau, 255]$, will not be included in the expanded background if they are

Fig. 1 Intensity range computation for bimodal background distribution of text embedded image (grey shade denotes background intensities)

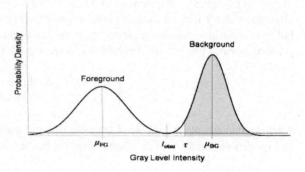

enclosed by object pixels $f(x, y) \in [0, \tau - 1]$. This leads to a great advantage in proper segmentation of foreground objects. Examples are shown in Figs. 2b and 3b, whereas if multiple seed points are chosen, foreground regions enclosed by object pixels are lost as shown in Fig. 2c. Therefore, single seed is considered for this stage.

Secondly, varying the value of k, background expansion strength may be controlled. If $k = 0$, $\tau = \mu_{BG}$ and if $k = 1$, $\tau = \mu_{otsu}$. So, selection of k plays an important role in object isolation.

Foreground dilation and feature map generation Now, morphological dilation is applied on foreground objects with $X \times Y$ mask for clear visibility by increasing the degree of discrimination from background. Here, dimension of the kernel is choosing in such a way that inter text line isolation will be high, but each character of text lines will be connected more preciously which leads to better segmentation of text components from background. The dilated image is considered as the feature map from which each foreground component is extracted and feature extraction is performed. It may be observed from Fig. 3c that in feature map, foreground objects are clearly distinguishable from background area and feature map contains both text and non-text components.

Fig. 2 Effect of region growing algorithm with varying number of seeds. **a** Sample image, **b** image with background suppression for single seed point, **c** image with background suppression for multiple seed points

Fig. 3 Foreground object isolation for a sample image: **a** normal text image, **b** isolated foreground objects, **c** dilated foreground objects

2.3 Component Feature Extraction

Foreground components extracted with connected component analysis are used for extraction of various statistical features that characterize the textness of objects and are potentially discriminating between text components (TC) and non-text components (NC). Sample components are shown in Fig. 4. It may be noted that each component is in greyscale at this stage.

However, there can be foreground components that contain both text and non-text. These mixed components (MC) cannot be used for training a binary classifier. Therefore, during trainings, MC's are not considered.

In this work, we have computed several texture-based, shape-based, region-based features for classifying these components as text component (TC) and non-text component (NC). Extracted features are described below:

Normalized Aspect Ratio Let height and width of a component be h and w, respectively. Then, normalized aspect ratio r is obtained as shown in Eq. 4. It has been observed that values of aspect ratio significantly differ from text components to non-text components in most of the cases.

$$\text{Aspect Ratio } (r) = \frac{\min(h, w)}{\max(h, w)} \tag{4}$$

where $\min()$ and $\max()$ functions return the minimum and maximum of the argument, respectively.

Circularity It is a shape-based feature and is defined as the ratio between component perimeter and area. Perimeter (P) of an object is the number of pixel lies in

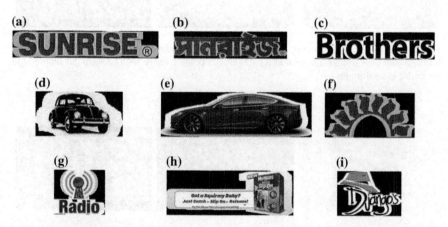

Fig. 4 Sample foreground component images: **a–c** text components, **d–f** non-text components, **g–i** mixed mode components

the boundary region called contour. Area (A) of a component is defined by the total number of pixels present in the image. Mathematical expression is shown in Eq. 5.

$$\text{Circularity} = \frac{P^2}{A} \tag{5}$$

Occupancy Ratio (OR) It is the ratio of total number of pixels present in the object, i.e. area to the bounding box area. This ratio gives a measure of how much space a given component occupies with respect to its bounding rectangle. For non-text components, this ratio is too small or too large many a times. Mathematical expression is shown in Eq. 6.

$$\text{Occupancy ratio (OR)} = \frac{\text{Area (Component)}}{\text{Area (Bounding Box)}} \tag{6}$$

Transiency Normally, text lines are self-contrasting and transient in nature compared to homogenous region. This feature signifies how transient text components are. Transiency is computed using Eq. 7 [26].

$$T_R = \sum |f(x, y) - f(x - 1, y)| + |f(x, y) - f(x, y - 1)| \tag{7}$$

It is one of the most important features for classifying TC and NC.

Horizontal Transition Density Transition from background to foreground or vice versa is more in text region compared to non-text region. So, here, number of horizontal transition along X-axis has been computed and divided by the X-dimension of the component.

Vertical Transition Density Similarly, transition along Y-axis from background to foreground and vice versa is also computed for component classification. It may be noted that Otsu threshold is used for deciding transition in both the cases.

Energy It is a very important and discriminating feature for component classification. Energy defines a measure of homogeneity of an image. Here, value of energy is 1 mean's all the pixels are having constant intensities. When energy value decreases, it signifies that images are having different pixel intensities. Equation 8 shows the mathematical expression.

$$\text{Energy} = \sum_{i=0}^{L-1} p[i]^2 \tag{8}$$

where L is grey level intensity, p[i] = Probability density.

Elongateness It is a shape-based feature. In this paper, it has been observed that text components are significantly different from non-text components with respect to elongation. Elongation is the ratio of component minor to the major axis that is shown in Eq. 9. Here, major axis and minor axis of any component are calculated using principle component analysis (PCA).

$$\text{Elongateness} = \frac{\text{Component minor axis}}{\text{Component major axis}} \tag{9}$$

2.4 Component Classification

After computing features of foreground components, the feature set is further fed into a multi-layer Perceptron (MLP) having three layers—input layer, output layer and hidden layer. The number of neurons at the input layer is the number of features used, the number of neurons at the output layer is 2 and that of the hidden layer is empirically chosen.

3 Experimental Result

Experiment has been carried out with a data set of 227 images captured by high resolution camera shown in Fig. 5. Size and aspect ratio of images are not same. A total 712 foreground components are extracted from these images, out of which 353 are text components, 262 are non-text components and 97 are mixed mode components. So, the object segmentation accuracy is 0.8638 (for $k = 0.65$).

To train the MLP for component classifications, a TC's and NC's are labelled as 0 and 1, respectively. Then, the labelled set of TC's and NC's are divided into training and test set in the ratio of 2:1. Mixed mode components are not considered for classification. As the text/non-text separation is a binary classification, the number of true positive (TP), true negative (TN), false positive (FP) and false negative (FN) is measured with the help of labelled components and classified components. Then, F-Measure (FM) is measured using Precision (P) and Recall (R) as shown in Eq. 10.

$$F\,M = \frac{2 \times P \times R}{P + R} \tag{10}$$

where $R = \frac{TP}{TP+FN}$ and $P = \frac{TP}{TP+FP}$.

Fig. 5 Successfully classified sample images (text components are marked by bounding box)

The accuracy of classification of text and non-text components is also computed by these parameters. The formula for computing accuracy is shown in Eq. 11.

$$accuracy = \frac{TP + TN}{TP + TN + FP + FN} \qquad (11)$$

The overall accuracy of component classification is measured as the product of object segmentation accuracy and component classification accuracy.

After calculating all the above statistical parameters, final quantitative result is shown in Table 1.

Precision defines how many classified text components are relevant, and Recall defines how many relevant components are correctly classified. It may be observed that the obtained result of Precision and Recall are quite high and Recall rate is higher than Precision rate.

It may be noticed from Fig. 6 that some components are wrongly classified.

Table 1 Quantitative analysis by statistical measurement

Recall (R)	Precision (P)	F-Measure (FM)	Accuracy	Overall accuracy
0.9801	0.9637	0.9718	0.9648	0.8332

Fig. 6 Images with partially misclassified components

4   Conclusion

In this paper, a text region extraction scheme has been presented for camera captured text images by foreground object segmentation and component classification using various statistical features. Present work mainly focussed on various types of camera captured images for detecting text components. We have obtained object segmentation accuracy of 0.8638 and component classification accuracy of 0.9648. So, the overall text extraction accuracy is 0.8332. It may be stated that although, component classification accuracy is quite high, overall accuracy comes down due to relatively less segmentation accuracy. While component features are very strong, segmentation needs to be improved. However, for images with near homogenous background, current system works reasonably satisfactory for practical applications.

Acknowledgements The authors are thankful to the Department of Computer Science and Engineering of Aliah University for providing every support for carrying out this work. The first author is also thankful to Aliah University for providing research fellowship.

References

1. Zhang, Z., Zhang, C., Shen, W., Yao, C., Liu, W., Bai, X.: Multi-oriented text detection with fully convolutional networks. In: IEEE Conference on Computer Vision and Pattern Recognition, pp. 4159–4167, (2016).
2. Chen, X., Yuille, A. L.: Detecting and reading text in natural scenes. In. IEEE Conference on Computer Vision and Pattern Recognition, Vol. 2, pp. II-II. (2004).
3. Yao, C., Bai, X., Liu, W., Ma, Y., Tu, Z.: Detecting texts of arbitrary orientations in natural images. In. IEEE Conference on Computer Vision and Pattern Recognition pp. 1083–1090, (2012).

4. Yi, C., Tian, Y.: Text string detection from natural scenes by structure-based partition and grouping. In. IEEE Transactions on Image Processing, pp. 2594–2605, (2011).
5. Neumann, L., Matas, J.: Real-time scene text localization and recognition., In. IEEE Conference on Computer Vision and Pattern Recognition, pp. 3538–3545. IEEE, (2012).
6. Huang, W., Lin, Z., Yang, J., Wang, J.: Text localization in natural images using stroke feature transform and text covariance descriptors. In. IEEE International Conference on Computer Vision, pp. 1241–1248, (2013).
7. Epshtein, B., Ofek, E., Wexler, Y.: Detecting text in natural scenes with stroke width transform. In. IEEE Conference on Computer Vision and Pattern Recognition (CVPR), pp. 2963–2970, (2010).
8. Zhao, Y., Lu, T. and Liao, W.: A robust color-independent text detection method from complex videos. In International Conference on Document Analysis and Recognition (ICDAR), (pp. 374–378). IEEE, (2011).
9. Kim, K. I., Jung, K., Kim, J. H.: Texture-based approach for text detection in images using support vector machines and continuously adaptive mean shift algorithm. In. IEEE Transactions on Pattern Analysis and Machine Intelligence, pp. 1631–1639 (2003).
10. Taravat, A., Del Frate, F., Cornaro, C., Vergari, S.: Neural networks and support vector machine algorithms for automatic cloud classification of whole-sky ground-based images. In. IEEE Geoscience and remote sensing letters, pp. 666–670 (2015).
11. Coates, A., Carpenter, B., Case, C., Satheesh, S., Suresh, B., Wang, T., Ng, A. Y.: Text detection and character recognition in scene images with unsupervised feature learning. In. IEEE International Conference on Document Analysis and recognition (ICDAR), pp. 440–445, (2011).
12. Shi, Z., Setlur, S., Govindaraju, V.: A steerable directional local profile technique for extraction of handwritten arabic text lines. In. IEEE 10th International Conference on Document Analysis and Recognition (ICDAR), pp. 176–180, IEEE, (2009).
13. Pan, Y. F., Hou, X., Liu, C. L.: A hybrid approach to detect and localize texts in natural scene images. In. IEEE Transactions on Image Processing, pp. 800–813, (2011).
14. Dalal, N. and Triggs, B.: Histograms of oriented gradients for human detection. In. IEEE Computer Society Conference on Computer Vision and Pattern Recognition, (Vol. 1, pp. 886–893). IEEE, (2005).
15. Minetto, R., Thome, N., Cord, M., Leite, N.J. and Stolfi, J.: T-HOG: An effective gradient-based descriptor for single line text regions. Pattern recognition, 46(3), pp. 1078–1090, (2013).
16. Tian, S., Bhattacharya, U., Lu, S., Su, B., Wang, Q., Wei, X., Lu, Y. and Tan, C.L.: Multilingual scene character recognition with co-occurrence of histogram of oriented gradients. Pattern Recognition, 51, pp. 125–134, (2016).
17. Ojala, T., Pietikäinen, M. and Harwood, D.: A comparative study of texture measures with classification based on featured distributions. Pattern recognition, 29(1), pp. 51–59, (1996).
18. Mäenpää, T. and Pietikäinen, M.: Multi-scale binary patterns for texture analysis. Image analysis, pp. 267–275, (2003).
19. Goto, H. and Tanaka, M.: Text-tracking wearable camera system for the blind. In 10th International Conference on Document Analysis and Recognition, ICDAR'09. (pp. 141–145). IEEE, (2009).
20. Ye, Q., Huang, Q., Gao, W. and Zhao, D.: Fast and robust text detection in images and video frames. Image and Vision Computing, 23(6), pp. 565–576, (2005).
21. Ye, Q. and Doermann, D.: Text detection and recognition in imagery: A survey. IEEE transactions on pattern analysis and machine intelligence, 37(7), pp. 1480–1500, (2015).
22. Liang, J., Doermann, D. and Li, H.: Camera-based analysis of text and documents: a survey. International journal on document analysis and recognition, 7(2), pp. 84–104, (2005).
23. Song, Y., Liu, A., Pang, L., Lin, S., Zhang, Y., Tang, S.: A novel image text extraction method based on k-means clustering. In. 7th IEEE/ACIS International Conference on Computer and Information Science, pp. 185–190, IEEE, (2008).

24. Lu, S., Chen, T., Tian, S., Lim, J. H., Tan, C. L.: Scene text extraction based on edges and support vector regression. In. International Journal on Document Analysis and Recognition (IJDAR), pp. 125–135, (2015).
25. Hsieh, J. W., Yu, S. H., Chen, Y. S.: Morphology-based license plate detection from complex scenes. In. 16th IEEE International Conference on Pattern Recognition, Vol. 3, pp. 176–179, (2002).
26. Mollah, A. F., Basu, S., Nasipuri, M.: Text detection from camera captured images using a novel fuzzy-based technique. In. 3rd IEEE International Conference on Emerging Applications of Information Technology (EAIT), pp. 291–294, (2012).
27. Otsu, N.: A threshold selection method from gray-level histograms. Automatica, pp. 23–27, (1979).

Video Inpainting Based on Re-weighted Tensor Decomposition

Anjali Ravindran, M. Baburaj and Sudhish N. George

Abstract Video inpainting is the process of improving the information content in a video by removing irrelevant video objects and restoring lost or deteriorated parts utilizing the spatiotemporal features that are available from adjacent frames. This paper proposes an effective video inpainting technique utilizing the multi-dimensional data decomposition technique. In Tensor Robust Principal Component Analysis (TRPCA), a multi-dimensional data corrupted by gross errors is decomposed into a low multi-rank component and a sparse component. The proposed method employs an improved version of TRPCA called Re-weighted low-rank Tensor Decomposition (RWTD) to separate the true information and the irrelevant sparse components in a video. Through this, manual identification of the components which have to be removed is avoided. Subsequent inpainting algorithm fills the region with appropriate and visually plausible data. The capabilities of the proposed method are validated by applying into videos having moving sparse outliers in it. The experimental results reveal that the proposed method performs well compared with other techniques.

Keywords Video inpainting · Tensor decomposition · Sparsity · Low-rank tensor recovery

A. Ravindran (✉)
Department of Electronics and Communication Engineering, Government College
of Engineering Kannur, Parassinikkadavu (P.O.), Kannur 670563, Kerala, India
e-mail: anjali93ravindran@gmail.com

M. Baburaj
Department of Electronics and Communication Engineering,
National Institute of Technology Calicut, Calicut, India
e-mail: baburajmadathil@gcek.ac.in

M. Baburaj
Present Address: Government College of Engineering Kannur,
Parassinikkadavu (P.O.), Kannur 670563, Kerala, India

S. N. George
Department of Electronics and Communication Engineering, National Institute
of Technology Calicut, NIT Campus P.O., Calicut 673601, Kerala, India
e-mail: sudhish@nitc.ac.in

© Springer Nature Singapore Pte Ltd. 2018
B. B. Chaudhuri et al. (eds.), *Proceedings of 2nd International Conference
on Computer Vision & Image Processing*, Advances in Intelligent Systems
and Computing 703, https://doi.org/10.1007/978-981-10-7895-8_21

1 Introduction

Inpainting is a major technique used for restoring missing or defected regions in data (images or videos), by utilizing the available information. Unlike images, in videos, temporal as well as spatial information is available for inpainting. The objective of video inpainting techniques is to maintain the visual coherence throughout the video while inpainting. Video inpainting [1] has a wide range of applications like video modification for privacy protection [2], film restoration [3], red-eye removal [4], multimedia editing and visualization [5]. Video inpainting approaches can generally be classified into patch-based and object-based approaches. Patch-based inpainting methods work well in images, but are less effective in videos, due to the difficulty in handling spatial and temporal continuity.

Bertalmio et al. [6] proposed an inpainting technique, which relies on Partial Differential Equations (PDEs), fluid dynamics, and Stoke's theorem to transport the inpainting information into the region. But it cannot reproduce large textured regions adequately. Yan et al. [7] proposed a texture synthesis-based inpainting approach for removing logos from video clips. But it cannot handle natural scenes effectively. Newson et al. [8] proposed a patch-based inpainting which searches for the nearest neighbors of the occluded pixel patches, and the aggregate information was used for inpainting. Umeda et al. in [3] introduced a video inpainting technique which utilizes directional median filtering along with spatiotemporal, exemplar-based inpainting in order to fill the missing areas in a more effective way. It requires a threshold value to be set to determine whether the median filter has to be applied or not. Timothy et al. [9] proposed a patch-based video inpainting method which considers only a single object at a time for inpainting process. This makes the process a time-consuming operation. Venkatesh et al. [10] proposed an object-based inpainting which relies on dictionary learning for the foreground–background segmentation. In [11], Kumar et al. proposed a moving text line detection and extraction technique based on edge and connected component detection.

Multi-dimensional data also called tensor affected by outliers can be decomposed into low-rank and sparse components [12]. Kilmer et al. [13] proposed a tensor decomposition scheme based on tensor SVD (t-SVD). Here, tubal rank for low-rank tensor recovery is replaced with the tensor nuclear norm. This makes the non-convex problem a convex one. Lu et al. [12] proposed a Tensor Robust Principal Component Analysis (TRPCA) technique which is an elegant tensor extension of RPCA. Baburaj et al. [14] proposed an improved form of TRPCA called Re-weighted low-rank Tensor Decomposition (RWTD) by applying re-weighted techniques.

In this paper, we propose a video inpainting technique assuming that the irrelevant component in it is sparse. Video is decomposed into sparse and low-rank components using RWTD [14], and irrelevant sparse components are identified. These identified sparse components are replaced by appropriate data from low-rank components. Effectiveness of the technique is illustrated by applying it for the moving sparse outlier removal in a set of synthetic videos.

The organization of the rest of the paper is as follows: Sect. 2 gives a brief introduction about the tensors and basic operations on it. In Sect. 3, the proposed method and decomposition algorithm are explained in detail. An evaluation of the performance of our method is given in Sect. 4. Conclusions are drawn in the Sect. 5.

2 Preliminaries on Tensor and Notations

Throughout this paper, we use Euler script, e.g., $\mathcal{A}$ to denote tensors; bold face capital letters, e.g., $\mathbf{M}$ for matrices; bold lower-case letters, e.g., $\mathbf{v}$ for vectors; and lower-case letters, e.g., k for scalars.

A tensor is a multi-dimensional array of data [15–17] in the field of real numbers, i.e., $\mathbb{R}^{n_1 \times n_2 \times \ldots n_N}$. Vectors are referred to as first-order tensors, matrices as second-order tensors, and multi-dimensional data of order three or above is called higher-order tensors.

Tensor slices are matrices obtained from the tensor by keeping all, except two indices constant [16]. For example, for an order-3 tensor, $\mathcal{A}(k, :, :)$ denotes kth horizontal, $\mathcal{A}(:, k, :)$ denotes kth lateral, and $\mathcal{A}(:, :, k)$ denotes kth frontal slices. The kth frontal slices of $\mathcal{A}$ can be compactly represented as $\mathbf{A}^{(k)}$. Tensor fibers are vectors obtained from the tensor by keeping all, except one index constant. The notations $\mathcal{A}(:, i, j)$, $\mathcal{A}(i, :, j)$, and $\mathcal{A}(i, j, :)$ are used to represent mode-1, mode-2 and mode-3 fibers, respectively. For $\mathcal{A} \in \mathbb{R}^{n_1 \times n_2 \times n_3}$, $\mathit{fft}(\cdot)$ of $\mathcal{A}$ along third dimension, denoted by $\mathcal{A}_f$, is given by $\mathcal{A}_f = \mathit{fft}(\mathcal{A}, 3)$. **Unfold** operation on $\mathcal{A} \in \mathbb{R}^{n_1 \times n_2 \times n_3}$ gives a block $n_1 n_3 \times n_2$ matrix, whereas the **fold** command does the reverse operation. As mentioned in [13] and [16], the five block-based operations needed to implement a third-order tensor multiplication are defined as follows,

$$bcirc(\mathcal{A}) = \begin{bmatrix} \mathcal{A}^{(1)} & \mathcal{A}^{(n_3)} & \ldots & \mathcal{A}^{(2)} \\ \mathcal{A}^{(2)} & \mathcal{A}^{(1)} & \ldots & \mathcal{A}^{(3)} \\ \vdots & \vdots & \vdots & \vdots \\ \mathcal{A}^{(n_3)} & \mathcal{A}^{(n_3-1)} & \ldots & \mathcal{A}^{(1)} \end{bmatrix} \tag{1}$$

$$unfold(\mathcal{A}) = \begin{bmatrix} \mathcal{A}^{(1)} \\ \mathcal{A}^{(2)} \\ \vdots \\ \mathcal{A}^{(n_3)} \end{bmatrix}, \qquad fold(unfold(\mathcal{A})) = \mathcal{A} \tag{2}$$

$$bdiag(\mathcal{A}) = \begin{bmatrix} \mathcal{A}^{(1)} & & \\ & \ddots & \\ & & \mathcal{A}^{(n_3)} \end{bmatrix}, \qquad fold(bdiag(\mathcal{A})) = \mathcal{A} \tag{3}$$

Definition 1 (*Tensor Transpose*) For a third-order tensor $\mathcal{A} \in \mathbb{R}^{n_1 \times n_2 \times n_3}$, to obtain the transpose $\mathcal{A}^T \in \mathbb{R}^{n_2 \times n_1 \times n_3}$, each frontal slice of $\mathcal{A}$ is transposed first and the order of the transposed frontal slices are reversed 2 through n_3. For example, let $\mathcal{B} = \mathcal{A}^T$ then

$$\mathbf{B}^{(1)} = \mathbf{A}^{(1)^T}$$
$$\mathbf{B}^{(k)} = \mathbf{A}^{(n_3-k+2)^T}, \quad k = 2, 3, ..., n_3 \tag{4}$$

Definition 2 (*Identity Tensor*) A tensor $\mathcal{I} \in \mathbb{R}^{n_1 \times n_1 \times n_3}$ is said to be an identity tensor if its first frontal slice is an $n_1 \times n_1$ identity matrix and all other slices are zeros.

Definition 3 (*f-diagonal Tensor*) If each frontal slice of a tensor is a diagonal matrix, then the tensor is said to be f-diagonal.

Definition 4 (*t-product*) The product of two tensors, $\mathcal{A} \in \mathbb{R}^{n_1 \times n_2 \times n_3}$ and $\mathcal{B} \in \mathbb{R}^{n_2 \times n_4 \times n_3}$, is defined as,

$$\mathcal{M} = \mathcal{A} * \mathcal{B} = fold(bcir(\mathcal{A})unfold(\mathcal{B})),$$
$$\mathcal{M} \in \mathbb{R}^{n_1 \times n_4 \times n_3} \tag{5}$$

$$\mathcal{M}(i,j,:) = \sum_{k=1}^{n_3} \mathcal{A}(i,k,:) \circledast \mathcal{B}(k,j,:) \tag{6}$$

$$\mathcal{M}_f^{(k)} = \mathcal{A}_f^{(k)} \mathcal{B}_f^{(k)}, \quad k = 1, ..., n_3 \tag{7}$$

Definition 5 (*Orthogonal Tensor*) The tensor $\mathcal{Q} \in \mathbb{R}^{n_1 \times n_1 \times n_3}$ is orthogonal if

$$\mathcal{Q}^T * \mathcal{Q} = \mathcal{Q} * \mathcal{Q}^T = \mathcal{I} \tag{8}$$

Definition 6 (*Unitary Tensor*) The tensor $\mathcal{U} \in \mathbb{R}^{n_1 \times n_2 \times n_3}$ is unitary if

$$\mathcal{U}^T * \mathcal{U} = \mathcal{U} * \mathcal{U}^T = n\mathcal{I} \tag{9}$$

where $n \in \mathbb{R}$

Definition 7 (*Tensor Singular Value Decomposition (t-SVD)*) The singular value decomposition of a tensor $\mathcal{A} \in \mathbb{R}^{n_1 \times n_2 \times n_3}$, is given by,

$$\mathcal{A} = \mathcal{U} * \Sigma * \mathcal{V}^T \tag{10}$$

Here, $\mathcal{U}$ is a unitary tensor of size $n_1 \times n_1 \times n_3$, $\mathcal{V}$ is another unitary tensor having size $n_2 \times n_2 \times n_3$, and Σ is f-diagonal tensor of size $n_1 \times n_2 \times n_3$. Computation of matrix SVDs in the Fourier domain provides the t-SVD [12].

Definition 8 (*Tensor Multi-rank and Tubal Rank*) A vector **r** in $\mathbb{R}^{n_3}$ whose ith element is equal to the rank of ith frontal slice of a tensor $\mathcal{A} \in \mathbb{R}^{n_1 \times n_2 \times n_3}$ is called the multi-rank of $\mathcal{A}$. The tubal rank r_t is defined as largest rank of all frontal slices of $\mathcal{A}$ or $r_t = max(\mathbf{r})$.

Definition 9 (*Tensor Nuclear Norm*) For a tensor $\mathcal{A} \in \mathbb{R}^{n_1 \times n_2 \times n_3}$, the tensor nuclear norm is given by,

$$\| \mathcal{A} \|_{\circledast} = \sum_{k=1}^{n_3} \sum_{i=1}^{min(n_1,n_2)} |\Sigma_f(i,i,k)| \tag{11}$$

Definition 10 (*Weighted Tensor Nuclear Norm*) Let $\mathcal{A} \in \mathbb{R}^{n_1 \times n_2 \times n_3}$ and let $\mathcal{W}_A \in \mathbb{R}^{n_1 \times n_2 \times n_3}$ be a weight tensor and $\Sigma_f(i,j,k) \in \mathbb{R}^{n_1 \times n_2 \times n_3}$ be the singular value tensor of $\mathcal{A}$, then the Weighted Tensor Nuclear Norm (WTNN) operator $\mathcal{F}_{WNN}(.)$: $\mathbb{R}^{n_1 \times n_2 \times n_3} \rightarrow \mathbb{R}$ is defined as,

$$\| \mathcal{A} \|_{\mathcal{W}_{\circledast}} = \sum_{k=1}^{n_3} \sum_{i=1}^{min(n_1,n_2)} \mathcal{W}_A(i,i,k) |\Sigma_f(i,i,k)| \tag{12}$$

3 Proposed Method

Video inpainting refers to the process of filling the undesired or removed parts of a video with appropriate data available from adjacent frames. Videos are generally considered as a multi-dimensional array (tensor) of order 3.

The proposed inpainting method is based on the assumption that the insignificant moving sparse outliers in the video, which has to be removed and inpainted, is sparse. A low-rank sparse decomposition decomposes such video into a low-rank component which consists of true data and a sparse component which consists of the moving sparse outliers. Since videos are third-order tensors, a matricization process has to be done prior to the matrix decomposition which may sometimes cause information loss. So, a tensor decomposition technique will be more effective compared to matrix decomposition.

Tensor decomposition, in general, refers to the recovery of a low multi-rank tensor from sparsely corrupted ones [12]. It assumes that if a tensor $\mathcal{M} \in \mathbb{R}^{n_1 \times n_2 \times n_3}$ having sparse noise is given, then it can be decomposed into low multi-rank $\mathcal{L} \in \mathbb{R}^{n_1 \times n_2 \times n_3}$ and sparse $\mathcal{S} \in \mathbb{R}^{n_1 \times n_2 \times n_3}$ components such that some incoherence conditions [18] are satisfied. Similar to matrix decomposition, tensor decomposition employs the Alternating Direction Method of Multipliers (ADMMs) [12] approach which solves the optimization problem serially. The non-convex optimization problem for tensor decomposition is given by,

$$\min_{\mathcal{L},\mathcal{S}} \quad rank(\mathcal{L}) + \lambda \| \mathcal{S} \|_0 \quad \text{such that} \quad \mathcal{M} = \mathcal{L} + \mathcal{S} \tag{13}$$

where λ is the regularization parameter and $\| \cdot \|_0$ is the l_0 norm. Non-convex problems do not possess a single local minimum. Hence, the above problem is NP hard. The convex optimization problem corresponding to (13) is given by,

$$\min_{\mathcal{L},S} \quad \| \mathcal{L} \|_\circledast + \lambda \| S \|_1 \qquad \text{such that} \quad \mathcal{M} = \mathcal{L} + S \qquad (14)$$

where $\| \cdot \|_\circledast$ denotes the tensor nuclear norm and $\| \cdot \|_1$ is the l_1 norm. But this method suffers a lot when the tensor becomes complicated or when too many error samples are present. To account this, an enhanced decomposition technique called Re-weighted Tensor Decomposition (RWTD) is used. RWTD encompasses a sparsity enhancement technique through re-weighted norms. Re-weighted tensor decomposition [14] can be expressed as,

$$\min_{\mathcal{L},S} \sum_{k=1}^{n_3} \sum_{i=1}^{min(n_1,n_2)} \mathcal{W}_{\mathcal{L}}(i,i,k) |\Sigma_f(i,i,k)| + \lambda \| \mathcal{W}_S \odot S \|_1$$
$$\text{such that} \quad \mathcal{M} = \mathcal{L} + S \qquad (15)$$

where $\mathcal{W}_{\mathcal{L}}$ and $\mathcal{W}_S$ are weights of the singular values of $\mathcal{L}$ and entries of S, respectively. $\Sigma_f(i,j,k)$ denotes the singular values of $\mathcal{L}$ and $\odot$ denotes standard Hadamard product. The weighted nuclear norm and weighted l_1 norm provide more closer approximations of rank and l_0 norm [14]. Using Definition 10, Eq. (15) can be expressed in a simplified form as,

$$\min_{\mathcal{L},S} \quad \| \mathcal{L} \|_{\mathcal{W}_\circledast} + \lambda \| \mathcal{W}_S \odot S \|_1$$
$$\text{such that} \quad \mathcal{M} = \mathcal{L} + S \qquad (16)$$

Alternating Direction Method of Multipliers (ADMM) [19, 20] can be applied to solve this optimization problem.

Given the original video with sparse outliers $\mathcal{M} \in \mathbb{R}^{n_1 \times n_2 \times n_3}$, RWTD [14] splits the video into a low-rank component $\mathcal{L} \in \mathbb{R}^{n_1 \times n_2 \times n_3}$ and a sparse component $S \in \mathbb{R}^{n_1 \times n_2 \times n_3}$. The moving sparse outliers are obtained here as the sparse component. The RWTD decomposition is illustrated in Fig. 1. This sparse component undergoes a thresholding operation to create an index set of insignificant pixels, Ω.

$$\Omega(i,j,k) = \begin{cases} 1, & \text{if } S(i,j,k) > \tau \\ 0, & \text{otherwise} \end{cases} \qquad (17)$$

where τ is an appropriate threshold, having value between 0 and 1.

The projection of $\mathcal{L}$ onto Ω, which is denoted by $\mathcal{P}_\Omega(\mathcal{L})$, in conjunction with the

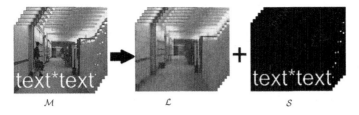

Fig. 1 Illustration of low-rank sparse decomposition of tensors

---

Algorithm 1: Proposed Video Inpainting Algorithm

Input: Observed Data $\mathcal{M} \in \mathbb{R}^{n_1 \times n_2 \times n_3}$
Output: Inpainted Data

1 $\mathcal{L}, \mathcal{S}, \mathcal{Y} \in \mathbb{R}^{n_1 \times n_2 \times n_3}$

2 $\min_{\mathcal{L}, \mathcal{S}} \| \mathcal{L} \|_{W_\circledast} + \lambda \| \mathcal{W}_S \odot \mathcal{S} \|_1$, such that $\mathcal{M} = \mathcal{L} + \mathcal{S}$

3 Find index set of unwanted pixels. $\Omega(i,j,k) = \begin{cases} 1, & \text{if } S(i,j,k) > \tau \\ 0, & \text{otherwise} \end{cases}$

4 Find $\mathcal{P}_\Omega(\mathcal{L}) = \begin{cases} \mathcal{L}(i,j,k) & \text{if } (i,j,k) \in \Omega \\ 0 & \text{if } (i,j,k) \notin \Omega \end{cases}$

5 Find $\mathcal{P}_{\Omega^\perp}(\mathcal{M}) = \begin{cases} 0 & \text{if } (i,j,k) \in \Omega \\ \mathcal{M}(i,j,k) & \text{if } (i,j,k) \notin \Omega \end{cases}$

6 Inpainted result, $\mathcal{Y} = \mathcal{P}_\Omega(\mathcal{L}) + \mathcal{P}_{\Omega^\perp}(\mathcal{M})$

7 **return** $\mathcal{Y}$

---

orthogonal projection of $\mathcal{M}$ onto Ω denoted by $\mathcal{P}_{\Omega^\perp}(\mathcal{M})$ is used for the inpainting process. The simple projection operator $\mathcal{P}_\Omega(\mathcal{L})$ is defined as,

$$\mathcal{P}_\Omega(\mathcal{L})(i,j,k) = \begin{cases} \mathcal{L}(i,j,k) & \text{if } (i,j,k) \in \Omega \\ 0 & \text{if } (i,j,k) \notin \Omega \end{cases} \tag{18}$$

and its complementary projection $\mathcal{P}_{\Omega^\perp}(\mathcal{M})$ is defined as,

$$\mathcal{P}_{\Omega^\perp}(\mathcal{M})(i,j,k) = \begin{cases} 0 & \text{if } (i,j,k) \in \Omega \\ \mathcal{M}(i,j,k) & \text{if } (i,j,k) \notin \Omega \end{cases} \tag{19}$$

Final inpainted tensor $\mathcal{Y}$ is obtained as,

$$\mathcal{Y} = \mathcal{P}_\Omega(\mathcal{L}) + \mathcal{P}_{\Omega^\perp}(\mathcal{M}) \tag{20}$$

Along with the decomposition step, inpainting accuracy and the resulting visual plausibility are greatly dependent on these projections. The whole inpainting algorithm is given in Algorithm 1.

4   Results

The performance of the proposed method has been evaluated in terms of its visual quality. Experiments were conducted on 64-bit processor having 2 GHz processor speed, and we have used R2015a version of MATLAB.

We have used a set of videos, each having 25 frames of frame size 144 × 176 which are available online [23]. Moving sparse outliers for different combinations of size and direction of motion were created separately in MATLAB and were added with the videos in the dataset which in turn causes an increment in the rank of the video. Diagonal, vertical, and horizontal motions were considered for each outlier size. Figure 2 shows the results for the proposed video inpainting method. The videos 'Hall' and 'Container' are having less background motions. So the proposed algorithm provides better results in these videos. The videos 'Highway' and 'Coastguard' with limited background motions also provide moderate results for all the sparse outlier-motion combinations. 'Soccer' and 'Bus' are the videos recorded by nonstationary cameras, so the results are lightly degraded.

An error metric was computed for each inpainted video to evaluate how much closer the inpainted result is to the original video without moving sparse outliers. The error metric used here for evaluation is,

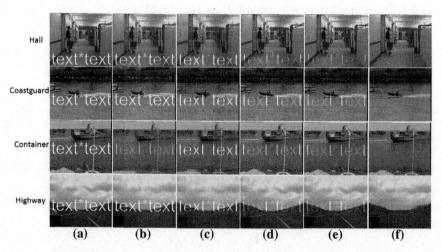

Fig. 2 Inpainting results of four videos. **a** Original video frames with moving sparse outliers, results obtained through **b** TRPCA, **c** NNTF, **d** LRR, **e** TD, and **f** our method

Table 1 Inverse Relative Squared Error (IRSE) in dB for the inpainted videos for sparse outliers of different sizes and movements

Video	Method used	Outlier size and movement					
		Small, vertical	Large, vertical	Small, diagonal	Large, diagonal	Small, random	Large, random
Hall	TRPCA [12]	27.56	27.68	27.69	25.97	27.67	28.01
	LRR [21]	30.77	27.14	30.31	29.62	30.41	30.21
	TD [15]	30.78	27.76	30.31	29.64	30.39	30.23
	NNTF [22]	23.49	16.40	23.80	22.68	23.74	22.78
	Proposed	**35.55**	**36.47**	**36.21**	**34.23**	**35.67**	**35.86**
Container	TRPCA [12]	21.76	22.82	22.83	22.48	22.81	21.96
	LRR [21]	29.35	26.27	29.57	29.06	29.51	28.98
	TD [15]	29.34	26.81	29.56	29.05	29.47	28.96
	NNTF [22]	23.62	15.70	23.95	22.43	23.97	22.10
	Proposed	**31.41**	**31.76**	**31.74**	**31.46**	**31.62**	**31.63**
Highway	TRPCA [12]	26.29	26.26	26.27	25.70	26.23	26.26
	LRR [21]	30.37	27.62	30.38	29.95	30.41	30.08
	TD [15]	28.37	27.96	30.39	29.97	30.42	28.10
	NNTF [22]	24.96	18.00	27.14	24.41	27.00	24.39
	Proposed	**30.95**	**31.92**	**31.28**	**31.11**	**31.75**	**30.62**
Coastguard	TRPCA [12]	21.96	21.95	21.95	21.74	21.87	21.80
	LRR [21]	23.49	22.63	23.38	23.39	23.29	23.50
	TD [15]	23.31	22.78	23.19	23.26	23.20	23.32
	NNTF [22]	22.14	15.34	23.06	20.93	23.19	20.91
	Proposed	**31.89**	**32.20**	**32.25**	**26.56**	**31.75**	**32.58**
Soccer	TRPCA [12]	19.64	19.71	19.30	19.58	19.56	21.58
	LRR [21]	18.78	18.86	18.58	18.59	18.60	18.66
	TD [15]	18.54	18.73	18.31	18.39	18.32	18.47
	NNTF [22]	19.87	14.68	19.96	19.42	20.05	19.49
	Proposed	**26.39**	**24.55**	**26.63**	**26.70**	**26.07**	**25.99**
Bus	TRPCA [12]	15.42	15.35	15.38	15.43	15.38	14.77
	LRR [21]	16.16	12.51	16.33	16.04	16.29	15.98
	TD [15]	12.41	12.65	12.31	12.51	12.30	12.49
	NNTF [22]	14.20	10.87	14.21	14.04	14.16	14.06
	Proposed	**18.58**	**18.00**	**18.59**	**20.17**	**18.62**	**18.58**

$$Error = \frac{\|\mathcal{A} - \mathcal{Y}\|_F}{\|\mathcal{A}\|_F} \tag{21}$$

where $\mathcal{Y}$ is the inpainted output, $\mathcal{A}$ is the original video without sparse outliers, and $\|\cdot\|_F$ denotes the Frobenius norm. It can be expressed as the Inverse Relative Squared Error (iRSE) in decibel as,

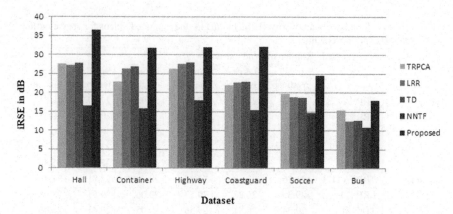

Fig. 3 Illustration of inverse relative squared error for inpainted videos with large, vertically moving sparse outliers

$$iRSE(in\ \ dB) = -20log_{10}\frac{\|\mathcal{A} - \mathcal{Y}\|_F}{\|\mathcal{A}\|_F} \tag{22}$$

In order to evaluate the effectiveness of the tensor decomposition step, we compared RWTD used here with some other tensor decomposition techniques [24] like Tensor Robust Principal Component Analysis (TRPCA), a Low-Rank Recovery (LRR) method using Linearized Alternating Direction Method with Adaptive Penalty [21], a Tensor Decomposition (TD) using Alternating Direction Augmented

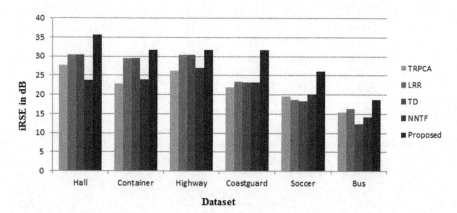

Fig. 4 Illustration of inverse relative squared error for inpainted videos with small, randomly moving sparse outliers

Lagrangian method [15], and a Nonnegative Tensor Factorisation (NNTF) [22] method. We present an evaluation of the results in Table 1. Figure 2 shows the result of inpainting through various tensor decomposition methods. It can be clearly observed that use of RWTD provides better inpainting results than other decomposition techniques. The inpainting results comparison for large, vertically moving sparse outliers and small, randomly moving sparse outliers are pictorially represented in Figs. 3 and 4 respectively.

5 Conclusion

In this paper, an efficient inpainting method for the moving sparse outliers removal in videos in a visually plausible way is implemented. Re-weighted tensor decomposition technique is used to obtain accurate video inpainting. This inpainting technique is applicable in the areas where irrelevant text lines scrolling through the frame in the TV channels can be removed. The proposed method was tested on a number of videos with various background and sparse outlier movements and has been found to outperform all other methods obtained by replacing RWTD with other decomposition techniques.

References

1. S. Moran, "Video inpainting," vol. 1, pp. 12–25, 2009.
2. W. Zhang, S. Cheung, and M. Chen, "Hiding privacy information in video surveillance system," in *Image Processing, 2005. ICIP 2005. IEEE International Conference on*, vol. 3. IEEE, 2005, pp. II–868.
3. Y. Umeda and K. Arakawa, "Removal of film scratches using exemplar-based inpainting with directional median filter," in *Communications and Information Technologies (ISCIT), 2012 International Symposium on*. IEEE, 2012, pp. 6–11.
4. S. Yoo and R.-H. Park, "Red-eye detection and correction using inpainting in digital photographs," *IEEE Transactions on Consumer Electronics*, vol. 55, no. 3, pp. 1006–1014, 2009.
5. V. V. Mahalingam, *Digital inpainting algorithms and evaluation*. University of Kentucky, 2010.
6. M. Bertalmio, A. L. Bertozzi, and G. Sapiro, "Navier-stokes, fluid dynamics, and image and video inpainting," in *Computer Vision and Pattern Recognition, 2001. CVPR 2001. Proceedings of the 2001 IEEE Computer Society Conference on*, vol. 1. IEEE, 2001, pp. I–I.
7. W.-Q. Yan and M. S. Kankanhalli, "Erasing video logos based on image inpainting," in *Multimedia and Expo, 2002. ICME'02. Proceedings. 2002 IEEE International Conference on*, vol. 2. IEEE, 2002, pp. 521–524.
8. A. Newson, A. Almansa, M. Fradet, Y. Gousseau, and P. Pérez, "Video inpainting of complex scenes," *SIAM Journal on Imaging Sciences*, vol. 7, no. 4, pp. 1993–2019, 2014.
9. T. K. Shih, N. C. Tang, and J.-N. Hwang, "Exemplar-based video inpainting without ghost shadow artifacts by maintaining temporal continuity," *IEEE transactions on circuits and systems for video technology*, vol. 19, no. 3, pp. 347–360, 2009.
10. M. V. Venkatesh, S.-c. S. Cheung, and J. Zhao, "Efficient object-based video inpainting," *Pattern Recognition Letters*, vol. 30, no. 2, pp. 168–179, 2009.

11. P. Kumar and P. Puttaswamy, "Moving text line detection and extraction in tv video frames," in *Advance Computing Conference (IACC), 2015 IEEE International.* IEEE, 2015, pp. 24–28.
12. C. Lu, J. Feng, Y. Chen, W. Liu, Z. Lin, and S. Yan, "Tensor robust principal component analysis: Exact recovery of corrupted low-rank tensors via convex optimization," in *Proceedings of the IEEE International Conference on Computer Vision and Pattern Recognition, (CVPR),* 2016.
13. M. E. Kilmer, K. Braman, N. Hao, and R. C. Hoover, "Third-order tensors as operators on matrices: A theoretical and computational framework with applications in imaging," *SIAM Journal on Matrix Analysis and Applications,* vol. 34, no. 1, pp. 148–172, 2013.
14. B. M. and S. N. George, "Reweighted low-rank tensor decomposition and its applications in video denoising," *CoRR,* vol. abs/1611.05963, 2016. [Online]. Available.
15. D. Goldfarb and Z. Qin, "Robust low-rank tensor recovery: Models and algorithms," *SIAM Journal on Matrix Analysis and Applications,* vol. 35, no. 1, pp. 225–253, 2014.
16. C. D. Martin, R. Shafer, and B. LaRue, "An order-p tensor factorization with applications in imaging," *SIAM Journal on Scientific Computing,* vol. 35, no. 1, pp. A474–A490, 2013. [Online]. Available.
17. T. G. Kolda and B. W. Bader, "Tensor decompositions and applications," *SIAM Review,* vol. 51, no. 3, pp. 455–500, 2009. [Online]. Available.
18. V. Chandrasekaran, S. Sanghavi, P. A. Parrilo, and A. S. Willsky, "Rank-sparsity incoherence for matrix decomposition," *SIAM Journal on Optimization,* vol. 21, no. 2, pp. 572–596, 2011.
19. M. Yan and W. Yin, "Self equivalence of the alternating direction method of multipliers," arXiv preprint arXiv:1407.7400, 2014.
20. X. Yuan, "Alternating direction methods for sparse covariance selection," *preprint,* 2009.
21. Z. Lin, R. Liu, and Z. Su, "Linearized alternating direction method with adaptive penalty for low-rank representation," in *Advances in neural information processing systems,* 2011, pp. 612–620.
22. Y. Xu and W. Yin, "A block coordinate descent method for regularized multiconvex optimization with applications to nonnegative tensor factorization and completion," *SIAM Journal on imaging sciences,* vol. 6, no. 3, pp. 1758–1789, 2013.
23. https://media.xiph.org/video/derf/. [Online]. Available.
24. A. Sobral, T. Bouwmans, and E.-h. Zahzah, "Lrslibrary: Low-rank and sparse tools for background modeling and subtraction in videos," in *Robust Low-Rank and Sparse Matrix Decomposition: Applications in Image and Video Processing.* CRC Press, Taylor and Francis Group.

Deep Convolutional Neural Network for Person Re-identification: A Comprehensive Review

Harendra Chahar and Neeta Nain

Abstract In video surveillance, person re-identification (re-id) is a popular technique to automatically finding whether a person has been already seen in a group of cameras. In the recent years, availability of large-scale datasets, the deep learning-based approaches have made significant improvement in the accuracy over the years as compared to hand-crafted approaches. In this paper, we have distinguished the person re-id approaches into two categories, i.e., image-based and video-based approaches; deep learning approaches are reviewed in both categories. This paper contains the brief survey of deep learning approaches on both image and video person re-id datasets. We have also presented the current ongoing works, issues, and future directions in large-scale datasets.

Keywords Person re-identification · Convolutional neural network · Open-world person re-identification

1 Introduction

The definition of re-identification is introduced in [1] as follows: "To re-identify a particular subject, then, is to identify it as numerically the same particular subject as one encountered on a previous instance". In video surveillance, person identification is defined as whether the same person has been already observed at another place by different cameras. This person re-identification task is used for the safety purpose at public place, distributed large networks of cameras in public-parks, streets and university campuses, etc. It is very strenuous for human to manually monitor video surveillance systems to accurately and efficiently finding a probe or to track a person

H. Chahar (✉) · N. Nain
Department of Computer Science and Engineering,
Malaviya National Institute of Technology, Jaipur, India
e-mail: hchahar616@gmail.com

N. Nain
e-mail: nnain.cse@mnit.ac.in

© Springer Nature Singapore Pte Ltd. 2018
B. B. Chaudhuri et al. (eds.), *Proceedings of 2nd International Conference on Computer Vision & Image Processing*, Advances in Intelligent Systems and Computing 703, https://doi.org/10.1007/978-981-10-7895-8_22

Fig. 1 Typical examples of pedestrians shot by different cameras. Each column corresponds to one person. Huge variations due to the light, pose, and viewpoint changes

across a group of cameras. A person re-id system can be divided into three parts, i.e., person detection, person tracking, and person retrieval. In this survey, person retrieval part is defined as person re-identification. In computer vision field, matching accurately two images of the same person under intensive appearance changes, such as lighting, pose, occlusion, background clutter, and viewpoint, is the most challenging problems for re-id system depicted in Fig. 1. Given its significance in research and real-world application problem, the re-id community is growing rapidly in recent years.

Few person re-id surveys already exist [2–5]. In this survey, we mainly discuss the vision part, which is also a focus in the computer vision community, another difference from previous surveys is that we focus on different re-id subtasks currently available or likely to be visible in the future, and special emphasis has been given to deep learning methods for person re-identification and issues on very large-scale person re-id datasets, which are currently popular topics or will be reflected in future trends.

This paper is organized as follows: Sect. 2 introduces a brief history of person re-id; Sect. 3 describes different kinds of deep learning approaches in image-based person re-id systems. Section 4 presents deep learning approaches in video-based person re-id systems. In Sect. 5, we present different open ongoing issues and future directions on large-scale datasets. Conclusions have drawn in Sect. 6.

2 History of Person Re-id Systems

Person re-id research problem has started with multi-camera tracking in [6]. Later, Huang and Russell [7] have proposed a Bayesian formulation to estimate the posterior of predicting the appearance of objects in one camera given evidence observed in other camera views. This appearance model combines multiple spatial-temporal features like color, velocity, vehicle length, height, and width. More details of multi-camera tracking are presented in [6].

In 2005, Wojciech Zajdel et al. [8] have proposed a method to re-identify, wherein a unique latent label is used for each person, and a dynamic Bayesian network is defined to encode the probabilistic relationship between the labels and features from the tracklets. Bayesian inference algorithm is used for determining the Id of incoming person by computing posterior label distributions.

In 2010, authors in [9, 10] have proposed technique for multi-shot person re-id. Color is a common feature used in both works, and in [10] authors additionally employ a segmentation model to detect the foreground. Minimum distance among bounding boxes in two image sets has been used for distance measurement, and authors in [9] also use the Bhattacharyya distance for the color and generic epitome features.

In 2014, Yi et al. [11] and Li et al. [12] have proposed a siamese neural network, which is used to find whether a pair of input images belong to same subject. Since then, this deep learning becomes a popular option in computer vision community for person re-id.

3 Deep Learning-Based Person Re-identification on Image Datasets

In 2006, Gheissari et al. [13] have proposed a method for person re-id based on using single images. Consider a closed-world model scenario, where G is set of N images, denoted as $\{g_i\}_{i=1}^{N}$ belongs to N different identities $1, 2, ..., N$. For a query image q, its identity is determined by:

$$i^* = argmax_{i \in 1,2,...,N} sim(q, g_i),$$ (1)

where sim (,) is a similarity function and i^* is the identity of query image q.

In 2012, Krizhevsky et al. [14] won the ILSVRC'12 competition with a large margin by using convolutional neural network (CNN)-based deep learning model, since then CNN-based deep learning models have been becoming popular. Two kinds of CNN model, i.e., the classification model used in image classification [14] and object detection [15], have been employed in the vision community. Since, these deep learning based CNN architecture requires the large number of training data. Therefore, currently most of the CNN-based re-id methods are using the siamese model [11]. In [12], authors have proposed a CNN model to jointly handle misalignment, photometric and geometric transforms, occlusions, and background clutter. In this model, a patch matching layer is added which multiplies the convolution responses of two images in different horizontal stripes and uses product to compute patch similarity in similar latitude.

Improved siamese model has proposed by Ahmed et al. [16], wherein the cross-input neighborhood dissimilarity features have computed, which are used to compare the features from one input image to features in neighboring locations of the other

Table 1 Statistics of image-based benchmark datasets for person re-id

Dataset	Time	#ID	#Image	#Camera	Label
VIPeR [27]	2007	632	1264	2	Hand
iLIDS [28]	2009	119	476	2	Hand
GRID [29]	2009	250	1275	8	Hand
CUHK01 [30]	2012	971	3884	2	Hand
CUHK02 [31]	2013	1816	7264	10	Hand
CUHK03 [12]	2014	1467	13164	2	Hand/DPM
PRID 450S [32]	2014	450	900	2	Hand
Market-1501 [33]	2015	1501	32668	6	Hand/DPM

image. Varior et al. [17] have proposed a system based on a siamese network, which uses long short-term memory (LSTM) modules. This module is used to store spatial connection to enhance the discriminative ability of the deep features by sequential access of image parts. In [18], authors have proposed a method to find effective subtle patterns in testing of paired images fedded into the network by inserting a gating function after each convolutional layer. In [19], siamese network has been integrated with a soft attention-based model to adaptively focus on the important local parts of paired input images. Cheng et al. [20] have proposed a triplet loss function, which takes three images as input. After the first convolutional layer, each image is partitioned into four overlapping body parts and fused with a global one in the fully connected layer.

In [21], authors have been proposed a three-stage learning process for attribute prediction based on an independent dataset and an attributes triplet loss function has trained on datasets with id labels. In [22], training set consists of identities from multiple datasets and a softmax loss is used in the classification. This method provides good accuracy on large datasets, such as PRW [23] and MARS [24] without careful training sample selection. In [25], authors have proposed a method, wherein a single Fisher vector [26] for each image has been constructed by using SIFT and color histograms aggregation. Based on the input Fisher vectors, a fully connected network has been build and linear discriminant analysis is used as an objective function which provides high inter-class variance and low intra-class variance.

3.1 Accuracy on Different Datasets Over the Years

Different kinds of datasets have been released for image-based person re-id such as VIPeR [27], GRID [29], iLIDS [28], CUHK01 [30], CUHK02 [31], CUHK03 [12], and Market-1501 [33]. The statistics about these datasets have been provided in Table 1. From this table, we have observed that the size of datasets has been increased over the years. As compared to earlier datasets, recent datasets, such as CUHK03 and Market-1501, have over the 1000 subjects which is good amount for training the deep learning models. Still, computer vision community is looking for large amount of datasets to train the models because deep learning models fully depend on datasets and provide good performance on larger datasets.

For the evaluation, the cumulative matching characteristics (CMC) curve and mean average precision (mAP) are usually used in both image and video datasets for person re-identification methods. CMC calculate the probability that a query image appears in gallery datasets. No matter how many ground truth matches in the gallery, only the first match is counted in the CMC calculation. If there exist multiple ground truths in the gallery, then mean average precision (mAP) is used for evaluation, which provides all the true matches belong in the gallery datasets to the query image.

From Table 2, we have observed that improvement in rank-1 accuracy on the different datasets VIPeR [27], CUHK01 [30], CUHK03 [12], PRID [32], iLIDS [28], and Market-1501 [33] over the years. We have observed highest rank-1 accuracy on

Table 2 Rank-1 accuracy of different image-based person re-identification approaches based on deep learning architecture on different datasets, i.e., (VIPeR, CUHK-01, CUHK-03, PRID, iLIDS, and Market-1501)

Authors/year	Evolution	VIPeR (%)	CUHK-01 (%)	CUHK-03 (%)	PRID (%)	iLIDS (%)	Market-1501 (%)
D Y [34] (2014)	CMC	28.23	–	–	–	–	–
Wei Li [12] (2014)	CMC	–	27.87	20.65	–	–	–
Ahmed [16] (2015)	CMC	34.81	65.0	54.74	–	–	–
Shi-Zhe Chen [35] (2016)	CMC	38.37	50.41	–	–	–	–
Lin Wu [36] (2016)	CMC/mAP	–	71.14	64.80	–	–	37.21
Xiao [22] (2016)	CMC	38.6	66.6	**75.33**	**64.0**	**64.6**	–
Chi-Su [21] (2016)	CMC/mAP	43.5	–	–	22.6	–	39.4
Cheng [20] (2016)	CMC	**47.8**	53.7	–	22.0	60.4	–
Hao Liu [19] (2016)	CMC/mAP	–	**81.04**	65.65	–	–	48.24
Varior [17] (2016)	CMC/mAP	42.4	–	57.3	–	–	61.6
Varior [18] (2016)	CMC/mAP	37.8	–	68.1	–	–	**65.88**
Wang [37] (2016)	CMC	35.76	71.80	52.17	–	–	–

these datasets 47.8%, 81.04%, 75.33%, 64.0%, 64.6%, and 65.88% from these works [18–20, 22], respectively. Except the VIPeR dataset, from the literature, we have observed that deep learning methods provided new state of the art on remaining five datasets as compared to hand-crafted person re-id systems. We have also observed overwhelming advantage of deep learning [18, 22] on largest datasets CUHK03 and Market-1501 so far. The improvement in object detection and image classification methods using deep learning in the next few years will also continuously dominate person re-id community. We have also observed that rank-1 accuracy is 65.88% and mAP is 39.55% , which is quite low, on Market-1501 dataset. This indicates that although it is relatively easy to find rank-1 accuracy, it is not trivial to locate the hard positives and thus achieve a high recall (mAP). Therefore, there is still much room for further improvement, especially when larger datasets are to be released and important breakthroughs are to be expected in image-based person re-id.

4 Deep Learning-Based Person Re-identification on Video Datasets

In recent years, video-based person re-id has become popular due to the increased data richness which induces more research possibilities. It shares a similar formulation to image-based person re-id as Eq. 1. Video-based person re-id replaces images q and g with two sets of bounding boxes $\{q_i\}_{i=1}^{n_q}$ and $\{g_j\}_{j=1}^{n_g}$, where n_q and n_g are the number of bounding boxes within each video sequence, respectively.

The common difference between video-based and image-based person re-id is that there are multiple images for each video sequence. Therefore, either a multi-match strategy or a single-match strategy should be employed after video pooling. In the previous works [9, 10], multi-match strategy has been used which requires higher computational cost. This may lead to be problematic on large datasets. Alternatively, a global vector has been constructed by aggregates frame-level features, which has better scalability called as pooling-based methods. As a consequence, recent video-based re-id methods generally use the pooling step. It can be either max/average pooling as [24, 38] or learned by a fully connected layer [39].

In [24], authors have proposed a system which does not require to capture the temporal information explicitly, wherein the images of subjects are used as its training samples to train a classification CNN model with softmax loss. Max pooling has been used to aggregate the frame features which provided the competitive accuracy on three datasets. Hence, these methods have been proven to be effective. Still, there is room for improvement at this stage, and the person re-id community is looking to take ideas from community of action/event recognition.

Fernando et al. [40] have proposed model which is used to capture frame features generated over the time in a video sequence. Wang et al. [41] have proposed a model, wherein CNN model is embedded with a multi-level encoding layer and provides video descriptors of different sequence lengths.

In recent works of [38, 39, 42], where appearance features such as color and LBP are used as the starting point into recurrent neural networks to capture the time flow between frames. In [38], authors have been proposed a model, wherein CNN is used to extract features from consecutive video frames, after that these features are fedded through a recurrent final layer. Max or average pooling is used to combine features to produce an appearance feature for the video. In [42], authors have used the similar architecture as [38] with miner difference. The special kind of recurrent neural network, the gated recurrent unit, and an identification loss are used, which provide loss convergence and improve the performance. Yan et al. [39] and Zheng et al. [24] have proposed models which use the identification model to classify each input video into their respective subjects, and hand-crafted low-level features (i.e., color and local binary pattern) are fed into many LSTMs. The output of these is connected to a softmax layer. Wu et al. [43] have proposed a model to extract both spatial-temporal and appearance features from a video. A hybrid network is build by fusing these two types of features. From this survey, we may conclude that spatial-temporal models and discriminative combination of appearance are efficient solution in future video person re-id research community.

There exist many video-based person re-id datasets such as ETHZ [44], PRID-2011 [46], 3DPES [45], iLIDS-VID [47], MARS [24]. The statistics about these datasets have been provided in Table 3. The MARS dataset [24] was recently released which is a large-scale video re-id dataset containing 1,261 identities in over 20,000 video sequences. From Table 4, we have observed highest Rank-1 accuracy on iLIDS-VID and PRID-2011 datasets 58%, 70% respectively. Deep learning

Table 3 Statistics of video-based benchmark datasets for person re-id

Dataset	Time	#ID	#Track	#Bbox	#Camera	Label
ETHZ [44]	2007	148	148	8580	1	Hand
3DPES [45]	2011	200	1000	200 k	8	Hand
PRID-2011 [46]	2011	200	400	40 k	2	Hand
iLIDS-VID [47]	2014	300	600	44 k	2	Hand
MARS [24]	2016	1261	20715	1 M	6	DPM&GMMCP

Table 4 Rank-1 accuracy of different video-based person re-identification approaches based on deep learning architecture on different datasets, i.e., iLIDS-VID and PRIQ-2011

Authors/year	Evaluation	iLIDS-VID (%)	PRIQ-2011 (%)
Wu [42]	CMC	46.1	69.0
Yan [39]	CMC	49.3	58.2
McLaughlin [38]	CMC	**58**	**70**

methods are producing overwhelmingly superior accuracy in video-based person re-id. On both the iLIDS-VID and PRID-2011 datasets, the best performing methods are based on the convolutional neural network with optional insertion of a recurrent neural network [38].

5 Currently Ongoing Underdeveloped Issues and Future Directions

Annotating large-scale datasets has always been a focus in the computer vision community. This problem is even more challenging in person re-id, because apart from drawing a bounding box of a pedestrian, one has to assign him an ID. ID assignment is not trivial since a pedestrian may reenter the fields of view (FOV) or enter another observation camera a long time after the pedestrians first appearance. In this survey, we believe two alternative strategies can help bypass the data issue.

First, how to use annotations from tracking and detection datasets remains under-explored. The second strategy is transfer learning that transfers a trained model from the source to the target domain. Transferring CNN models to other re-id datasets can be more difficult because the deep model provides a good fit to the source. Xiao et al. [22] gather a number of source re-id datasets and jointly train a recognition model for the target dataset. Hence, unsupervised transfer learning is still an open issue for the deeply learned models.

The re-identification process can be viewed as a retrieval task, in which re-ranking is an important step to improve the retrieval accuracy. It refers to the reordering of the initial ranking result from which re-ranking knowledge can be discovered. For a detailed survey of search re-ranking methods, re-ranking is still an open direction in person re-id, while it has been extensively studied in instance retrieval.

We have observed that existing re-id works can be viewed as an identification task described in Eq. 1. In the identification task, query subjects are assumed to exist in the dataset and our aim is to determine the id of the query subject. On the other side, study of open-world person re-id systems is person verification task. This verification task is based on identification task described in Eq. 1 with one more constraint $sim(q, g_i) > h$, where h is the threshold. If this condition satisfies, then query subject q belongs to identity i^*; otherwise, subject q is determined as an outlier subject which is not presented in the dataset, although i^* is the first ranked subject in the identification phase.

Only few works have been done on open-world person re-id systems. Zheng et al. [48] have been designed a system which has dataset of several known subjects and a number of probes. The aim of this work is to achieve low false target recognition rate and high true target recognition. Liao et al. [49] have proposed a method which has two phases, i.e., detection and identification. In the first phase, it finds whether a probe subject is present in the dataset or not. In the second phase, it assigns an id to the accepted probe subject.

Open-world re-id still remains a challenging task as evidenced by the low recognition rate under low false accept rate, as shown in [48, 49].

5.1 Person Re-id in Very Large Datasets

In recent years, the size of data has increased significantly in the re-id community, which gives rise to community for use of deep learning approaches. However, it is evident that available datasets are still far from a real-world problem. We have observed that the largest dataset used in survey is 500 k [33], and evidence suggests that mAP drops over 7% compared to Market-1501 with a 19 k dataset. Moreover, in [33], approximate nearest neighbor search has used for fast retrieval with low accuracy.

From both a research and an application perspective, person re-id in very large datasets should be a critical direction in the future. There is also a need to design a person re-id systems for highly crowded scenes, e.g., in a public rally or a traffic jam. Therefore, there is a need to design an efficient method to improve both accuracy and efficiency of the person re-id systems. We also have to design a person re-id system which is robust and large-scale learning of descriptors and distance metrics. As a consequence, training a global person re-id model with adaptation to various illumination condition and camera location is a priority.

6 Conclusion

Person re-identification is gaining extensive interest in the modern scientific community. We have presented a history of person re-id systems. Then, deep learning approaches have been discussed in both images and video-based datasets. We also highlight some important open issues that may attract further attention from the community. They include solving the data volume issue, re-id re-ranking methods, and open-world person re-id systems. The integration of discriminative feature learning, detector/tracking optimization, and efficient data structures will lead to a successful person re-identification system which we believe are necessary steps toward practical systems.

References

1. Plantinga, A.: Things and persons. The Review of Metaphysics, pp. 493–519 (1961)
2. D'Orazio, T., Grazia C.: People re-identification and tracking from multiple cameras: A review., In 19th IEEE International Conference on Image Processing (ICIP), pp. 1601–1604 (2012)

3. Bedagkar, G., Apurva, Shishir K.S.: A survey of approaches and trends in person re-identification, Image and Vision Computing, Vol. 32 no. 4, pp. 270–286 (2014)
4. Gong, S., Cristani, M., Yan, S., Loy, C. C.: Person re-identification, Springer, Vol. 1 (2014)
5. Satta, R.: Appearance descriptors for person re-identification: a comprehensive review, arXiv preprint arXiv:1307.5748 (2013)
6. Wang, X.: Intelligent multi-camera video surveillance: A review, Pattern recognition letters, Vol. 34 no. 1, pp. 3–19 (2013)
7. Huang, T., Russell, S.: Object identification in a bayesian context, In IJCAI, Vol. 97, pp. 1276–1282 (1997)
8. Zajdel, W., Zivkovic, Z., Krose, B. J. A.: Keeping track of humans: Have I seen this person before?, In Proceedings of the IEEE International Conference on Robotics and Automation, pp. 2081–2086, IEEE (2005)
9. Bazzani, L., Cristani, M., Perina, A., Farenzena, M., Murino, V.: Multiple-shot person re-identification by hpe signature, In 20th International Conference on Pattern Recognition (ICPR), pp. 1413–1416, IEEE (2010)
10. Farenzena, M., Bazzani, L., Perina, A., Murino, V., Cristani, M.: Person re-identification by symmetry-driven accumulation of local features, In IEEE Conference on Computer Vision and Pattern Recognition (CVPR), pp. 2360–2367, IEEE (2010)
11. Yi, D., Lei, Z., Liao, S., Li, S. Z.: Deep metric learning for person re-identification, In 22nd International Conference on Pattern Recognition (ICPR), pp. 34–39, IEEE (2014)
12. Li, W., Zhao, R., Xiao, T., Wang, X.: Deepreid: Deep filter pairing neural network for person re-identification, In Proceedings of the IEEE Conference on Computer Vision and Pattern Recognition, pp. 152–159, (2014)
13. Gheissari, N., Sebastian, T. B., Hartley, R.: Person reidentification using spatiotemporal appearance. In IEEE Computer Society Conference on Computer Vision and Pattern Recognition, Vol. 2, pp. 1528–1535, IEEE (2006)
14. Krizhevsky, A., Sutskever, I., Hinton, G. E.: Imagenet classification with deep convolutional neural networks, In Advances in neural information processing systems, pp. 1097–1105 (2012)
15. Girshick, R., Donahue, J., Darrell, T., Malik, J.: Rich feature hierarchies for accurate object detection and semantic segmentation, In Proceedings of the IEEE conference on computer vision and pattern recognition, pp. 580–587 (2014)
16. Ahmed, E., Jones, M., Marks, T. K.: An improved deep learning architecture for person re-identification, In Proceedings of the IEEE Conference on Computer Vision and Pattern Recognition, pp. 3908–3916 (2015)
17. Varior, R. R., Shuai, B., Lu, J., Xu, D., Wang, G.: A siamese long short-term memory architecture for human re-identification, In European Conference on Computer Vision, Springer International Publishing, pp. 135–153 (2016)
18. Varior, R. R., Haloi, M., Wang, G.: Gated siamese convolutional neural network architecture for human re-identification, In European Conference on Computer Vision, Springer International Publishing, pp. 791–808 (2016)
19. Liu, H., Feng, J., Qi, M., Jiang, J., Yan, S.: End-to-end comparative attention networks for person re-identification, arXiv preprint arXiv:1606.04404 (2016)
20. Cheng, D., Gong, Y., Zhou, S., Wang, J., Zheng, N.: Person re-identification by multi-channel parts-based CNN with improved triplet loss function, In Proceedings of the IEEE Conference on Computer Vision and Pattern Recognition, pp. 1335–1344 (2016)
21. Su, C., Zhang, S., Xing, J., Gao, W., Tian, Q.: Deep attributes driven multi-camera person re-identification, In European Conference on Computer Vision, Springer International Publishing, pp. 475–491 (2016)
22. Xiao, T., Li, H., Ouyang, W., Wang, X.: Learning deep feature representations with domain guided dropout for person re-identification, In Proceedings of the IEEE Conference on Computer Vision and Pattern Recognition, pp. 1249–1258 (2016)
23. Zheng, L., Zhang, H., Sun, S., Chandraker, M., Tian, Q.: Person re-identification in the wild, arXiv preprint arXiv:1604.02531 (2016)

24. Zheng, L., Bie, Z., Sun, Y., Wang, J., Su, C., Wang, S., Tian, Q.: Mars: A video benchmark for large-scale person re-identification, In European Conference on Computer Vision, Springer International Publishing, pp. 868–884 (2016)

25. Wu, L., Shen, C., van den Hengel, A.: Deep linear discriminant analysis on fisher networks: A hybrid architecture for person re-identification, Pattern Recognition (2016)

26. Perronnin, F., Snchez, J., Mensink, T.: Improving the fisher kernel for large-scale image classification, In European conference on computer vision, Springer Berlin Heidelberg, pp. 143–156 (2010)

27. Gray, D., Tao, H.: Viewpoint invariant pedestrian recognition with an ensemble of localized features, In European conference on computer vision, Springer Berlin Heidelberg, pp. 262–275 (2008)

28. Wei-Shi, Z., Shaogang, G., Tao, X.,: Associating groups of people, In Proceedings of the British Machine Vision Conference, pp. 23.1–23.11 (2009)

29. Loy, C. C., Xiang, T., Gong, S.: Multi-camera activity correlation analysis, In IEEE Conference on Computer Vision and Pattern Recognition, pp. 1988–1995, IEEE (2009)

30. Li, W., Zhao, R., Wang, X.: Human reidentification with transferred metric learning, In Asian Conference on Computer Vision, Springer Berlin Heidelberg, pp. 31–44 (2012)

31. Li, W., Wang, X.: Locally aligned feature transforms across views, In Proceedings of the IEEE Conference on Computer Vision and Pattern Recognition, pp. 3594–3601 (2013)

32. Roth, P. M., Hirzer, M., Kstinger, M., Beleznai, C., Bischof, H.: Mahalanobis distance learning for person re-identification, In Person Re-Identification, pp. 247–267, Springer (2014)

33. Zheng, L., Shen, L., Tian, L., Wang, S., Wang, J., Tian, Q.: Scalable person re-identification: A benchmark, In Proceedings of the IEEE International Conference on Computer Vision, pp. 1116–1124 (2015)

34. Yi, D., Lei, Z., Liao, S., Li, S. Z.: Deep metric learning for person re-identification, in Proceedings of International Conference on Pattern Recognition, pp. 2666–2672 (2014)

35. Chen, S. Z., Guo, C. C., Lai, J. H.: Deep ranking for person re-identification via joint representation learning, IEEE Transactions on Image Processing, Vol. 25 no.5, pp. 2353–2367 (2016)

36. Wu, L., Shen, C., Hengel, A. V. D.: Personnet: person re-identification with deep convolutional neural networks. arXiv preprint arXiv:1601.07255 (2016)

37. Wang, F., Zuo, W., Lin, L., Zhang, D., Zhang, L.: Joint learning of single-image and cross-image representations for person re-identification, In Proceedings of the IEEE Conference on Computer Vision and Pattern Recognition, pp. 1288–1296 (2016)

38. McLaughlin, N., Martinez del Rincon, J., Miller, P.: Recurrent convolutional network for video-based person re-identification, In Proceedings of the IEEE Conference on Computer Vision and Pattern Recognition, pp. 1325–1334 (2016)

39. Yan, Y., Ni, B., Song, Z., Ma, C., Yan, Y., Yang, X.: Person re-identification via recurrent feature aggregation, In European Conference on Computer Vision, Springer International Publishing, pp. 701–716 (2016)

40. Fernando, B., Gavves, E., Oramas, J., Ghodrati, A., Tuytelaars, T.: Rank pooling for action recognition, IEEE transactions on pattern analysis and machine intelligence (2016)

41. Wang, P., Cao, Y., Shen, C., Liu, L., Shen, H. T.: Temporal pyramid pooling based convolutional neural networks for action recognition, arXiv preprint arXiv:1503.0122 (2015)

42. Wu, L., Shen, C., Hengel, A. V. D.: Deep recurrent convolutional networks for video-based person re-identification: An end-to-end approach, arXiv preprint arXiv:1606.01609 (2016)

43. Wu, Z., Wang, X., Jiang, Y. G., Ye, H., Xue, X.: Modeling spatial-temporal clues in a hybrid deep learning framework for video classification, In Proceedings of the 23rd ACM international conference on Multimedia, pp. 461–470 (2015)

44. Ess, A., Leibe, B., Van Gool, L.: Depth and appearance for mobile scene analysis, In IEEE 11th International Conference on Computer Vision, pp. 1–8 (2007)

45. Baltieri, D., Vezzani, R., Cucchiara, R.: 3dpes: 3d people dataset for surveillance and forensics, In Proceedings of the 2011 joint ACM workshop on Human gesture and behavior understanding, pp. 59–64 (2011)

46. Hirzer, M., Beleznai, C., Roth, P. M., Bischof, H.: Person re-identification by descriptive and discriminative classification, In Scandinavian conference on Image analysis, Springer Berlin Heidelberg, pp. 91–102 (2011)
47. Wang, T., Gong, S., Zhu, X., Wang, S.: Person re-identification by video ranking, In European Conference on Computer Vision, Springer International Publishing, pp. 688–703 (2014)
48. Zheng, W. S., Gong, S., Xiang, T.: Towards open-world person re-identification by one-shot group-based verification, IEEE transactions on pattern analysis and machine intelligence, Vol. 38 no. 3, pp. 591–606 (2016)
49. Liao, S., Mo, Z., Zhu, J., Hu, Y., Li, S. Z.: Open-set person re-identification, arXiv preprint arXiv:1408.0872 (2014)

Flexible Threshold Visual Odometry Algorithm Using Fuzzy Logics

Rahul Mahajan, P. Vivekananda Shanmuganathan, Vinod Karar and Shashi Poddar

Abstract Visual odometry is a widely known art in the field of computer vision used for the task of estimating rotation and translation between two consecutive time instants. The RANSAC scheme used for outlier rejection incorporates a constant threshold for selecting inliers. The selection of an optimum number of inliers dispersed over the entire image is very important for accurate pose estimation and is decided on the basis of inlier threshold. In this paper, the threshold for inlier classification is adapted with the help of fuzzy logic scheme and varies with the data dynamics. The fuzzy logic is designed with an assumption about the maximum possible camera rotation that can be observed between consequent frames. The proposed methodology has been applied on KITTI dataset, and a comparison has been laid forth between adaptive RANSAC with and without using fuzzy logic with an aim of imparting flexibility to visual odometry algorithm.

Keywords Visual odometry · RANSAC · Fuzzy logic · Navigation

1 Introduction

With the advancement in automation and mobile robotics, machine vision has become an integral part of several industrial and commercial applications. Visual odometry (VO) is one of the most important aspects of machine vision that aims at computing the motion vector of any moving vehicle. The current vision-based navigation scheme has its origin from the works done by Moravec [1], Matthies-Shafer [2], and Lounguet-Higgins [3] in 1980s and has several new

R. Mahajan (✉)
Ajay Kumar Garg Engineering College, Ghaziabad, India
e-mail: rahulmahajan1704@gmail.com

P. Vivekananda Shanmuganathan
VIT University, Vellore, India

V. Karar · S. Poddar
CSIR-Central Scientific Instruments Organisation, Chandigarh, India

© Springer Nature Singapore Pte Ltd. 2018
B. B. Chaudhuri et al. (eds.), *Proceedings of 2nd International Conference on Computer Vision & Image Processing*, Advances in Intelligent Systems and Computing 703, https://doi.org/10.1007/978-981-10-7895-8_23

dimensions added to it since then. VO is a subset of the complex localization problem solved in the widely known simultaneous localization and mapping (SLAM) framework [4] and does not require to store features over frames. Several approaches for performing VO have been presented over the years and are either dense [5] or sparse [6, 7] in nature. The VO can be also classified on the kind of camera used for capturing a scene [8, 9] or on the number of views used, that is, monocular or stereo [10] or multi-view cameras. RGB-D cameras are also being used for indoor odometry purposes [11] or where the scene is not very far away from the camera [12].

In this work, a sparse feature-based technique has been used for estimating motion through a grayscale stereo camera-based setup. A general VO pipeline includes feature extraction, matching, triangulation, pose estimation, and optionally refinement. A comparison among various feature descriptors used for feature extraction, and matching techniques are given in [13, 14]. There are several triangulation techniques of which the one proposed by Hartley is very popular and used in this work [6]. Pose estimation (rotation and translation) in VO is performed using two prominent schemes, that is, absolute orientation [15–17] and relative orientation schemes [18]. In an absolute orientation scheme, the pose is estimated by fitting an appropriate rotation and translation, which minimizes the corresponding feature position error across stereo frames in 3D. The relative orientation scheme does not require feature triangulation, but carries out matching over 2D correspondences, estimating poses in a framework of perspective from n-projections. At times, the pose estimated from either of the pose estimation framework is refined using optimization subroutine, popularly known as bundle adjustment techniques [19, 20]. The above framework of pose estimation has several subroutines in which the error gets accumulated, which either may be in the feature extraction, feature matching, or 3D triangulation process. These 3D triangulated points are inclusive of outliers, which need to be segregated for achieving accurate pose estimate.

Several outlier rejection schemes have been proposed in the literature over the past few years like M-estimation [21], least median of squares [22], RANSAC [23], and their several modifications. One of the most widely used approaches for handling such uncertainties is referred as Random Sampling and Consensus (RANSAC) and has been studied in much detail [24]. RANSAC is an outlier rejection scheme which is used to improve the accuracy of VO [10, 13] or Visual SLAM [13]. Various modifications have been proposed in the RANSAC algorithms to improve its efficiency and comparisons of different classes have been provided in [24, 25]. In this paper, the RANSAC scheme has been modified by applying fuzzy logic to choose its inlier threshold. The RANSAC scheme used here requires three different points for pose estimation and incorporates the absolute orientation scheme. The paper does not claim any improvement in the working efficiency of RANSAC but shows an improvement in the initial pose estimation made using the proposed method. The later part of the paper is divided into different sections. In Sect. 2, the theoretical background required for understanding the visual odometry pipeline is described for ready reference. Section 3 contains the proposed methodology along with the need for using fuzzy logic and the entire fuzzy system design procedure. Section 4 presents the results and analysis of the proposed scheme, and the accuracy comparisons are made with respect to the simple adaptive RANSAC scheme. Finally, Sect. 5 concludes the paper along with its future scope.

2   Theoretical Background

Visual odometry is a widely growing art that has started finding varying applications ranging from terrestrial navigation to unknown environment mapping. The proposed scheme of fuzzy-based adaptive RANSAC for improving navigation estimates requires several subroutines to be studied and implemented. This section describes few of the mathematical preliminaries in brief for better and comprehensive understanding of the paper and is presented under following four subsections.

2.1   Feature Extraction, Matching, and Selection

Feature-based visual odometry technique requires an extraction of features from the 2D image which are then converted into descriptors. Traditional feature detectors like the ones proposed by Moravec [1] and Harris [26] were simple corner detectors but did not ensure scale invariance. In order to overcome these problems, Lowe et al. [27] proposed a scale-invariant feature descriptor (SIFT) which has undergone several changes since then. Present-day feature detectors are described with the help of descriptor which takes cues from the region around the feature point. An improvement over the 128-length SIFT vector was proposed as speeded-up robust feature (SURF) [28] and is used widely for computer vision tasks. These feature descriptors are used to match features between two image frames. Grid-based technique divides the complete 2D image into a grid of subimages from which a finite number of features are selected, if available. This helps in improving the robustness of the estimated pose and has been advocated in [29, 30].

2.2   Triangulation

The pose estimation task using absolute orientation scheme requires the matched feature points to be triangulated to 3D space. Triangulation is a process of obtaining the 3D position for a given 2D image point using the camera calibration (K) and the projection matrix (P). One of the traditional mechanisms to obtain 3D point is to back-propagate the rays originating from the 2D-matched feature points in the stereo image and obtain the position where these two corresponding rays meet in

3D space. However, owing to the presence of noise and error in previous steps, these rays originating from the 2D image points generally do not intersect at a point. This uncertainty has several reasons such as lens distortion, wrong correspondence matching. There exist several mechanisms to reduce these effects of which the direct linear transformation (DLT) technique proposed by Hartley and Sturm [31] is very popular. Although several other optimal schemes for triangulation do exist, the DLT has been used here for its simplicity.

2.3 Outlier Rejection

RANSAC is one of the most widely used outlier rejection scheme used in computer vision applications. It works on hypothesized and verified manner by selecting a subset of points and generating a model hypothesis. It then verifies the list of points which are consistent in the entire dataset with the hypothesized model and classifies them as inliers and the rest of the points as outliers. It hypothesizes for a few numbers of times and chooses the hypothesis with maximum number of inliers. The minimization function used for RANSAC is the feature position error given as

$$E = \sum_{i=1}^{n} ||(RX_i + t) - Y_i||^2 \tag{1}$$

Here, R and t denote rotation and translation, respectively, and, X and Y are the n 3D point correspondences in consequent frames. In order to apply this RANSAC scheme, the correct estimation of outlier percentage is needed beforehand. This helps in calculating the maximum number of iterations for which the RANSAC algorithm should be repeated in order to obtain a set with maximum inliers. However, this outlier percentage is not known previously and cannot be predicted accurately. Adaptive RANSAC [6] is one of the techniques which adapts this parameter on the basis of varying outlier population and calculates the maximum number of iterations required. The adaptive RANSAC has some added advantages over the simple RANSAC algorithm for which it is preferred in this paper and are as follows. Firstly, there is no need of assuming a rigid outlier probability for all the frames beforehand. Secondly, more robust solution is obtained when the outlier probability is more than assumed and has lesser computation when outlier probability is lesser than the assumed.

2.4 Pose Estimation

The 3D matched feature correspondences are passed through an outlier rejection scheme, thus improving the robustness of estimated rotation and translation in pose estimation scheme. This paper uses the absolute orientation scheme for pose estimation and incorporates the promising technique proposed by Umeyama [17]. The main goal of this scheme is to obtain the six degrees of freedom (3 rotation, 3 translation) between two subsequent 3D image point clouds. Umeyama's method is an improvement over the pose estimation scheme proposed by Horn [15] and Arun et al. [16]. The pose estimation scheme applied here is briefly described in the following steps and initiates with finding R which minimizes

$$\epsilon^2 = \frac{1}{n}\sum_{i=1}^{n}\|Y - RX\|^2. \tag{2}$$

This is followed by estimating translation t as:

$$t = \mu_y - R\mu_x. \tag{3}$$

Here, μ_x and μ_y are the centroids of the two 3D point clouds. The minimization problem of the Eq. (2) is solved using the singular value decomposition. Calculate centroids of the 3D point clouds

$$\mu_x = \frac{1}{n}\sum_{i=1}^{n} X_i, \quad \mu_y = \frac{1}{n}\sum_{i=1}^{n} Y_i. \tag{4}$$

- Calculate the correlation matrix using

$$\sum_{xy} = \frac{1}{n}\sum_{i=1}^{n} (Y_i - \mu_y)(X_i - \mu_x)T. \tag{5}$$

- Determining UDV^T by the singular value decomposition of $\sum_{xy}$

-
$$S = \begin{cases} I, & \text{if } \det(U).\det(V) = 1 \\ diag(1, 1, \ldots, -1), & \text{if } \det(U).\det(V) = -1 \end{cases}$$

- Rotation, R is obtained as

$$R = USV^T. \tag{6}$$

- And, translation, t is obtained as

$$t = \mu_y - R\mu_x. \tag{7}$$

The rotation and translation obtained using the above scheme are then used to obtain the transformation matrix as:

$$T = \begin{bmatrix} r_{11} & r_{12} & r_{13} & t_x \\ r_{21} & r_{22} & r_{23} & t_y \\ r_{31} & r_{32} & r_{33} & t_z \\ 0 & 0 & 0 & 1 \end{bmatrix} \tag{8}$$

Here r_{ij} and t_i are the components of rotation matrix and translation matrix. The rotation and translation are estimated in the camera coordinate frame and is then converted into a world coordinate frame. The pose is simply a 3×4 matrix stacking rotation and translation elements in the world coordinate frame given as $P = [R_w | t_w]$. This transformation matrix is concatenated for generating path and refining pose in some of the schemes. The bundle adjustment technique is not discussed in detail here as it is not within the scope of this paper.

3 Proposed Methodology

The proposed scheme of fuzzy-aided visual odometry scheme has been applied to the online available KITTI dataset. The KITTI dataset has been generated by capturing consequent images from a stereo camera mounted on a car along with other aiding sensors which help in generating the ground truth. In this work, the absolute orientation scheme for pose estimation is used which minimizes the position error of 3D triangulated point clouds. The feature extraction methodology used here, that is, speeded-up robust features (SURF) scheme, is rotation and affine invariant, robust to noise and has good repeatability in terms of extracted features. Unlike traditional schemes, the proposed scheme applies grid-based feature selection for providing spatial variance to the feature space and improves pose estimation [8, 30]. Figures 1 and 2 show the feature selection with and without the use of the grid-based technique. It can be observed from Fig. 1 that the features are concentrated in specific regions of the image while in Fig. 2, only a few of the features are selected from each grid, providing importance to the weak features lying in sparse regions as well. The features selected from different grids are then triangulated using a direct linear transformation technique [6] and sent for pose estimation framework.

Many a times developer faces situations where parameters such as inlier threshold, the similarity matching threshold for feature point selection, etc., are chosen by the developer and is solely a matter of developer's experience. Hence, the selection of these parameters and their optimization is a task of utmost importance to provide robustness to the RANSAC algorithm [32]. A novice may

Fig. 1 Feature selection without grid-based feature selection

Fig. 2 Feature selection with grid-based feature selection

experience great difficulty in selecting these parameter values, unless recommended or obtained based on the environment. However, there always exists an ambiguity in the parameter selection as there is no fine line that advocates selection of a defined value. The main aim of this work is to induce flexibility in one of the many parameter selection tasks, that is, the inlier selection threshold in RANSAC for the visual odometry pipeline. The proposed scheme monitors rotation over consecutive frames and triggers threshold selection on the basis of these inputs to the fuzzy logic.

The proposed scheme of using fuzzy logic to estimate RANSAC threshold is a novel approach and is found to improve the estimated motion accuracy. The number of inliers increases with the increasing RANSAC threshold, allowing more feature points to enter into the pose estimation framework. These 3D point clouds have several outliers between two consecutive frames and need to be removed using outlier rejection schemes as discussed in Sect. 2.3. The outlier rejection scheme, RANSAC, needs a threshold for selecting inliers [6, 23, 33] which varies as per the situation. It is important to have sufficient number of inliers dispersed over the complete image to obtain better pose estimate.

It has been found empirically that a considerable variation in the number of inliers is seen while the vehicle takes a turn. As the yaw angle of the vehicle changes while the vehicle turns, the number of feature points extracted get reduced, which leads to lesser inliers with the same threshold. It is thus hypothesized here to adapt the inlier threshold with the varying dynamics of the vehicle. Mamdani-type fuzzy logic [34] is incorporated here to regulate this threshold which is used for classifying correspondences as inliers. This fuzzy logic is a multiple-input single-output (MISO) system for which the previous frame's rotation and change in rotation are the inputs and the output is the inlier threshold value. Both inputs, 'rotations' and 'change in rotations,' are defined using the fuzzy sets as: big negative (BN), small negative (SN), zero (ZE), small positive (SP), and big positive (BP). The output, 'inlier threshold,' is classified as small (S), medium (M), large (LA), larger (LR), and largest (LT) while the membership function used is triangular. The fuzzy logic design for input parameters is shown in Figs. 3 and 4, and for the output in Fig. 5. There are certain observations that were made before developing this fuzzy logic system. Some of the observations for the current problem are as follows: (i) A variation of inliers is observed when RANSAC is performed over the same data points in different runs, owing to its probabilistic nature; (ii) previous frame's rotation has direct relation to image blurriness and inversely proportional to the number of feature matches; and (iii) structured environment leads to better matching of features across frames than the unstructured scenes of tree and vegetation. The above observations have thus inspired certain assumptions for the fuzzy logic design and are as follows: (i) At least 25 inliers are required to obtain a reliable pose estimate; (ii) the vehicle tends to move in a direction in which it is traveling and would not change its direction of travel abruptly from one frame to another; (iii) higher the absolute rotation of the previous frame, higher is the threshold value for RANSAC. The above observations and analysis have been used to define limits for different fuzzy input sets and are defined separately for rotation (R) and change in rotation (R). The ranges for the fuzzy sets shown in Figs. 3, 4, and 5 are selected empirically and can be modified by the designer. The main aim of this paper is to demonstrate the effectiveness of using fuzzy logic to improve estimated motion by modifying the RANSAC threshold.

The generalized rules for the fuzzy logic system designed here are as follows: (i) If change in rotation is positive and the previous rotation is positive, threshold increases, (ii) if change in rotation is negative and the previous rotation is positive, threshold decreases, (iii) if change in rotation is positive and the previous rotation is

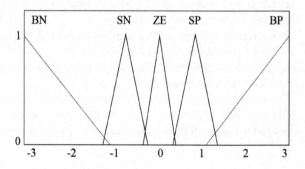

Fig. 3 Membership functions for input 'change in rotation'

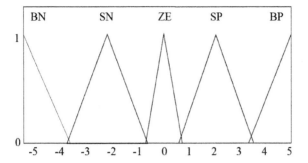

Fig. 4 Membership functions for input 'previous frame's rotation'

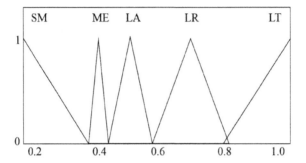

Fig. 5 Membership functions for output 'threshold for RANSAC'

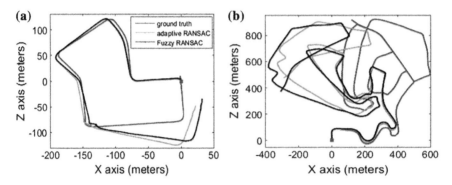

Fig. 6 Visual odometry results for sequence **a** and **b** in left and right, respectively

negative, threshold decreases, and (iv) if change in rotation is negative and the previous rotation is negative, threshold increases. Figure 6 below depicts mapping of changing threshold with the fuzzy set selection, as per the input.

4   Results and Discussion

The algorithm developed here is used to estimate pose for the online available KITTI visual odometry dataset. The odometry results obtained from the proposed methodology are compared with the ground truth data for testing and analysis. As seen, there is an improvement in the pose estimation by applying fuzzy logic scheme to vary RANSAC threshold. The fuzzy adaptive RANSAC for visual odometry (FARVO) pipeline is compared to simple adaptive RANSAC. The odometry results of two of the sequences along with their average translational and rotational errors are shown in Fig. 6. Figure 6 depicts the estimated path using the proposed methodology (key: black) along with the ground truth (key: blue) and an adaptive version of RANSAC without fuzzy logic (key: red). The comparison between the estimated and ground truth path is numerically carried out on the basis of average rotational and translational errors. The evaluation metric [33] for error calculation is given as:

$$E_{rot}(F) = \frac{1}{|F|} \sum_{(i,j) \in F} \angle[(\hat{p}_j \ominus \hat{p}_l) \ominus (p_j \ominus p_i)]. \tag{9}$$

$$E_{trans}(F) = \frac{1}{|F|} \sum_{(i,j) \in F} \left\| (\hat{p}_j \ominus \hat{p}_l) \ominus (p_j \ominus p_i) \right\|_2. \tag{10}$$

Here, F is a set of frames (i, j), $\hat{p}$ and p are estimated and true camera poses, respectively, $\ominus$ denotes the inverse of standard motion composition operator, and $\angle[.]$ denotes the rotation angle. The two datasets are selected such that one of them shows several rotational changes in between while the other has fewer rotation changes comparatively.

As seen, the trajectory obtained using FARVO pipeline could not be directly commented regarding its performance as compared to the results obtained using simple VO pipeline. Figures 7 and 8 provide numerical comparison between the

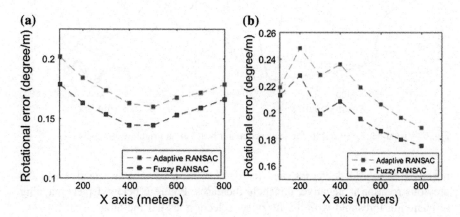

Fig. 7 Average rotational error for sequence **a** and **b** in left and right, respectively

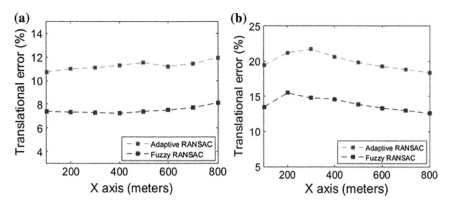

Fig. 8 Average translational error for sequence **a** and **b** in left and right, respectively

rotational and translational errors for sequence A and B as compared to ground truth. It can be observed that the FARVO scheme performs better than the simple adaptive RANSAC scheme which has constant inlier threshold at all the times (Tables 1 and 2).

A further investigation has been carried out by applying the FARVO scheme on different sequences provided in KITTI benchmark and is presented in Table 3. Though an improvement in the pose estimation is observed using the fuzzy-aided scheme, it does not necessarily improve the pose estimate at every instance. There do exist some sequences which showed increase in average rotational error using fuzzy-RANSAC algorithm also but a reduction in average translational error is

Table 1 Mapping of threshold change with fuzzy rule selection

	Increase
⟶	
Small	Medium	Large	Larger	Largest
Decrease	⟵			

Table 2 Fuzzy rule base for adaptive RANSAC

ΔR	Rotation (R)				
	Big negative	Small negative	Zero	Small positive	Big positive
Big negative	Largest	larger	Larger	Large	Large
Small negative	Larger	Large	Medium	Medium	Large
Zero	Larger	Medium	Small	Medium	Larger
Small positive	Large	Medium	Medium	Large	Larger
Big positive	Large	Large	Larger	Larger	Largest

Table 3 Average rotational and translational errors

Sl. no.	Average rotational error using adaptive RANSAC (degree/m)	Average rotational error using FARVO (degree/m)	Average translational error using adaptive RANSAC (% error)	Average translational error using FARVO (% error)
1	0.190029	0.183659	14.08384	9.45367
2	0.216545	0.202576	19.51388	14.52276
3	0.16172	0.151493	13.18257	8.268767
4	0.144733	0.147414	31.33495	18.65981
5	0.172694	0.16538	14.97984	11.14408
6	0.150165	0.14178	18.18112	10.9517
7	0.186957	0.177638	11.97847	8.323461
8	0.190529	0.187746	16.13237	11.79259
9	0.167662	0.188933	21.69584	17.71804
10	0.237451	0.233538	16.68511	11.03329

observed for all the training sequences. Though the results are not very accurate as compared to the ground truth, this paper embarks upon the usage of fuzzy logic for design of a flexible visual odometry algorithm which is free of rigid threshold and adapts according to the previous motion history.

5 Conclusion and Future Scope

RANSAC is a widely used outlier rejection scheme in computer vision applications like homography computation and pose estimation. It is used to segregate inliers and outliers from a set of data based on mathematical hypothesis. In this paper, the fuzzy logic design has been proposed to reduce dependency on the rigid threshold for extracting inliers in the RANSAC loop. The paper has implemented an adaptive RANSAC and a 3D-to-3D motion estimation scheme to estimate rotation and translation between two image frames. The proposed methodology of fuzzy-aided RANSAC is found to reduce the translation error for most of the cases and as compared to the simple adaptive RANSAC scheme. The rationale of choosing fuzzy logic for threshold adaptation is its ability to provide a solution in a lesser known environment. It has also been found in the literature that there does not exist any well-defined relation for the yaw movement undergone by the agent and its effect on the inliers. Thus, this research gap has been addressed in this work with the help of a non-learning-type control mechanism. In future, the proposed scheme would be applied for 2D-3D and 2D-2D motion estimation schemes along with terrain information to improve threshold adaptation mechanism for known environments.

References

1. Moravec, H.P., Obstacle avoidance and navigation in the real world by a seeing robot rover. 1980, DTIC Document.
2. Matthies, L. and S. Shafer, Error modeling in stereo navigation. IEEE Journal on Robotics and Automation, 1987. 3(3): p. 239–248.
3. Longuet-Higgins, H.C., A computer algorithm for reconstructing a scene from two projections. Readings in Computer Vision: Issues, Problems, Principles, and Paradigms, MA Fischler and O. Firschein, eds, 1987: p. 61–62.
4. Mouragnon, E., et al. Real time localization and 3d reconstruction. in 2006 IEEE Computer Society Conference on Computer Vision and Pattern Recognition (CVPR'06). 2006.
5. Kerl, C., J. Sturm, and D. Cremers. Dense visual SLAM for RGB-D cameras. in 2013 IEEE/RSJ International Conference on Intelligent Robots and Systems. 2013.
6. Hartley, R. and A. Zisserman, Multiple view geometry in computer vision. 2003: Cambridge university press.
7. Armangué, X. and J. Salvi, Overall view regarding fundamental matrix estimation. Image and vision computing, 2003. 21(2): p. 205–220.
8. Corke, P., D. Strelow, and S. Singh. Omnidirectional visual odometry for a planetary rover. in Intelligent Robots and Systems, 2004. (IROS 2004). Proceedings. 2004 IEEE/RSJ International Conference on. 2004. IEEE.
9. Scaramuzza, D. and R. Siegwart, Appearance-guided monocular omnidirectional visual odometry for outdoor ground vehicles. IEEE transactions on robotics, 2008. 24(5).
10. Nistér, D., O. Naroditsky, and J. Bergen. Visual odometry. in Computer Vision and Pattern Recognition, 2004. CVPR 2004.
11. Fang, Z. and Zhang, Y., 2015. Experimental evaluation of RGB-D visual odometry methods. *International Journal of Advanced Robotic Systems*, 12(3), p. 26.
12. Huang, A.S., Bachrach, A., Henry, P., Krainin, M., Maturana, D., Fox, D. and Roy, N., 2017. Visual odometry and mapping for autonomous flight using an RGB-D camera. In *Robotics Research* (pp. 235–252). Springer International Publishing.
13. Fraundorfer, F. and D. Scaramuzza, Visual odometry: Part i: The first 30 years and fundamentals. IEEE Robotics and Automation Magazine, 2011. 18(4): p. 80–92.
14. Giachetti, A., Matching techniques to compute image motion. Image and Vision Computing, 2000. 18(3): p. 247–260.
15. Horn, B.K., H.M. Hilden, and S. Negahdaripour, Closed-form solution of absolute orientation using orthonormal matrices. JOSA A, 1988. 5(7): p. 1127–1135.
16. Arun, K.S., T.S. Huang, and S.D. Blostein, Least-squares fitting of two 3-D point sets. IEEE Transactions on pattern analysis and machine intelligence, 1987(5): p. 698–700.
17. Umeyama, S., Least-squares estimation of transformation parameters between two point patterns. IEEE Transactions on pattern analysis and machine intelligence, 1991.
18. Haralick, B.M., et al., Review and analysis of solutions of the three point perspective pose estimation problem. International journal of computer vision, 1994. 13(3): p. 331–356.
19. Lourakis, M. and A. Argyros, The design and implementation of a generic sparse bundle adjustment software package based on the levenberg-marquardt algorithm. 2004, Technical Report 340, Institute of Computer Science-FORTH, Heraklion, Crete, Greece.
20. Sünderhauf, N., et al., Visual odometry using sparse bundle adjustment on an autonomous outdoor vehicle, in Autonome Mobile Systeme 2005. 2006, Springer. p. 157–163.
21. Torr, P.H. and A. Zisserman, MLESAC: A new robust estimator with application to estimating image geometry. Computer Vision and Image Understanding, 2000.
22. Rousseeuw, P.J., Least median of squares regression. Journal of the American statistical association, 1984. 79(388): p. 871–880.
23. Fischler, M.A. and R.C. Bolles, Random sample consensus: a paradigm for model fitting with applications to image analysis and automated cartography. Communications of the ACM, 1981. 24(6): p. 381–395.

24. Raguram, R., J.-M. Frahm, and M. Pollefeys. A comparative analysis of RANSAC techniques leading to adaptive real-time random sample consensus. in European Conference on Computer Vision. 2008. Springer Berlin Heidelberg.
25. Choi, S., T. Kim, and W. Yu, Performance evaluation of RANSAC family. Journal of Computer Vision, 1997. 24(3): p. 271–300.
26. Harris, C.G. and J. Pike, 3D positional integration from image sequences. Image and Vision Computing, 1988. 6(2): p. 87–90.
27. 26. Lowe, D.G., Distinctive image features from scale-invariant keypoints. International journal of computer vision, 2004. 60(2): p. 91–110.
28. Bay, H., et al., Speeded-up robust features (SURF). Computer vision and image understanding, 2008. 110(3): p. 346–359.
29. Kitt, B., A. Geiger, and H. Lategahn. Visual odometry based on stereo image sequences with RANSAC-based outlier rejection scheme. in Intelligent Vehicles Symposium. 2010.
30. Nannen, V. and G. Oliver. Grid-based Spatial Keypoint Selection for Real Time Visual Odometry. in ICPRAM. 2013.
31. Hartley, R.I. and P. Sturm, Triangulation. Computer vision and image understanding, 1997. 68(2): p. 146–157.
32. Carrasco, P.L.N. and G.O. Codina, Visual Odometry Parameters Optimization for Autonomous Underwater Vehicles. Instrumentation viewpoint, 2013(15).
33. Geiger, A., P. Lenz, and R. Urtasun. Are we ready for autonomous driving? the kitti vision benchmark suite. in Computer Vision and Pattern Recognition (CVPR), 2012.
34. Mamdani, E.H. and Assilian, S., 1975. An experiment in linguistic synthesis with a fuzzy logic controller. International journal of man-machine studies, 7(1), pp. 1–13.

Fast Single Image Learning-Based Super Resolution of Medical Images Using a New Analytical Solution for Reconstruction Problem

K. Mariyambi, E. Saritha and M. Baburaj

Abstract The process of retrieving images with high resolution using its low-resolution version is refereed to as super resolution. This paper proposes a fast and efficient algorithm that performs resolution enhancement and denoising of medical images. By using the patch pairs of high- and low-resolution images as database, the super-resolved image is recovered from their decimated, blurred and noise-added version. In this paper, the high-resolution patch to be estimated can be expressed as a sparse linear combination of HR patches over the database. Such linear combination of patches can be modelled as nonnegative quadratic problem. The computational cost of proposed method is reduced by finding closed form solution to the associated image reconstruction problem. Instead of traditional splitting strategy of decimation and convolution process, we decided to use the decimation and blurring operator's frequency domain properties simultaneously. Simulation result conducted on several images with various noise level shows the potency of our SR approach compared with existing super-resolution techniques.

1 Introduction

Most of the electronic imaging applications like medical imaging, video, surveillance, etc. high resolution (HR) images are desired. Despite the advances in medical image acquiring systems, the image resolution is decreased by several factors due to distinct technological and physical applications. The quality of image analysis and

K. Mariyambi (✉) · E. Saritha · M. Baburaj
Department of Electronics and Communication Engineering,
Government College of Engineering Kannur, Parassinikkadavu (P.O.),
Kannur 670563, Kerala, India
e-mail: mambikhader12@gmail.com

E. Saritha
e-mail: sarithae@gcek.ac.in

M. Baburaj
e-mail: baburajmadathil@gcek.ac.in

© Springer Nature Singapore Pte Ltd. 2018
B. B. Chaudhuri et al. (eds.), *Proceedings of 2nd International Conference on Computer Vision & Image Processing*, Advances in Intelligent Systems and Computing 703, https://doi.org/10.1007/978-981-10-7895-8_24

processing algorithm are affected by a series of artefacts introduced by these factors. Hence, image resolution enhancement is very challenging in case of medical image processing. The most direct method to obtain the HR image is to improve imaging system hardware. But it should be restricted because hardware improvement results many other technical costs. There comes the importance of a signal processing technique called super resolution (SR), and it is the process of retrieving of images of HR from its low-resolution (LR) version. Image reconstruction by super resolution (SR) is an active research area, because of its ability to overcome the resolution limitations that often found in low-cost imaging sensors as well as digital imaging technology.

Single image SR [1, 2] and multiple image SR [3, 4] are the two broad categories of SR method. In multiple image SR, HR images are constructed from a several LR images. Here information contained in a multiple low-resolution frames that are non-redundant is combined for generating the corresponding high-resolution image. But in single image SR, a single image is used to construct the HR image. In interpolation-based single image SR method such as bicubic interpolation [5], the unknown pixels are interpolated based on surrounding pixels to generate HR image. Despite their simplicity and easy implementation, the interpolation-based algorithm in case of LR image generated by anti-aliasing operation of HR image, blur the high-frequency details. It generates smooth image with ringing and jagged artefact. In reconstruction-based single image SR approach [6–9], reconstruction problem is formulated by integrating priors or by using regularisation in an inverse reconstruction problem.

Another category of single image SR is learning-based (or example-based) method. Some example-based SR is based on nearest neighbour search [10]. In this method, the nearest neighbours of given LR image patch are computed from the database, then corresponding HR patch is obtained by learning the function that map the HR and LR patches correctly. The disadvantage of this method is that their performance mostly depends on parameter and quality of nearest neighbours. Recently, after the introduction of sparse coding technique Yang et al. [11] proposed example-based SR via sparse representation with promising performance. In [11], HR image is obtained by using the pre-learned dictionary of HR and LR images. The drawback of this method is that their quality depends on pre-learned dictionary and learning dictionary is a time-consuming process.

Most of existing example-based SR methods consider the input image as noiseless. But in case of medical images, this assumption is not valid. Trinh et al. [2] have been achieved promising performance in case of high level of noise by using sparse coding technique. We focus on developing example-based SR and denoising of medical using fast algorithm. In case of medical images, it is easy to obtain the standard images of same organ at approximately same location, which helps to obtain good database. In [2], Trinh et al. proposed a sparse weight model to find the HR patch. We propose further development and contribution mainly in image reconstruction optimisation. In this work, we derived a closed form solution to the reconstruction problem by employing the decimation and blurring operator's intrinsic frequency domain properties simultaneously. These properties were already used in [1, 12, 13].

The structure of this paper is as follows. In Sect. 2, detailed explanation of proposed method is described. Results and comparison with several existing SR methods are recorded in Sect. 3. Finally, conclusion is reported in Sect. 4.

2 Proposed Method

In this paper, we introduce a fast sparse weight learning-based SR algorithm to enhance the resolution and to perform denoising of medical images. The key idea of this method is finding nonnegative sparse coefficient for an input patch $\mathbf{y}_m^l \in \mathbb{R}^m$ using database $\mathbf{P}_l = \{\mathbf{u}_i^l, i \in \mathcal{I}\}$, here example patches $\mathbf{u}_i^l$ which are consistent to $\mathbf{y}_m^l$ assigns nonzero coefficient. This SR approach estimates the super-resolved image by first finding HR patch using the given LR patch and database. Then these HR patches are fused to recover the coarse estimate of corresponding HR image. The finer estimate of corresponding HR image is obtained by finding the solution to the reconstruction optimisation problem using a new analytical solution with l_2 regularisation. Before going in detail, we first explain the image degradation model and the properties of blurring and decimation matrix.

2.1 Image Degradation Model

In SR problem, an input LR image is considered as noisy added HR image with blurring and decimation and mathematically,

$$\mathbf{Y} = \mathbf{DBX} + \mathbf{n} \tag{1}$$

where, vector $\mathbf{Y} \in \mathbb{R}^{N_l \times 1}$ is the given LR image and vector $\mathbf{X} \in \mathbb{R}^{N_h \times 1}$ is the HR image to be estimated. For the sake of simplicity, $\mathbf{n}$ is considered as zero mean additive white Gaussian noise (AWGN) with independent identically distribution (i.i.d.). The matrix $\mathbf{B}$ is blurring matrix, i.e. convolution with some blurring kernal (point spread function (PSF)) and $\mathbf{D}$ is decimation matrix and it models the functionality of camera sensor.

The LR image can be arranged as a set of image patches, where each patch is used to perform SR. Thus, we can represent the LR image $\mathbf{Y}$ as a set of N image patches as,

$$\mathbf{Y} = \{\mathbf{y}_m^l, m = 1, 2, \ldots N\} \tag{2}$$

where the vector $\mathbf{y}_m^l \in \mathcal{R}^{l \times 1}$ is the LR image patch ($l = \sqrt{l} \times \sqrt{l}$). Similarly, the HR image is also be the combination of same number $\mathbf{N}$ of image patches

$\{\mathbf{x}_m^h, m = 1, 2, \ldots N\}$, where $\mathbf{x}_m^h \in \mathcal{R}^{h \times 1}$ $(h = \sqrt{h} \times \sqrt{h})$. Where $\sqrt{h} = d\sqrt{l}$, d is magnification (decimation) factor. The relation between HR and LR patch is given by

$$\mathbf{y}_m^l = \mathbf{DBx}_m^h + \mathbf{n}_m \tag{3}$$

where $\mathbf{n}_m$ is the noise in the mth patch.

The basic assumption about blurring and decimation operators is stated as follows.

Assumption 1 The blurring kernal is shift invariant, and it represents cyclic convolution operator.

It has been regularly used in image processing literature [12–14] as blurring matrix is a block circulant with circulant blocks (BCCB) matrix. Using this assumption, decomposition of blurring and its conjugate transpose matrix is given by

$$\mathbf{B} = \mathbf{F}^H \Lambda \mathbf{F} \tag{4}$$

$$\mathbf{B}^H = \mathbf{F}^H \Lambda^H \mathbf{F} \tag{5}$$

where $\mathbf{F}$ and $\mathbf{F}^H$ denote Fourier and inverse Fourier transforms, and Λ is a diagonal matrix.

Assumption 2 The decimation operator D is similar to down-sampling operator and its conjugate transpose D^H is an interpolation operator.

This assumption also widely used in many research work [11–13]. The decimation matrix satisfies the relation $\mathbf{DD}^H = \mathbf{I}$

The super-resolved image is obtained by performing the following three phases: database construction phase and image patch SR phase and HR image reconstruction phase.

2.2 Database Construction

In this paper, database is considered as normalised patch pairs constructed from example images. A set $\{\mathbf{P}_i^h, i \in \mathcal{I}\}$ of vectorised image patches is first derived from example images. Then, vectorised patch $\mathbf{P}_i^l$ corresponding to $\mathbf{P}_i^h$ is determined by

$$\mathbf{p}_i^l = \mathbf{DBP}_i^h \tag{6}$$

where $\mathbf{P}_i^h$ corresponds to HR patch and $\mathbf{P}_i^l$ corresponds to LR patch. During database construction noise is not considered, i.e. LR image is noise free. Finally, the high-resolution/low-resolution patch pairs are used as database and are given by

$$(\mathbf{P}_l, \mathbf{P}_h) = \left((\mathbf{u}_i^l, \mathbf{u}_i^h) = \left(\frac{\mathbf{P}_i^l}{||\mathbf{P}_i^l||}, \frac{\mathbf{P}_i^h}{||\mathbf{P}_i^h||} \right), i \in \mathcal{I} \right) \tag{7}$$

Here $||\mathbf{u}_i^l||_2 = 1$ and $\mathbf{DBu}_i^h = \mathbf{u}_i^l$ for all $i \in \mathcal{I}$

2.3 Image Patch SR

In this work, sparse weight model is used for patch SR. In this phase, we try to find the estimate of the HR patch $\mathbf{x}_m^h$, denoted as $\widehat{\mathbf{x}}_m^h$ from input LR patch y_m^l using the database (P_l, P_h), where $\mathbf{y}_m^l = \mathbf{DBx}_m^h + \mathbf{n}_i$. In case of medical images due to the local structures repetition, it is easy to find subset of patches $\mathbf{u}_i^h \in P_h$ that have similar structures of $\mathbf{x}_m^h$. These patches are then used to find the estimate $\widehat{\mathbf{x}}_m^h$.

Linear combination of the HR patches in the database $\mathbf{P}_h$ is used to find HR patch $\mathbf{x}_m^h$ as,

$$\mathbf{x}_m^h = \sum_{i \in I} \alpha_{mi} \mathbf{u}_i^h \tag{8}$$

where the vector $\alpha^m = [\alpha_{m1}, \alpha_{m2}, \dots, \alpha_{mi}, \dots]^T \geq 0$ represents the sparse coefficient. Equation (8) can be written as,

$$\mathbf{x}_m^h = \mathbf{P}_h \alpha^m \tag{9}$$

Both $\mathbf{x}_m^h$ and $\mathbf{P}_h$ are nonnegative. The corresponding LR patch $\mathbf{y}_m^l$ of $\mathbf{x}_m^h$ is obtained by multiplying Eq. (9) by $\mathbf{DB}$ gives

$$\mathbf{DBx}_m^h = \mathbf{DBP}_h \alpha^m = \mathbf{P}_l \alpha^m \tag{10}$$

Using Eq. (3), we obtain the relation,

$$\mathbf{P}_l \alpha^m = \mathbf{y}_m^l - \mathbf{n}_m \tag{11}$$

This gives that

$$\mathbf{y}_m^l - \mathbf{P}_l \alpha^m = \mathbf{n}_m \tag{12}$$

Then sparse coefficient vector α^m is obtained by minimising the following optimisation problem.

$$\alpha^m = \underset{\alpha \geq 0}{argmin} ||\alpha||_0 + \sum_{i \in I} \omega_{mi} \alpha_{mi}$$

$$subject \quad to \quad ||\mathbf{y}_m^l - \sum_{i \in I} \alpha_{mi} \mathbf{u}_i^l||_2^2 \leq \rho \sigma_m^2 \tag{13}$$

where $\alpha = [\alpha_{m1}, \alpha_{m2}, \ldots, \alpha_{mi}, \ldots]^T$, ρ denotes a given positive number, σ_m represents the standard deviation of the noise in the mth patch, the ω_{mi} is a penalty coefficients that lean on the dissimilarity between x_m^h and u_i^h.

For high dissimilarity between x_m^h and u_i^h, the value of penalty coefficient ω_{mi} is high, while the value of α_{mi} is small. In other words, α_{mi} is large for high similarity between x_m^h and u_i^h. However, the dissimilarity (or similarity) between x_m^h and u_i^h is computed from its LR versions y_i^l and u_k^l. With coefficient α_{mi} is small, the objective function in (14) can be minimised. Here, the penalty coefficients w_{ik} are defined as

$$\omega_{mi} = \psi(d(\mathbf{y}_m^l, \mathbf{u}_i^l)) \tag{14}$$

where d is a dissimilarity criterion finding dissimilarity (or similarity) between $\mathbf{y}_m^l$ and $\mathbf{u}_i^l$, ψ is a nonnegative increasing function.

In order to define dissimilarity criterion between $\mathbf{y}_m^l$ and $\mathbf{u}_i^l$, we use the explanation of consistency of patches. If two patches $\mathbf{x}_1$ and $\mathbf{x}_2$ are consistent, then there exists a constant $c \in \mathbb{R}$ such that $\mathbf{x}_1 = c\mathbf{x}_2$. If $\mathbf{u}_i^l$ and $\mathbf{DBx}_m^h$ are consistent, then the patch $\mathbf{u}_i^l$ and $\mathbf{y}_m^l$ are ideally similar and are related as

$$\mathbf{y}_m^l = \mu_{mi}\mathbf{u}_i^l + \mathbf{n}_m \tag{15}$$

Since noise component $\mathbf{n}_m \sim N(0, \sigma_m^2)$, the mean of n_m, $\mathbf{E}(\mathbf{n}_m) \approx 0$. Then, the constant μ_{mk} is approximately given as:

$$\mathbf{E}(\mathbf{y}_m^l) = \mu_{mi}\mathbf{E}(\mathbf{u}_i^l) + \mathbf{E}(\mathbf{n}_m) \implies \mu_{mi} = \frac{\mathbf{E}(\mathbf{y}_m^l)}{\mathbf{E}(\mathbf{u}_i^l)} \tag{16}$$

Statistical property of residual patch noise is evaluated using the parameter a_{mi} given by

$$a_{mi} = |\mathbf{E}(\mathbf{y}_m^l - \mu_{mi}\mathbf{u}_i^l)| + |\mathbf{Var}(\mathbf{y}_m^l - \mu_{mi}\mathbf{u}_i^l) - \sigma_m^2| \simeq 0 \tag{17}$$

Then dissimilarity criterion is then given by

$$d(\mathbf{y}_m^l, \mathbf{u}_i^l) = ||\mathbf{y}_m^l - \mu_{mi}\mathbf{u}_i^l||_2^2 + a_{mi} \tag{18}$$

Then the function ψ_i is computed as

$$\psi_i(t) = \begin{cases} e^t & \text{if} \quad t > \gamma(m\sigma_m^2) \\ t & \text{if} \quad t \leq \gamma(m\sigma_m^2) \end{cases} \tag{19}$$

where γ is a constant and m is the total number of elements in the vector y_m^l. Then sparse decomposition problem becomes

$$\alpha^m = \underset{\alpha \geq 0}{argmin} ||\alpha||_1 + \sum_{i \in I} \omega_{mi} \alpha_{mi}$$

$$subject \quad to \quad ||y_i^l - \sum_{k \in I} \alpha_{ik} u_k^l||_2^2 \leq \epsilon \sigma_i^2 \tag{20}$$

By using Lagrange multiplier method, Eq. (21) become

$$\alpha^m = \underset{\alpha \geq 0}{argmin} \frac{1}{2} ||\mathbf{y}_m^l - \sum_{i \in I} \alpha_{mi} \mathbf{u}_i^l||_2^2 + \lambda \sum_{i \in I} (1 + \omega_{mi}) \alpha_{mi} \tag{21}$$

where the parameter λ is a sparsity regularisation parameter. Equation (22) can also be written as:

$$\alpha^m = \underset{\alpha \geq 0}{argmin} \frac{1}{2} ||\mathbf{y}_m^l - \mathbf{U}_m \alpha||_2^2 + \mathbf{w}_m^T \alpha \tag{22}$$

where U_m is the matrix which is created by concatenating all columns vector $\mathbf{u}_i^l$ and $\mathbf{w}_m$ is formed by concatenating all the coefficients $\lambda(1 + \omega_{mi})$, here $i \in I$.

The problem in Eq. (23) is a quadratic nonnegative programming (QNP) which can be effectively solved by using many algorithms. In this work, multiplicative updates algorithm proposed by Hoyer in [15] is used to solve the above problem and is shown in Algorithm 1.

Algorithm 1: Multiplicative updates algorithm for QNP [15]

Input: $\alpha = \alpha_0 > 0$, iteration number K
Update: k=0
1   While $k < K$ and $||\mathbf{y}_m^l - \mathbf{U}_m \alpha_t||_2^2 > m\sigma_m^2$
2   $\alpha_{t+1} = \alpha_t . * (\mathbf{U}_m^T \mathbf{y}_m^l)./(\mathbf{U}_m^T \mathbf{U}_m \alpha_t + w_m)$;
3   $k = k + 1$;
4   End
Output: $\alpha^i = \alpha_t$

Once α^m is obtained, then desired HR patch $\hat{\mathbf{x}}_m^h$ can be computed as

$$\hat{\mathbf{x}}_m^h = \sum_{i \in I} \alpha_m^i \mathbf{u}_i^h \tag{23}$$

Similarly, the denoised version of LR image $\hat{\mathbf{y}}_m^l$ can be computed as

$$\hat{\mathbf{y}}_m^l = \sum_{i \in I_i} \alpha_m^i \mathbf{u}_i^l \tag{24}$$

The entire patch SR process is summarised as Algorithm 2.

Algorithm 2: Fast Sparse Weight Super Resolution algorithm

Input: LR image **Y**, magnification factor d, database $(\mathbf{P}_l, \mathbf{P}_h)$, sparsity
        regularization parameter λ
Output: Final estimate of HR $\widehat{X}^{final}$
Begin:
- Split **Y** into arranged set patches $\mathbf{y}_m^l$
- **For** each patch $\mathbf{y}_m^l$

  - Compute $d(\mathbf{y}_m^l, \mathbf{u}_i^l)$ using equation (17), (18) and (19).
  - Compute w_{mi} using equation (15) and (20).
  - Find α^m using Algoritm 1.
  - Estimate HR patch $\widehat{\mathbf{x}}_m^h$ using equation (24) and denoised LR patch $\mathbf{y}_m^l$ using equation (25).

- **End**
- **Fusion:** Find coarse estimate of HR image $\widehat{X}^{coarse}$
  and the denoised LR image $\mathbf{Y}^{denoise}$.
- **Reconstruction** of final HR image.

  - Decompose blurring matrix **B**,
    $\mathbf{B} = \mathbf{F}^H \Lambda \mathbf{F}$
  - Compute Λ using equation (28).
  - Calculate fourier transform of **r** denoted as **Fr**,
    $\mathbf{Fr} = \mathbf{F}(\mathbf{B}^H \mathbf{D}^H \mathbf{Y}^{denoise} + 2\tau \widehat{\mathbf{X}}^{coarse})$
  - Compute the final HR image $\widehat{\mathbf{X}}^{final}$ using equation (27).

End:

2.4 HR Image Reconstruction

The obtained HR patches are then fused to obtain the coarse estimate of HR image, $\widehat{\mathbf{X}}^{coarse}$. Similarly denoised LR image $\mathbf{Y}^{denoise}$ is obtained by fusing denoised LR patches. Finally, the finer estimate of HR image, $\widehat{\mathbf{X}}^{final}$ is computed by minimising the following optimisation problem.

$$\min_x \frac{1}{2}||\mathbf{Y}^{denoise} - \mathbf{DBX}||_2^2 + \tau||\mathbf{X} - \widehat{\mathbf{X}}^{coarse}||_2^2 \tag{25}$$

where τ is the reconstruction regularisation parameter. In order to reduce the computational time required to find optimisation problem (26), we find an analytical solution to the problem (26) by considering the Assumptions 1 and 2. The final solution to the reconstruction problem (26) is given by

$$\widehat{\mathbf{X}}^{final} = \frac{1}{2\tau}\left(r - \mathbf{F}^H\left(\underline{\Lambda}^H\left(2\tau d\mathbf{I}_{N_l} + \underline{\Lambda}\underline{\Lambda}^H\right)^{-1}\underline{\Lambda}\right)\mathbf{F}_r\right) \tag{26}$$

where $\mathbf{r} = \mathbf{B}^H \mathbf{D}^H \mathbf{Y}^{denoise} + 2\tau \widehat{\mathbf{X}}^{coarse}$ and the matrix $\underline{\Lambda}$ is defined by,

$$\underline{\Lambda} = [\Lambda_1, \Lambda_2, \ldots \Lambda_d] \tag{27}$$

where d is the magnification factor and the blocks Λ_i satisfy the relationship,

$$diag\{\Lambda_1, \Lambda_2, \ldots, \Lambda_d\} = \Lambda \tag{28}$$

The Algorithm 2 summarises the implementation of proposed SR approach called Fast Sparse Weight SR (FSWSR).

3 Experimental Results

This section demonstrates the experimental results performed on both noiseless images and noisy images with different magnification factor. The experimental test is performed on five HR test images as shown in Fig. 1. The proposed method called Fast Sparse Weight SR (FSWSR) is compared with existing SR method such as bicubic interpolation (Bb), Sparse coding-based SR (ScSR) [11] and SR by Sparse Weight (SRSW) [2]. The objective quality of super-resolved image is measured using Peak Signal to Noise Ratio (PSNR) and Structural SIMilarity index (SSIM). The experimental values of PSNR, SSIM and computational time with various noise level is shown in Table 1. From the table, it is clear that our method out performs the existing SR method (Figs. 2 and 3).

The performance of proposed method is also evaluated with various experimental parameters such as noise level, magnification factor as a function of reconstruction regularisation parameter. The obtained PSNR and SSIM curve are depicted in Fig. 4. From the figure, it is observed the PSNR and SSIM curve increase with reconstruction regularisation parameter τ.

For subjective evaluation of our method, results of SR of MRI image of ankle with decimation factor, d = 2 is shown in Fig. 2. For further illustrating the denoising effectiveness of our method, the result of MRI of ankle with heavy noise level ($\sigma = 20$) is shown in Fig. 3.

(a)   **(b)**   **(c)**   **(d)**   **(e)**

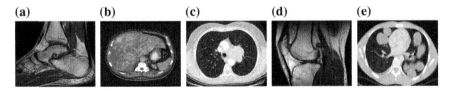

Fig. 1 Test images. **a** MRI of ankle. **b** CT of abdomen. **c** CT of chest. **d** MRI of knee. **e** CT of thorax

Table 1 Performance analysis of proposed SR method

Image	σ	PSNR				SSIM			
		Bicubic	ScSR	SRSW	Proposed	Bicubic	ScSR	SRSW	Proposed
(a)	$\sigma = 0$	17.9	27.7	29.25	35.52	0.7375	0.8367	0.8965	0.9895
	$\sigma = 5$	17.74	26.7	29.13	34.82	0.5651	0.7197	0.8905	0.9742
	$\sigma = 10$	17.1	24.18	28.87	33.30	0.3913	0.5591	0.8771	0.9642
	$\sigma = 20$	16.9	19.71	27.90	32.35	0.3872	0.3221	0.8322	0.9331
(b)	$\sigma = 0$	17.9	27.7	28.7	33.52	0.7375	0.8367	0.8495	0.9860
	$\sigma = 5$	17.74	26.7	27.8	33.05	0.5651	0.7197	0.8861	0.9712
	$\sigma = 10$	17.1	24.8	25.71	29.18	0.3913	0.5592	0.7998	0.9518
	$\sigma = 20$	16.9	19.71	24.65	28.15	0.3872	0.4512	0.7489	0.9199
(c)	$\sigma = 0$	17.5	27.16	28.13	33.25	0.7299	0.8311	0.8425	0.9816
	$\sigma = 5$	17.29	26.53	27.89	33.00	0.5610	0.7178	0.8818	0.9699
	$\sigma = 10$	17.1	24.65	25.69	29.17	0.3968	0.5585	0.7989	0.9510
	$\sigma = 20$	16.73	19.67	24.27	28.10	0.3871	0.4499	0.7479	0.9178

Time (S)

Bicubic	ScSR	SRSW	Proposed
0.47	25.8	12.12	3.85

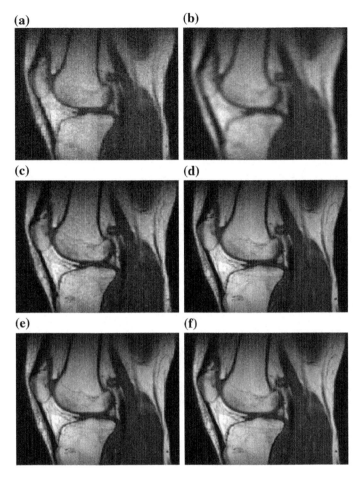

Fig. 2 Results of SR of MRI image of ankle with decimation factor $d = 2$. **a** LR image with $\sigma = 5$. **b** Bicubic interpolation. **c** Result of ScSR [11]. **d** Result of SRSW [2]. **e** Proposed FSRSW method. **f** Original image

4 Conclusion

This paper proposes fast learning-based SR for medical images to enhance the resolution and to perform denoising. This method uses the idea that both the standard images and LR image are taken at approximately same location. The coarse measurement of HR image is obtained as sparse decomposition problem; finally, the finer measurement of HR image is obtained via fast algorithm using intrinsic properties of decimation blurring operator in frequency domain. The results show that performance of our method has improved results over other existing SR methods.

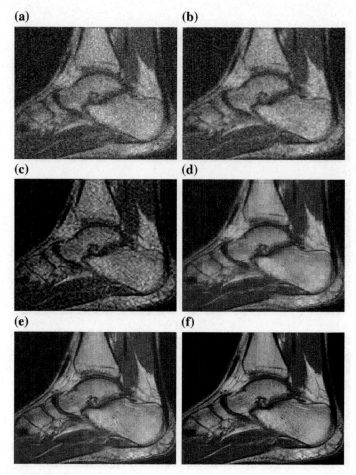

Fig. 3 Results of SR of MRI image of ankle with decimation factor $d = 2$. **a** LR image with $\sigma = 20$. **b** Result of bicubic interpolation. **c** Result of SCSR [11]. **d** Result of SRSW [2]. **e** Proposed FSRSW method. **f** Original image

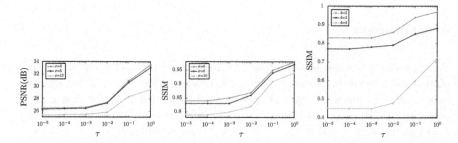

Fig. 4 PSNR and SSIM curve for different parameters as a function of reconstruction regularization parameter

References

1. N. Zhao, Q. Wei, A. Basarab, N. Dobigeon, D. Kouam, and J. Y. Tourneret, "Fast single image super-resolution using a new analytical solution for l2-l2 problems," *IEEE Transactions on Image Processing*, vol. 25, no. 8, pp. 3683–3697, 2016.
2. D.-H. Trinh, M. Luong, F. Dibos, J.-M. Rocchisani, C.-D. Pham, and T. Q. Nguyen, "Novel example-based method for super-resolution and denoising of medical images," *IEEE Transactions on Image processing*, vol. 23, no. 4, pp. 1882–1895, 2014.
3. T. Huang, "Multi-frame image restoration and registration," *Advances in computer vision and Image Processing*, vol. 1, pp. 317–339, 1984.
4. M. Irani and S. Peleg, "Improving resolution by image registration," *CVGIP: Graphical models and image processing*, vol. 53, no. 3, pp. 231–239, 1991.
5. P. Thévenaz, T. Blu, and M. Unser, "Image interpolation and resampling," *Handbook of medical imaging, processing and analysis*, pp. 393–420, 2000.
6. J. Sun, Z. Xu, and H.-Y. Shum, "Image super-resolution using gradient profile prior," in *Computer Vision and Pattern Recognition, 2008. CVPR 2008. IEEE Conference on*. IEEE, 2008, pp. 1–8.
7. Y.-W. Tai, S. Liu, M. S. Brown, and S. Lin, "Super resolution using edge prior and single image detail synthesis," in *Computer Vision and Pattern Recognition (CVPR), 2010 IEEE Conference on*. IEEE, 2010, pp. 2400–2407.
8. J. Sun, J. Sun, Z. Xu, and H. Y. Shum, "Gradient profile prior and its applications in image super-resolution and enhancement," *IEEE Transactions on Image Processing*, vol. 20, no. 6, pp. 1529–1542, 2011.
9. M. K. Ng, P. Weiss, and X. Yuan, "Solving constrained total-variation image restoration and reconstruction problems via alternating direction methods," *SIAM journal on Scientific Computing*, vol. 32, no. 5, pp. 2710–2736, 2010.
10. A. Rueda, N. Malpica, and E. Romero, "Single-image super-resolution of brain mr images using overcomplete dictionaries," *Medical image analysis*, vol. 17, no. 1, pp. 113–132, 2013.
11. J. Yang, J. Wright, T. S. Huang, and Y. Ma, "Image super-resolution via sparse representation," *IEEE Transactions on Image Processing*, vol. 19, no. 11, pp. 2861–2873, 2010.
12. M. D. Robinson, C. A. Toth, J. Y. Lo, and S. Farsiu, "Efficient fourier-wavelet super-resolution," *IEEE Transactions on Image Processing*, vol. 19, no. 10, pp. 2669–2681, 2010.
13. F. Šroubek, J. Kamenický, and P. Milanfar, "Superfast superresolution," in *Image Processing (ICIP), 2011 18th IEEE International Conference on*. IEEE, 2011, pp. 1153–1156.
14. Z. Lin and H.-Y. Shum, "Fundamental limits of reconstruction-based superresolution algorithms under local translation," *IEEE Transactions on Pattern Analysis and Machine Intelligence*, vol. 26, no. 1, pp. 83–97, 2004.
15. P. O. Hoyer, "Non-negative sparse coding," in *Neural Networks for Signal Processing, 2002. Proceedings of the 2002 12th IEEE Workshop on*. IEEE, 2002, pp. 557–565.
16. H. Chang, D.-Y. Yeung, and Y. Xiong, "Super-resolution through neighbor embedding," in *Computer Vision and Pattern Recognition, 2004. CVPR 2004. Proceedings of the 2004 IEEE Computer Society Conference on*, vol. 1. IEEE, 2004, pp. I–I.
17. A. Gilman, D. G. Bailey, and S. R. Marsland, "Interpolation models for image super-resolution," in *Electronic Design, Test and Applications, 2008. DELTA 2008. 4th IEEE International Symposium on*, Jan 2008, pp. 55–60.
18. A. Beck and M. Teboulle, "A fast iterative shrinkage-thresholding algorithm for linear inverse problems," *SIAM journal on imaging sciences*, vol. 2, no. 1, pp. 183–202, 2009.
19. S. C. Park, M. K. Park, and M. G. Kang, "Super-resolution image reconstruction: a technical overview," *IEEE signal processing magazine*, vol. 20, no. 3, pp. 21–36, 2003.

Analyzing ConvNets Depth for Deep Face Recognition

Mohan Raj, I. Gogul, M. Deepan Raj, V. Sathiesh Kumar, V. Vaidehi
and S. Sibi Chakkaravarthy

Abstract Deep convolutional neural networks are becoming increasingly popular in large-scale image recognition, classification, localization, and detection. In this paper, the performance of state-of-the-art convolution neural networks (ConvNets) models of the ImageNet challenge (ILSVRC), namely VGG16, VGG19, OverFeat, ResNet50, and Inception-v3 which achieved top-5 error rates up to 4.2% are analyzed in the context of face recognition. Instead of using handcrafted feature extraction techniques which requires a domain-level understanding, ConvNets have the advantages of automatically learning complex features, more training time, and less evaluation time. These models are benchmarked on AR and Extended Yale B face dataset with five performance metrics, namely Precision, Recall, F1-score, Rank-1 accuracy, and Rank-5 accuracy. It is found that GoogleNet ConvNets model with Inception-v3 architecture outperforms than other four architectures with a Rank-1 accuracy of 98.46% on AR face dataset and 97.94% accuracy on Extended Yale B face dataset. It confirms that deep CNN architectures are suitable for real-time face recognition in the future.

M. Raj (✉) · I. Gogul · M. Deepan Raj · V. Sathiesh Kumar · S. Sibi Chakkaravarthy
Department of Electronics Engineering, Madras Institute of Technology Campus,
Anna University, Chennai, India
e-mail: mohanraj4072@gmail.com

I. Gogul
e-mail: gogulilangoswami@gmail.com

M. Deepan Raj
e-mail: deepanraj.18@gmail.com

V. Sathiesh Kumar
e-mail: sathiieesh@gmail.com

S. Sibi Chakkaravarthy
e-mail: sb.sibi@gmail.com

V. Vaidehi
School of Computing Science and Engineering, VIT University, Chennai,
Tamil Nadu, India
e-mail: vaidehimitauc@gmail.com

© Springer Nature Singapore Pte Ltd. 2018
B. B. Chaudhuri et al. (eds.), *Proceedings of 2nd International Conference
on Computer Vision & Image Processing*, Advances in Intelligent Systems
and Computing 703, https://doi.org/10.1007/978-981-10-7895-8_25

Keywords Deep learning · Face recognition · Convolutional neural networks Computer vision

1 Introduction

In recent years, machine learning has reached its pinnacle in automation and deep learning started achieving success in numerous research areas of Computer Vision. As Data became Big Data, traditional CPUs are getting replaced by powerful GPUs for computationally intensive applications. The need to use deep learning systems is of utmost importance in various domains such as medical image analysis, face recognition, robotics, self-driving cars to achieve better results. Before a decade, traditional feature extraction methods such as Local Binary Pattern (LBP) by Ojala et al. [1], Scale-Invariant Feature Transform (SIFT) by David Lowe [2], Histogram of Oriented Gradients (HoG) by Dalal and Triggs [3] achieved good results in publicly available datasets. It was mainly due to the use of Bag of Visual Words approach along with a machine learning classifier such as linear SVM as shown by Yang et al. [4]. Recent research works in the machine learning community showed that automatic learning of these features in raw images is possible if multiple layers of nonlinear activation functions are used. This led to the introduction of a neural networks, namely the CNN which was first applied on a larger dataset (ILSVRC ImageNet challenge 2012) by Krizhevsky et al. [5] to identify the label of an unknown image among 1000 categories. After AlexNet [5], VGGNet [6], and GoogleNet [7] models showed that even more deeper architectures could be built to improve the accuracy of recognition. Although deeper architectures such as VGGNet [6], GoogleNet [7], OverFeat [8], and ResNet [9] achieve good results, these models could take longer computation time to train on a single GPU (more than 5–6 months). To evaluate such models with increased depth, data and computational resources such as multi-GPU clusters are needed.

Face recognition in unconstrained environment is a challenging task. In digital revolution, face recognition is found to be very useful in the near future, as it leads to many applications like surveillance, security, emotion recognition and home automation. Humongous researches and implementation for face recognition are being done, to meet the current state-of-the-art requirements [10–12]. Recent deep learning models to perform face recognition proposed by Google [13] and Facebook [14] used their own private datasets which are huge. Google used 200 million images of 8 million unique identities to train a CNN, and Facebook used 4.4 million images of 4030 unique identities to train a CNN. Also, publicly available datasets such as AR dataset, Extended Yale B dataset, and Labeled Faces in the Wild (LFW) dataset used face recognition by the research community.

In this paper, a methodology is presented to analyze the performance of five deep convent architectures utilized in the ILSVRC ImageNet challenge. The ILSVRC ImageNet challenge is conducted every year for image recognition, classification, localization, and detection. In this paper, VGG16, VGG19, OverFeat, Inception-v3, and ResNet50 are the Deep ConvNets architectures considered for analysis. AR and Extended Yale B face dataset are considered to benchmark these deep ConvNets models. By performing this analysis, it is possible to get actionable insights into the use of Deep ConvNets architectures for face recognition using different publicly available datasets.

This paper is organized as follows: Sect. 2 presents the deep learning technique for face recognition, Sect. 3 discusses the effect of ConvNets depth for face recognition, Sect. 4 analyzes the performance of state-of-the-art ConvNets model, and Sect. 5 concludes the paper.

2 Deep Learning for Face Recognition

As deep learning ConvNets achieves good accuracy in ImageNet challenges (around 1 million training images with 1000 classes), the need to use these models for other vision related applications is increasing worldwide in Computer Vision research. An increase in the amount of digital data that is available in the Internet and the recent advancements in Graphics Processing Unit (GPU) has made Deep ConvNets a default choice when it comes to large-scale image recognition tasks. This is feasible mainly due to the characteristics of ConvNets which possess automatic feature learning capability, higher training time, and very less evaluation time. Instead of designing and training a ConvNets from scratch, which requires more computational resources such as multi-GPU clusters (huge cost), it is decided to go for feature extraction using ConvNets approach. Some of the existing pre-trained neural network architectures which are made publicly available are VGG16, VGG19, OverFeat, Inception-v3, and ResNet50 which are the popular ConvNets considered for model selection, feature extraction, and performance analysis. The overall methodology for analyzing the performance of these models in the context of face recognition is shown in Fig. 1.

Supervised learning is carried out when the training database holds images along with its labels. The overall methodology to perform deep learning for face recognition is divided into two phases.

1. Feature extraction using ConvNets
2. Training and evaluation

2.1 Feature Extraction Using ConvNets

Instead of using handcrafted feature descriptors such as Local Binary Pattern (LBP), Scale-Invariant Feature Transform (SIFT), Speeded-Up Robust Features (SURF), or Histogram of Oriented Gradients (HoG), which require domain-level understanding of the face recognition (FR) problem, pre-trained ConvNets are used

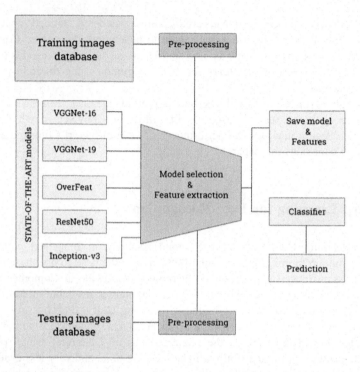

Fig. 1 Overall methodology involved in analyzing the performance of state-of-the-art ConvNets in the context of face recognition

as feature extractors. These feature extracted vectors are also referred as "Neural Codes". This is possible because of the pre-trained weights and architectures of these state-of-the-art models publicly made available for Computer Vision research. The model weights and architectures of the models considered are loaded locally prior to the training phase. Based on the model selected (1 out of 5 models— VGG16, VGG19, OverFeat, ResNet50, Inception-v3), the images in the training database are resized to a fixed dimension as shown in Table 1. After pre-processing, each of the images in the database is sent to the selected model architecture and features are extracted by removing the top fully connected layers. The extracted feature for that single image is flattened and stored in a list. Its corresponding label

Table 1 Fixed dimension to resize the images for state-of-the-art models considered	ConvNets model	Fixed dimension (for pre-processing)
	VGG16	224 × 224
	VGG19	224 × 224
	OverFeat	231 × 231
	Inception-v3	299 × 299
	ResNet50	224 × 224

is also stored in another list. This process is repeated for all the images in the training database. The extracted image features and labels are stored locally in an HDF5 file format as NumPy arrays. This feature extraction process is carried out for different benchmark face datasets such as AR and Extended Yale B.

2.2 Training and Evaluation

After extracting features from the training phase, the stored features and labels are loaded and split into training and testing data based on a parameter "train_test_split". If "train_test_split" is chosen as 0.2, then it means 80% of the overall data is used for training and 20% of the overall data is used to evaluate the trained model. Based on parameter tuning and grid-search methodology, it is found that logistic regression (LR) outperformed all the other machine learning classifiers such as Random Forests (RF), Support Vector Machine (SVM), and K-Nearest Neighbors (KNN). Two performance metrics are chosen to evaluate the trained model, namely Rank-1 accuracy and Rank-5 accuracy. Rank-1 accuracy gives the accuracy of the trained model when tested with an unseen test data on the first chance. Rank-5 accuracy gives the accuracy of the trained model when tested it with an unseen data given five chances.

3 Effect of ConvNets Depth on Face Recognition

The pre-trained models chosen share a common trait of having a deep architecture, meaning these networks have increased convolutional depth or increased number of layers. Based on the work by Romero et al. [15] and Ba and Caruana [16], it became evident that training a deeper neural network achieved better performance than a shallow network. But they have also specified that depth might make learning easier but it is not an essential factor to be considered. Thus, in the context of face recognition, how far does depth in ConvNets affect accuracy is analyzed. Figure 2 shows the architecture for five state-of-the-art Deep ConvNets considered for performance analysis.

3.1 VGGNet

VGGNet created by Simonyan et al. [6] obtained 1st place in the localization task with 25.3% error and 2nd place in classification task with 7.3% error. The major objective behind VGGNet is to obtain better accuracy in ImageNet challenge 2014 by implementing different configurations of deep convolutional neural network architectures that use a sequential stack of convolutional layers with millions of parameters to learn, thereby increasing the overall depth of the network (16–19

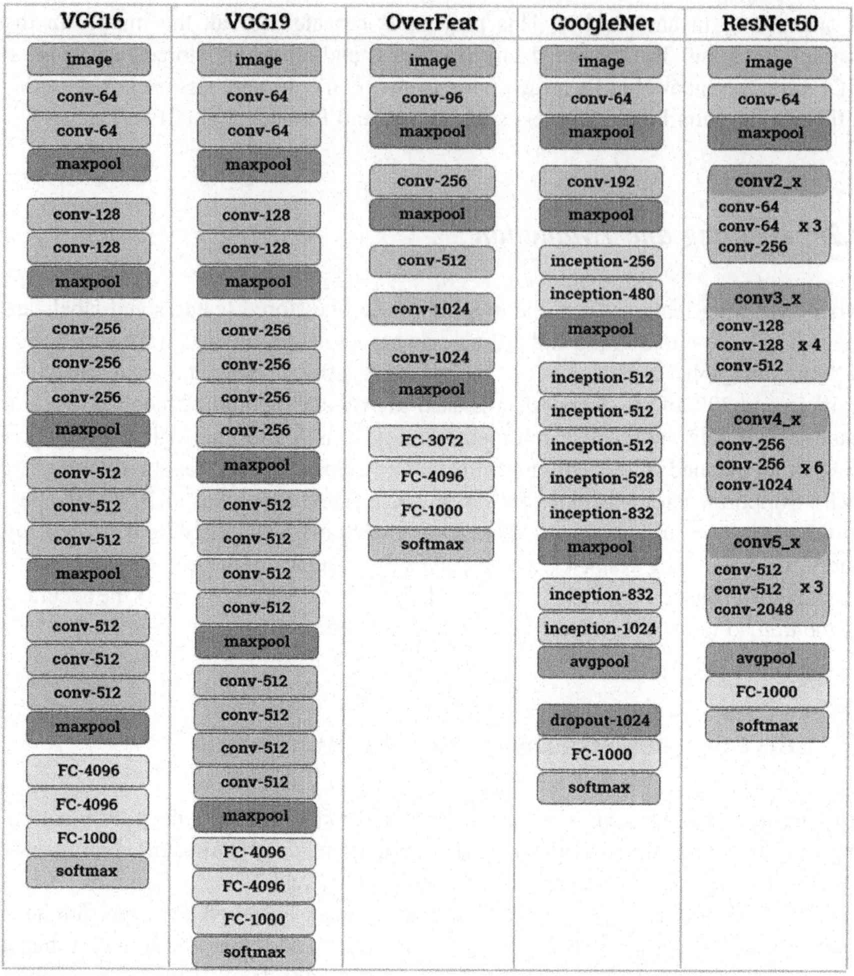

VGG16	VGG19	OverFeat	GoogleNet	ResNet50
image	image	image	image	image
conv-64	conv-64	conv-96	conv-64	conv-64
conv-64	conv-64	maxpool	maxpool	maxpool
maxpool	maxpool	conv-256	conv-192	conv2_x conv-64 conv-64 x 3 conv-256
conv-128	conv-128	maxpool	maxpool	
conv-128	conv-128	conv-512	inception-256	
maxpool	maxpool	conv-1024	inception-480	conv3_x conv-128 conv-128 x 4 conv-512
conv-256	conv-256	conv-1024	maxpool	
conv-256	conv-256	maxpool	inception-512	
conv-256	conv-256	FC-3072	inception-512	conv4_x conv-256 conv-256 x 6 conv-1024
maxpool	maxpool	FC-4096	inception-512	
conv-512	conv-512	FC-1000	inception-528	
conv-512	conv-512	softmax	inception-832	
conv-512	conv-512		maxpool	conv5_x conv-512 conv-512 x 3 conv-2048
maxpool	conv-512		inception-832	
conv-512	conv-512		inception-1024	
conv-512	maxpool		avgpool	avgpool
conv-512	conv-512			FC-1000
maxpool	conv-512		dropout-1024	softmax
FC-4096	conv-512		FC-1000	
FC-4096	conv-512		softmax	
FC-1000	maxpool			
softmax	FC-4096			
	FC-4096			
	FC-1000			
	softmax			

Fig. 2 State-of-the-art deep convolutional neural network models chosen for performance analysis in the context of face recognition

weight layers). Unlike other top-performing architectures of ILSVRC-2012 such as Krizhevsky et al. [5], and ILSVRC-2013 such as Zeiler and Fergus [17], and Sermanet et al. [8] (which used larger receptive fields in the first convolutional layer), the VGGNet uses smaller [3 × 3] receptive fields or convolution kernels throughout the entire network with a convolution stride size of 1 (i.e., no loss of information). The network architecture starts with an input image of fixed size [224 × 224], followed by multiple convolutional layers and max-pooling layers with different input and output dimensions. Six different configurations are presented such as A (11 weight layers), A-LRN (11 weight layers with Local Response Normalization), B (13 weight layers), C (16 weight layers), D (16 weight layers), and E (19 weight layers). The convolutional layers are activated using ReLU nonlinear activation function. [2 × 2] window size is chosen to perform max-pooling over 5 max-pool layers giving (×2 reduction). Three fully connected (FC) layers are involved for the final classification followed by a softmax classifier. Configurations D (VGG16) and E (VGG19) were chosen for performance analysis as these models are publicly available.

3.2 GoogLeNet

GoogLeNet created by Szegedy et al. [7] is a very deep convolutional neural network, achieved better accuracy for classification and detection in ILSVRC 2014 ImageNet challenge. GoogLeNet, code named as Inception, has an architecture that has increased depth and width while keeping a constant computational budget. It achieved a top-5 error rate of 6.67% on the ILSVRC 2014 challenge. The overall architecture of GoogLeNet contains four sections, namely stem, inception modules, auxillary, and output classifier. Stem contains a chain of convolutions, max-pooling, and Local Response Normalization (LRN) operations. The inception module in GoogLeNet is a unique approach where a set of convolution and pooling is performed at different scales on the input volume, computed in parallel and concatenated together to produce the output volume (DepthConcat). There are 9 such inception modules with two max-pooling layers in between to reduce spatial dimensions. In the recent variants of GoogleNet, auxillary classifiers are ignored after the introduction of batch normalization. Prior to the output classifier, average pooling is performed, followed by a fully connected layer with a softmax activation function. When compared with VGGNet (which has around 180 million parameters), GoogLeNet has less parameters to learn (around 5 million). An improved variant of GoogLeNet called "Inception-v3" is considered for performance analysis, which adds factorized convolutions and aggressive regularization.

3.3 OverFeat

OverFeat developed by Sermanet et al. [8] is another deep convolutional neural network that achieved state-of-the-art accuracy in ILSVRC 2013 ImageNet challenge with Rank 4 in classification and Rank 1 in localization and detection. This network is not only for the purpose of image classification, but also demonstrated the novel approach for localization and detection using a single ConvNets. It uses a multiscale and sliding window approach in a ConvNets as well as a novel approach to localization by learning to predict object boundaries. Although this network architecture is similar to that of Krizhevsky et al. [5] (best in ILSVRC 2012 ImageNet challenge), major improvements were contributed toward network design and inference step. The input image to the OverFeat network is resized to a fixed dimension of [231 × 231]. There are a total of 8 layers in the network with convolution, max-pooling, and fully connected layers having varied number of channels, filter size, stride size, pooling size, and pooling stride size. ReLU non-linear activation function is used from layers 1–5 (similar to Krizhevsky et al. [5]), but with no contrast normalization, with nonoverlapping pooling regions and with larger feature maps in layers 1 and 2. Before the output classifier, there are three fully connected layers with 3072, 4096, and 1000 nodes followed by a softmax classifier. The publicly released OverFeat feature extractor is considered for performance analysis in the context of face recognition.

3.4 ResNet50

ResNet developed by Kaiming He et al. [9] are deep convolutional neural networks that achieved good accuracy in ILSVRC 2015 ImageNet challenge with Rank 1 in classification task (ensemble of these residual nets achieved 3.57% error rate). These networks demonstrated a unique approach, where instead of learning unreferenced functions in the network, they explicitly reformulate the layers as learning residual functions with reference to the layer inputs. The authors also evaluated residual nets with a depth of up to 152 layers (which is 8 times that of VGGNet) while still maintaining a lower complexity. Five types of configurations are presented such as ResNet18 (18-layer), ResNet34 (34-layer), ResNet50 (50 layer), ResNet101 (101 layer), and ResNet152 (152 layer). Each of these configurations has different input/output dimensions, filter size, stride size, pooling size, and pooling stride size. The input image to the network is resized to a fixed dimension of [224 × 224]. ResNets perform all the standard operations of a ConvNets such as convolution, max-pooling, and batch normalization. After each convolution and before activation, batch normalization is performed. Stochastic Gradient Descent (SGD) is used as the optimizer with a batch size of 256. Based on the error that is getting accumulated, learning rate is adjusted accordingly (from an initial value of 0.1), momentum is chosen as 0.9, and weight decay is chosen as 0.0001. The overall network is trained for 6,00,000 epochs. The publicly released ResNet50 (50-layer) is chosen for performance analysis.

3.5 Error Rates

All the five models considered for performance analysis were evaluated based on two performance metrics, namely "top-1 validation error" (in %) and "top-5 validation error" (in %) taken during ILSVRC challenge. The ImageNet challenge normally consists of three data split: training data, validation data, and testing data. Table 2 shows the performance metrics of the state-of-the-art Deep ConvNets

Table 2 Error rates (in %) of state-of-the-art Deep ConvNets models on ILSVRC challenge

ConvNets model	Evaluation method	Top-1 val.error (%)	Top-5 val.error (%)
VGG16	Dense	24.8	7.5
	Multi-crop	24.6	7.5
	Multi-crop and dense	24.4	7.2
VGG19	Dense	24.8	7.5
	Multi-crop	24.6	7.4
	Multi-crop and dense	24.4	7.1
OverFeat	7 accurate models, 4 scales, fine stride	33.96	13.24
Inception-v3	144 crops evaluated	18.77	4.2
ResNet50	Single model	20.74	5.25

models on validation data. Each model used different evaluation method to give these results.

4 Results and Discussions

4.1 Software Requirements

The experimental setup for the proposed methodology is carried out using Intel Xeon processor with NVIDIA Quadro K2000 GPU and 28 GB RAM. Python programming language is used for the overall experiment from data processing, feature extraction, model training till model evaluation. A developed, efficient, and modular deep learning library for Python called Keras created by François Chollet is used for the overall experiment. The entire experiment is carried out on Windows 7 Operating System (OS) with Theano as backend for Keras. Other scientific computational python packages used are NumPy, SciPy, matplotlib, h5py, scikit-learn, and OpenCV 3.1.

4.2 Dataset

Two publicly available face datasets are considered for analyzing the five ConvNets architectures. The AR [18] face dataset contains 4000 color images of 126 people (50 men and 76 women) with different facial expressions, illumination condition, and occlusions in two different sessions per person. The Extended Yale B database [19] contains 2432 face images of 38 subjects under 64 different illumination

AR face dataset   Extended Yale B face dataset

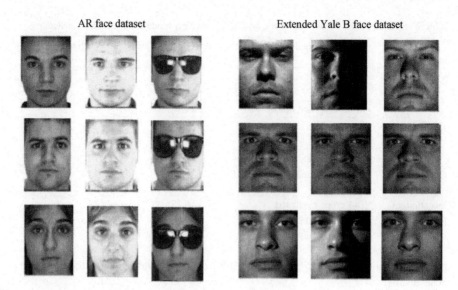

Fig. 3 Sample face images from AR face dataset and Extended Yale B face dataset

Table 3 Face recognition dataset summary

Dataset	No. of. unique identities	Training images	Testing images
AR	100	2340	260
Extended Yale B	38	2182	243

conditions. Figure 3 shows some sample images from AR face dataset and Extended Yale B face dataset. Table 3 shows the summary of these two datasets.

Feature extraction using ConvNets is carried out after performing pre-processing on the training images. The feature vector dimension for each image is taken from these deep ConvNets which is shown in Table 4.

All the five Deep ConvNets considered are used as feature extractors by using the activations of top fully connected layers which learn higher-level features from the training images. Tables 5 and 6 show the recognition report of the five Deep ConvNets models considered for face recognition on AR and Extended Yale B face dataset, respectively. It could be seen that VGG16 and VGG19 both achieved 100%

Table 4 Activation nodes present in the chosen layer of ConvNets for feature extraction

Model	Feature vector shape	Activation layer
VGG16	(4096)	FC-4096
VGG19	(4096)	FC-4096
OverFeat	(3072)	FC-3072
Inception-v3	(2048)	FC-2048
ResNet50	(4096)	FC-4096

Table 5 Recognition report for AR dataset on different ConvNets architectures

Model	Precision (%)	Recall (%)	F1-score (%)	Rank-1 (%)	Rank-5 (%)
VGG16	100	100	100	99.62	100
VGG19	100	100	100	99.62	100
OverFeat	98	96	96	96.15	99.23
Inception-v3	99	98	98	98.46	100
ResNet50	64	56	56	55.77	76.15

Table 6 Recognition report for Extended Yale B dataset on different ConvNets architectures

Model	Precision (%)	Recall (%)	F1-score (%)	Rank-1 (%)	Rank-5 (%)
VGG16	95	93	93	93.42	97.94
VGG19	95	93	93	93	96.71
OverFeat	95	94	94	94.24	97.53
Inception-v3	98	98	98	97.94	98.77
ResNet50	75	72	72	77.02	87.65

Rank-5 testing accuracy, which means that if the trained model is given a chance to guess the label of an unknown image five times, it will correctly guess it with a probability of 1. Google's Inception-v3 architecture outperformed all the other deep architectures by a substantial amount on a much challenging dataset, namely Extended Yale B dataset. This performance could mainly be contributed due to the usage of the novel inception module. The inception module of GoogLeNet is shown in Fig. 4. Table 7 shows the face recognition accuracy between ConvNets model and handcrafted feature extraction technique.

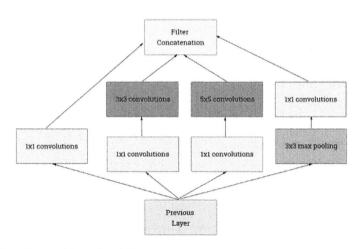

Fig. 4 Inception module of GoogleNet

Table 7 Face recognition accuracy between ConvNets and handcrafted feature extraction technique

Dataset	ConvNets accuracy %	Handcrafted features (SIFT) accuracy %
AR	98.46	94.78
Extended Yale B	97.94	93.61

During the selection of the convolution filter size of a convolution layer, typical choices include [1 × 1], [3 × 3], [5 × 5], and [7 × 7]. The inception module in GoogleNet proposed a new way to perform convolution on the previous layer by performing multiple convolutions (which are computed in parallel over the input volume) which gets concatenated at the output volume. Instead of using [3 × 3] and [5 × 5] filter sizes directly on the input volume (which greatly increases the depth of output volume), dimensionality reduction (which could also be thought of as pooling of features) is performed using [1 × 1] convolution filters, so that the overall depth at the output volume does not get increased at a higher rate. Thus, the inception module holds a smaller filter convolution, a medium filter convolution, a large filter convolution, and a pooling operation performed in parallel on the input volume which learns extremely fine grain details, higher-level details, and combats overfitting (due to the presence of pooling operation). In addition to this, the presence of Rectifier Linear Unit (ReLu) nonlinear activation after each convolutional layer enhances the performance.

The incepton-v3 architecture does not have fully connected layers at the top, instead it uses "average pool" operation which greatly reduces the learnable parameters involved. Thus, instead of stacking layers in a CNN sequentially (in the case of VGGNet and OverFeat network), GoogLeNet showed a different type of deep architecture such as the "inception" module (network in a network) which highly contributes to achieve better results. Although VGG16, VGG19, OverFeat, and ResNet50 achieve good results on AR dataset as well as Extended Yale B dataset, these deep architectures have sequentially stacked layers of convolution, max-pooling, and fully connected layers. Thus, it is inferred that the presence of different deep architectures such as the "inception" module in GoogLeNet contributes more in achieving better results. Tables 8 and 9 show the feature sizes stored locally after feature extraction using each of the Deep ConvNets model as well as the

Table 8 Feature size and timing details for feature extraction and training for Extended Yale B dataset on different ConvNets models

Model	Feature size (MB)	Feature extraction time (mins)	Training time (mins)
VGG16	40.6	7	7
VGG19	40.6	7	7
OverFeat	61.7	8	8
Inception-v3	20.3	8	4
ResNet50	20.3	6	1

Table 9 Feature size and timing details for feature extraction and training for AR dataset on different ConvNets models

Model	Feature size (MB)	Feature extraction time (mins)	Training time (mins)
VGG16	37.8	7	2
VGG19	37.8	8	2
OverFeat	57.2	2	3
Inception-v3	18.9	7	2
ResNet50	18.9	6	2

timing details of feature extraction and training. From the Tables 8 and 9, it is observed that Inception-v3 has lesser feature size when compared to other four ConvNets models.

5 Conclusion and Future Work

This paper evaluates the performance of the Deep ConvNets architectures such as VGG16, VGG19, OverFeat, ResNet50, and Inception-v3 for the face dataset (AR and Extended Yale B). From the experimentation, it is found that the Inception-v3 ConvNets model outperforms than other four ConvNets model with an accuracy of 98.7%. From the performance analysis, it is confirmed that the ConvNets model outperforms than traditional feature extraction-based technique (SIFT) in terms of recognition accuracy. It is found that instead of making a single CNN model to train and make predictions, an ensemble of Deep ConvNets could be used to increase the overall accuracy.

Acknowledgements This research project is supported by DAE-BRNS, Department of Atomic Energy, Government of India. The authors would like to extend their sincere thanks to DAE-BRNS for their support.

References

1. T. Ojala, M. Pietikäinen and T. Mäenpää, "Multiresolution Gray Scale and Rotation Invariant Texture Classification with Local Binary Patterns", IEEE Transactions on Pattern Analysis and Machine Intelligence, vol. 24, no. 7, pp. 971–987, 2002.
2. D.G. Lowe, "Distinctive Image Features from Scale-Invariant Keypoints", International Journal of Computer Vision, 91–110. https://doi.org/10.1023/b:visi.0000029664.99615.94, 2004.
3. N. Dalal and B. Triggs, "Histograms of Oriented Gradients for Human Detection", IEEE Computer Society Conference on Computer Vision and Pattern Recognition (CVPR), Vol. 1, 886–893, 2005.

4. J. Yang, Y.G. Jiang, A.G. Hauptmann, and C.W. Ngo, "Evaluating Bag-of-Visual Words Representations in Scene Classification", Proceedings of the International Workshop on Multimedia Information Retrieval (ACM), 197–206, 2007.

5. A. Krizhevsky, I. Sutskever, and G.E. Hinton, "ImageNet Classification with Deep Convolutional Neural Networks", Conference on Neural Information Processing Systems (NIPS), pp. 1106–1114, 2012.

6. K. Simonyan and A. Zisserman, "Very Deep Convolutional Networks for Large-Scale Image Recognition", International Conference on Learning Representations (ICLR), 2015.

7. C. Szegedy, W. Liu, Y. Jia, P. Sermanet, S. Reed, D. Anguelov, D. Erhan, V. Vanhoucke, and A. Rabinovich, "Going Deeper with Convolutions", IEEE Conference on Computer Vision and Pattern Recognition (CVPR), 2015.

8. P. Sermanet, D. Eigen, X. Zhang, M. Mathieu, R. Fergus, and Y. LeCun, "Overfeat: Integrated Recognition, Localization and Detection using Convolutional Networks", International Conference on Learning Representations (ICLR), 2014.

9. K. He, X. Zhang, S. Ren, J. Sun, "Deep Residual Learning for Image Recognition", IEEE Conference on Computer Vision and Pattern Recognition (CVPR), 2016.

10. Z. Cao, Q. Yin, X. Tang, and J. Sun, "Face Recognition with Learning based Descriptor", Proceedings of Computer Vision and Pattern Recognition (CVPR), 2010.

11. P. Li, S. Prince, Y. Fu, U. Mohammed, and J. Elder. "Probabilistic Models for Inference about Identity", IEEE Transactions on Pattern Analysis and Machine Intelligence (PAMI), 2012.

12. T. Berg and P. Belhumeur, "Tom-vs-Pete Classifiers and Identity preserving Alignment for Face Verification", Proceedings of British Machine Vision Conference (BMVC), 2012.

13. F. Schroff, D. Kalenichenko, and J. Philbin, "Facenet: A Unified Embedding for Face Recognition and Clustering", IEEE Conference on Computer Vision and Pattern Recognition (CVPR), 2015.

14. Y. Taigman, M. Yang, M. Ranzato, and L. Wolf, "Deep-Face: Closing the Gap to Human Level Performance in Face Verification", IEEE Conference on Computer Vision and Pattern Recognition (CVPR), 2014.

15. A. Romero, N. Ballas, S.E. Kahou, A. Chassang, C. GattaandY. Bengio, "Fitnets: Hints for Thin Deep Nets", arXiv:1412.6550, 2014.

16. J. Ba and R. Caruana, "Do deep nets really need to be deep?", Advances in Neural Information Processing Systems 27, arXiv:1312.6184, 2014.

17. M.D. Zeiler, and R. Fergus, "Visualizing and Understanding Convolutional Networks", European Conference on Computer Vision (ECCV), 2014.

18. A.M. Martinez and R. Benavente, "The AR Face Database", CVC Technical Report #24, June 1998.

19. "The Extended Yale Face Database B", available online: http://vision.ucsd.edu/~leekc/ExtYaleDatabase/ExtYaleB.html.

Use of High-Speed Photography to Track and Analyze Melt Pool Quality in Selective Laser Sintering

Sourin Ghosh, Priti P. Rege and Manoj J. Rathod

Abstract Manufacturing industry is moving toward a process model involving rapid and frequent product deliveries to increase their consumer base. Employing laser sintering methods in product fabrication provides a superior quality, low cost and high-fidelity solution to support this. Effective monitoring and diagnostics of laser sintering process become a critical task in this regard. This paper focuses on analyzing spatters and plume generated during continuous laser sintering for fabrication of circular rings. Analysis using high-speed photography and subsequent image processing was undertaken. By varying laser parameters, the generated spatter and plume was tracked and features such as spatter count, spatter size, and plume area were examined. Results show that spatter count and plume size are related to the variations in laser power intensity. Optimal power settings are shown to produce best quality product. The proposed analysis method could be used to monitor the stability of laser sintering process.

Keywords Metallurgy · Selective laser sintering · Spatter tracking
Plume tracking · High-speed camera · Frame rate

1 Introduction

Engineering parts made of Iron (Fe) alloys are widely used in the manufacturing sector in industries like automobiles, aerospace, construction. Fabrication using traditional methods like forging, powder metallurgy involves a high turnaround time and results in a sub-par product quality with poor mechanical properties.

S. Ghosh (✉) · P. P. Rege
Department of Electronics and Telecommunication, College of Engineering Pune,
Pune, India
e-mail: souringhosh1@live.com

M. J. Rathod
Department of Metallurgy and Material Science, College of Engineering Pune,
Pune, India

© Springer Nature Singapore Pte Ltd. 2018
B. B. Chaudhuri et al. (eds.), *Proceedings of 2nd International Conference on Computer Vision & Image Processing*, Advances in Intelligent Systems and Computing 703, https://doi.org/10.1007/978-981-10-7895-8_26

Hence, to hasten the process of bringing their products to market, companies are turning to selective laser sintering (SLS) [1, 2]. SLS is a process where a continuous laser beam scans the powder bed layer by layer. Each layer is created by melting, sintering, and bonding the particles in a thin lamina [3]. The size of the layer can be as thin as the particle size of powder used. The experiments described in this paper use powdered iron for synthesizing circular metal rings of thickness 3 mm with an inner diameter of 8 mm.

The process of SLS is highly dynamic and involves unstable heat transfer from the laser beam to the metal substrate [4]. So, monitoring of the entire activity becomes a necessity toward ensuring highest quality builds [5, 6]. An effective method of quality measurements involves image processing techniques for object tracking and shape measurements. It is already being used in high-power laser welding [7] where techniques have been developed to observe and provide measurements for ejected spatters and laser-induced plume. A combined occurrence of these two phenomena is undesirable as it results in inadequate transfer of heat to the melt surface. Application of image processing for understanding this phenomenon in SLS and providing quality measures has not been explored much in industry.

This paper presents a simple method for monitoring laser sintering process and introduces an image processing algorithm for feature extraction of ejected spatters and plume growth. Through the use of high-speed photography image sequences of plume and spatter, filtered through a spectral UV filter, have been captured during the fiber laser sintering. When these images are input to the developed algorithm, it tracks plume growth and ejected spatters by comparing consecutive frames. It is able to provide statistics based on features including plume area, spatter count, spatter area. The correlation of sintering quality with plume and spatter features is investigated.

The paper is organized as follows. Section 2 describes the experimental setup of the fiber laser assembly for laser sintering. Section 3 presents the methodology of image processing and subsequent feature extraction with a detailing of the spatter and plume tracking algorithm. Section 4 highlights the achieved results and discusses about the relation between sintering quality, spatter, and plume features.

2 Experimental Setup

An SEM analysis of the Iron (Fe) particles to be used during the process is shown in Fig. 1. Particle size varies in the range of 13–31 μm captured at 1000x resolution. They are loosely bound with no visible compaction.

This was followed by experimentation using the setup shown in Fig. 2. The iron powder is spread as a uniform layer of 1 mm thickness over a steel substrate. This assembly is kept on a mechanical turntable with horizontal and vertical movement capability (x-y direction). The assembly is covered by a transparent cubical acrylic chamber with a spherical opening at the top for laser firing. Argon gas is fed into the chamber for protecting sintered area from oxidation. A 400 W Yb-doped fiber laser fitted with a beam collimator is used in continuous mode with a constant scanning

Fig. 1 SEM image of iron powder under consideration with marked particle sizes of 13.26 and 30.36 μm

speed of 3 mm/s. Phantom Flex v311 high-speed camera [9] is set up at an incidence angle of 20° with maximum aperture and exposure setting. Additionally, for suppressing saturation and blooming of CCD sensor of camera, a spectral band-pass filter with passband wavelength (180–534 nm) and stopband wavelength (740–1070 nm) is placed in front of the camera. The resulting difference in contrast on applying the filter is visible in Fig. 3a, b. The captured video sequences are transferred from the camera's flash memory to the monitoring computer for further processing. The fiber laser power is varied from 80 to 400 W. The shielding gas flow is set at 36 L/min and nozzle angle is 0°. The turntable is driven by a precise motor assembly which can be moved in x and y directions. The laser power and turntable movement are programmable through computer software pre-installed with the laser assembly. Experiments are performed with three power settings, low power (80 W), medium power (200 W), and high power (400 W). The frame rate for capturing video sequences is set at 3000 fps with an image resolution of 512 pixel by 512 pixel and 8-bit color depth.

2.1 Inclusion Criteria

1. Laser firing action in continuous mode.
2. Power variation between 80–400 W.
3. 4 s video per experiment
4. Observation of 90 frames (Frame no. 1233–1322) out of 12342 frames.

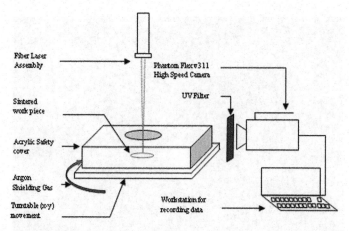

Fig. 2 Experimental setup of laser sintering process

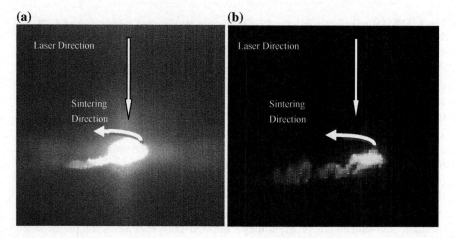

Fig. 3 **a** Image captured without use of UV filter showing saturation (power: 80 W). **b** Image captured through a UV filter showing high contrast well-defined melt pool (power: 80 W)

2.2 *Exclusion Criteria*

1. Laser firing action in pulsing mode.
2. Videos with multiple saturations and blooming.

3 Methodology

3.1 Sample Preparation

Powdered iron is spread as a thin layer (1 mm thick) over a steel metal base. The preparation is covered with a cube-shaped transparent acrylic box with a spherical opening at the top. The covering is supplied with Argon shielding gas during laser firing. Shielding is required to prevent particulate oxide formations in the final product due to interaction of melt pool surface with surrounding air. Apart from shielding, it aids cooling the collimator lens [8].

3.2 Laser Assembly Preparation

The fiber laser is set to fire in continuous mode. Laser scanning speed is set at 3 mm/s. Laser beam collimator is set for 2 mm beam diameter. Experiments are conducted by setting power to 80, 200, and 400 W.

3.3 Image Acquisition and Preprocessing

During each experiment, high-speed video sequences are recorded by the specialized camera [9]. The process of laser action over the loosely bound particles is an event where rapid spatter movement can be tracked only through high-speed imaging with frame rates greater than 1000 frames per second and not an ordinary CCD camera.

The raw data is transferred to a monitoring workstation where image files in TIFF format are extracted. TIFF is the best format as compared to JPEG or BMP as it preserves information. Ninety frames in this format were extracted from the raw video file generated by the Phantom Camera Control Software with each frame is activity at 1/3000th of a second. The algorithm designed in MATLAB v2012a considers these images as input.

No preprocessing was applied to extracted images and the developed algorithm was tested directly on them.

3.4   Spatter Tracking and Feature Extraction

Algorithm

1: Initialization
2: Read two successive color frames
3: Color to Gray scale conversion
4: Perform Background Subtraction
5: Perform binarisation
6: Compute Region properties: Filled Area, Centroid, Bounding Box
7: If Filled Area is more than or equal to 2 pixels then
8:          Mark with blue bounding box
9: Else
10:          Mark with red bounding box
11: End If
12: If not last frame, go to 2
13: End

The algorithm picks the nth and n−1th color frames $I\_color(n)$ and $I\_color(n-1)$ as shown in Fig. 4a and 4b, respectively. Spatter movement is marked with red circle. It converts them to grayscale images $I\_gray(n)$ and $I\_gray(n-1)$, shown in Fig. 5a and 5b, respectively.

(a)   **(b)**

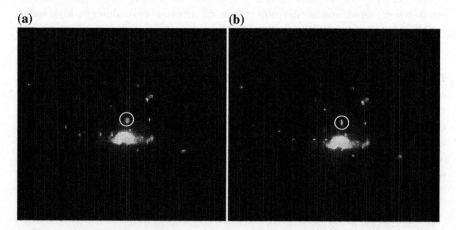

Fig. 4 a nth color frame $I\_color(n)$. **b** n−1th color frame $I\_color(n-1)$ (*power: 80 W, prominent spatter movement is circled*)

(a) **(b)**

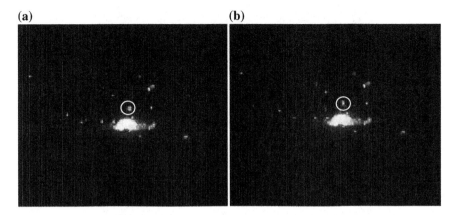

Fig. 5 **a** nth gray frame *I_gray(n)*. **b** n−1th gray frame *I_gray(n−1)* (*prominent spatter movement is circled*)

This step is followed by background subtraction using the frame difference method:

$$P(I\_diff(x, y, n)) = P(I\_gray(x, y, n)) - P(I\_gray(x, y, n-1)) \tag{1}$$

where $P(I\_gray(x, y, n))$ is the pixel intensity value at location (x, y) in the nth frame and $P(I\_gray(x, y, n-1))$ is the pixel intensity value at the same location in the previous frame.

Background subtraction is a widely used method to detect object motion [10]. Algorithms using a varied mix of background models are compared in [11, 12] and includes the frame difference approach. Frame difference method allows the tracking algorithm to access inter-frame variations in the least possible time required. Its choice is best suited here due to continuous rapid motion in every frame.

Fig. 6 Binary image of background subtracted frame *I_diff*

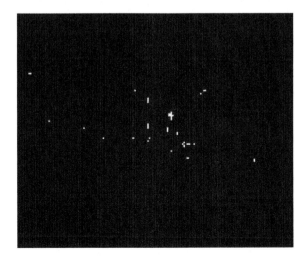

Fig. 7 Detected spatters with superimposed markers on input frame (*green for area < 2 and blue for area >2*)

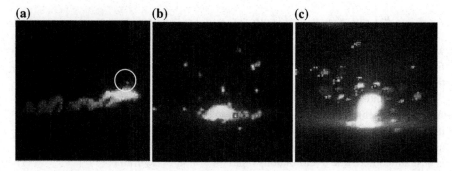

Fig. 8 a–c Detected spatters with superimposed markers on input frame at 80, 200 and 400 W (*green for area < 2 and blue for area ≥ 2*)

Next step is the binarization of *I_diff* with an optimally selected threshold as shown in Fig. 6. This step is essential to compute the region properties of *I_diff*.

Next, region properties including filled area, bounding box, and centroid are calculated for all closed objects in I_diff. All the identified spatters with Filled Area ≥ 2 are bounded with a blue marker and Filled Area < 2 pixels with a green marker as shown in Fig. 7. This classification threshold of 2 pixels is concluded after observing individual spatter size features divided over 270 frames. Figure 8a–c show the tracked spatters for power settings of 80 W, 200 W, and 400 W, respectively. Figure 8a shows a single ejected spatter encircled with a white marker.

3.5 Plume Tracking and Feature Extraction

Algorithm

1: Initialization

2: Read current frame

3: Color to Gray scale conversion

4: Perform Binarisation

5: Compute region properties: Filled Area, Centroid and Bounding Box

6: PlumeArea = max (Filled Area)

7: Mark with red bounding box

8: If not last frame, go to 2

9: End

The algorithm takes image I_gray_n and converts into binary image based on an optimally selected threshold. Then, it computes the region properties and segments the plume region based on plume centroid and filled area. The extracted plume shape is bounded with a red marker box. The tracked plume shape for power of 80, 200, and 400 W is shown in Figs. 9, 10, and 11, respectively.

(a)   **(b)**

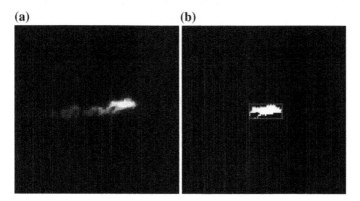

Fig. 9 a, b Weld pool image with marker on detected plume (80 W)

(a)   (b)

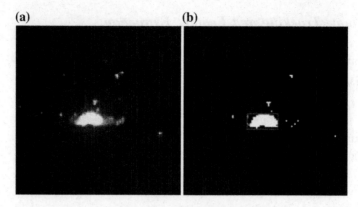

Fig. 10 a, b Weld pool image with marker on detected plume (200 W)

(a)   (b)

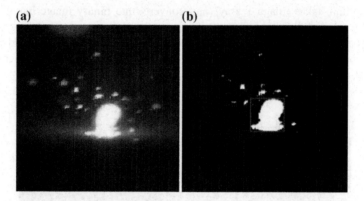

Fig. 11 a, b Weld pool image with marker on detected plume (400 W)

4 Results and Discussion

4.1 Spatter Tracking

Figure 12 shows the plot of total ejected spatter count against the frames (1–90) for three power combinations. Due to the back pressure exerted by the laser firing on the loosely compacted iron particles, significant spatters are observed for 400 W power setting. Also, a dip in spatter count is observed for frame number 60–62 due to iron particles sticking to each other due to extreme heat.

Plots of spatter count with Filled Area < 2 pixels and Filled Area $\geq$ 2 pixels against respective frame numbers shown in Figs. 13 and 14 suggest a proportional rise in high density and low-density spatter count.

Figure 15a–c shows the sintered product with power settings of 80, 200, and 400 W, respectively. It can be observed that the ring structure remains intact in

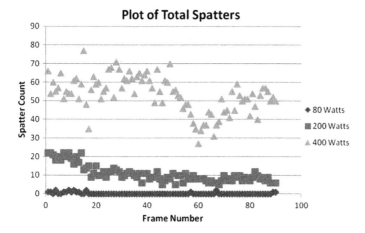

Fig. 12 Plot of total spatters against frame progression

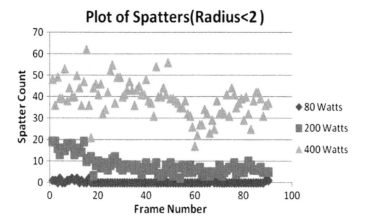

Fig. 13 Plot of spatters (radius < 2) against frame progression

Fig. 15b only. The ring quality in Fig. 15a is inferior and iron particles have started unbinding right after sintering. Also, the ring can be seen to stick to the base metal in Fig. 15c and is undesirable.

4.2 Plume Tracking

The results of the plume tracking algorithm were plotted with plume area against frame number. Figure 16 highlights that setting the power to 400 W results in rapid expansion of plume area as compared to 80 and 200 W. Larger plume size results in inefficient transfer of heat from the laser to melt pool surface.

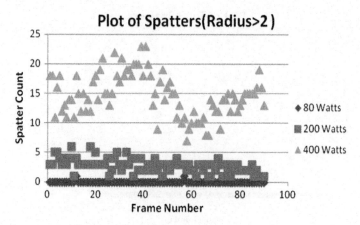

Fig. 14 Plot of spatters (radius > 2) against frame progression

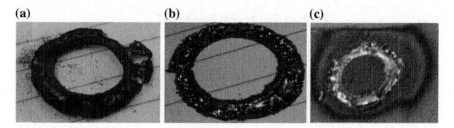

Fig. 15 **a–c** Manufactured product with Power: 80, 200, and 400 W, respectively

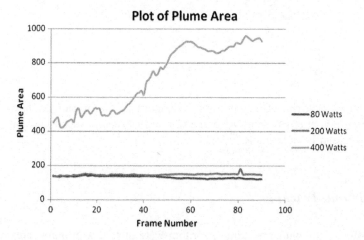

Fig. 16 Plot of plume area against frame progression

5　Conclusion

Application of high-speed photography in monitoring fiber laser sintering of circular iron rings has been suggested. Spatter and plume generation during the process has been captured through high contrast videos recorded at 3000 frames per second. An image processing algorithm based on background subtraction for extracting the laser-induced spatter and plume features has been proposed. Accurate tracking of spatter and plume growth is accomplished. Results show that the algorithm can accurately track spatter movement over the considered dataset. It is observed that there is a 200% increase in spatter count of non-compacted iron powder on doubling the laser power. There is a corresponding increase in average plume size by 400% in high-power conditions leading to undesirable sintered product quality.

The proposed algorithm can be used in diagnostics of rapid spatter and plume growth during selective laser sintering. Evidently, experimentation with laser power of 200 W produces a high-quality sintered product with ejected spatters and plume having small sizes with growth in the same direction of sintering. It prevents energy absorption and beam scattering ensuring optimal quality of synthesized rings.

A GUI-based tool has been developed to provide frame by frame tracking of spatter and plume to manufacturing technicians. Plume growth monitoring is a critical issue for them and the proposed use of image processing is an effective approach for ensuring high-quality sintered products.

Acknowledgements The authors are grateful to Department of Electronics and Telecommunication, College of Engineering, Pune, for permitting the usage of the high-speed camera to capture the entire laser sintering process.

References

1. Morgan, R, Papworth, A, Sutcliffe, C, Fox, P, O'Neill, W.: High density net shape components by direct laser re-melting of single-phase powders. J Mater Sci.; 37:3093–3100 (2002).
2. Craeghs, Tom and Clijsters, Stijn and Yasa, Evren and Kruth, Jean-Pierre: Online quality control of selective laser melting, Proceedings of the Solid Freeform Fabrication Symposium, Austin, TX, pp. 212–226 (2011).
3. Olakanmi, E.O., Cochrane, R.F., Dalgarno, K.W.: A review on selective laser sintering/melting (SLS/SLM) of aluminium alloy powders: Processing, microstructure, and properties, Progress in Materials Science, 74, pp. 401–477 (2015).
4. Kruth, J.P., Froyen, L., Van Vaerenbergh, J., Mercelis, P., Rombouts, M. and Lauwers, B.: Selective laser melting of iron-based powder, Journal of Materials Processing Technology, 149(1), pp. 616–622 (2004).
5. Berumen, S., Bechmann, F., Lindner, S., Kruth, J. P., Craeghs, T.: Quality control of laser-and powder bed-based Additive Manufacturing (AM) technologies, Physics procedia, 5, pp. 617–622 (2010).

6. Yasa, E., Deckers, J., Kruth, J. P.: The investigation of the influence of laser re-melting on density, surface quality and microstructure of selective laser melting parts, Rapid Prototyping Journal, 17(5), pp. 312–327 (2011).
7. You, D, Gao, X, Katayama, S: Monitoring of high-power laser welding using high-speed photographing and image processing. J Mech Sys and Sig Proc., 49(1–2), pp. 39–52 (2014).
8. Kar, A., Sankaranarayanan, S, Kahlen, F.J.: One-step rapid manufacturing of metal and composite parts, U.S. Patent 6,526,327 (2003).
9. Vision Research, www.visionresearch.com.
10. Piccardi M: Background subtraction techniques: A review, IEEE International Conference on Systems, Man and Cybernetics (2004).
11. Sobral, A., Vacavant A.: A comprehensive review of background subtraction algorithms evaluated with synthetic and real videos, Computer Vision and Image Understanding, 122, 4–21 (2014).
12. Prasad, DK., Rajan, D, Rachmawati, L, Rajabally, E, Quek, C: Video Processing From Electro-Optical Sensors for Object Detection and Tracking in a Maritime Environment: A Survey, IEEE Transactions on Intelligent Transportation Systems (2017).

Multi-Scale Directional Mask Pattern for Medical Image Classification and Retrieval

Akshay A. Dudhane and Sanjay N. Talbar

Abstract This paper presents a classification scheme for interstitial lung disease (ILD) pattern using patch-based approach and artificial neural network (ANN) classifier. A new feature descriptor, Multi-Scale Directional Mask Pattern (MSDMP), is proposed for feature extraction. Proposed MSDMP extracts microstructure information from a (31×31) size patches of the region of interest (ROI) which were marked by the radiologists. A two-layer feed-forward neural network is used for classification of ILD patterns. Also, proposed MSDMP feature descriptor has been tested on medical image retrieval system to check its robustness. Two benchmark medical datasets are used to evaluate the proposed descriptor. Performance analysis shows that the proposed feature descriptor outperforms the other existing state-of-the-art methods in terms of average recognition rate (ARR) and F-score.

Keyword ILD artificial neural network feature descriptor

1 Introduction

The interstitial lung disease (ILD) is broadly categorized having different conditions of the lung and collectively represents more than 130 different categories. Usually, all these types of ILD majorly cause thickening of the interstitium, a part of the lung anatomic structure. The interstitium provides support to the alveoli/air sacs; also, small blood vessels find their path through the interstitium, which allows oxygen exchange. On the other hand, ILD causes scarring of the interstitium which turns into a decrease in strength of the air sacs to store and carry oxygen and eventually

A. A. Dudhane (✉) · S. N. Talbar
Department of Electronics & Telecommunication Engineering, SGGSIE&T, Nanded, India
e-mail: akshay.aad16@gmail.com

S. N. Talbar
e-mail: sntalbar@sggs.ac.in

© Springer Nature Singapore Pte Ltd. 2018
B. B. Chaudhuri et al. (eds.), *Proceedings of 2nd International Conference on Computer Vision & Image Processing*, Advances in Intelligent Systems and Computing 703, https://doi.org/10.1007/978-981-10-7895-8_27

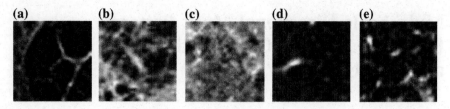

Fig. 1 Sample ILD patches of size 31 × 31; one from each class. **a** Tissue emphysema, **b** tissue fibrosis **c** tissue ground glass, **d** tissue normal, **e** tissue micronodule

patient may lose the ability to breathe. Commonly seen symptoms of the ILDs are shortness of breath, weakness, weight loss [1]. Over to this, many times cause of the ILD is unknown and which is known as idiopathic interstitial pneumonia (IIP).

It has been observed that growth of the ILD varies from person to person as well as with the ILD category [1] due to which diagnosis of ILD becomes an initial step to cure the disease. ILD diagnosis comprises of various physical examinations, pulmonary function test, as well as various scans (X-ray, computed tomography (CT) scans) to understand the tissue patterns. High-resolution computed tomography (HRCT) scans show sufficient variation in the tissue patterns of different ILDs due to which HRCT scans become popular modality to diagnose the disease. However, interpreting the type of ILD with the help of HRCT scans is challenging even for experienced radiologist because different ILDs may have different combinations of the tissue patterns. This property of the ILD motivates us to have a computer-aided diagnosis (CAD) system which helps the radiologists to get "second opinion" to increase their accuracy to diagnose the ILD pattern. A CAD system for lung CT scans includes three phases, namely lung segmentation, ILD quantification, and differential diagnosis. In this study, we are focusing on ILD quantification (specifically, ILD pattern classification). Among various ILD patterns, we are specifically focusing on five frequently seen ILDs namely fibrosis, ground glass, emphysema, micronodules, and healthy patterns which are shown in Fig. 1.

2 Related Work

Since HRCT scans of ILD show the appropriate difference in tissue patterns, a lot of research has been carried out in ILD pattern classification to build the CAD system. Various researchers have proposed different methods based on spatial as well as transform domain image analysis to classify ILD patterns [2–9], some of them are briefly enlightened as follows.

Initially, Mir et al. [2] analyzed CT images based on second-order statistics. Further, Renuka uppaluri et al. [3] have proposed adaptive multiple feature method (AMFM) for pulmonary parenchyma classification from computed tomography (CT) scans using regional approach. AMFM comprises of twenty-four features which were a gold standard since the use of multi-scale filter bank [5]. Multi-scale filter bank for ILD pattern classification was proposed by sluimer et al. [5]. Uchiyama et al. [6] built CAD system in which they have segmented lung field using morphological operations, and further, they have used six measures to classify the abnormal lung patterns. Moreover, Xu et al. used the ability of MDCT scanners to achieve 3D sub-millimeter resolution and extracted 3D AMFM features by extending 2D AMFM [4] to three dimensions. On the other hand, local operators can extract the regional information which helps to analyze the frequency of local structures [10, 11]. Inspired by volumetric local binary pattern (LBP), Murala et al. [12] proposed a multi-resolution analysis of LBP. Also, to incorporate the directional information in texture analysis, lot of research has been carried out using wavelet transform and Gabor filters [7–9, 13, 14]. Gabor filters are more analogous to the human visual system than wavelet analysis because it collects information at multiple scales and multiple orientations. Using Gabor filters, majunath et al. [13] extracted first-order moments from image texture to increase the accuracy of image retrieval system. However, these features are not rotation invariant. To address this problem, Han et al. [15] proposed scale- and rotational-invariant Gabor filter (RIGF). However, Gabor filters are computationally more complex. To overcome this problem, Depeursinge et al. [7] have used wavelet frames and gray level histograms for classification of five frequently seen ILD patterns (emphysema, fibrosis, ground glass, healthy, and micronodules). They have used B-spline wavelet basis functions and extracted first-order moments for each ROI. Moreover, Talbar et al. [16] proposed texture classification based on wavelet features. Also, isotropic wavelet frames [8] and optimized steerable wavelets [9] are used to extract information from HRCT scan of lungs.

However, the accuracy of any classification system depends upon its discriminative feature map and effective classifier. So, to improve the accuracy of ILD classification, many researchers have used different classifiers [14, 17–22]. Michinobu et al. [17] proposed histogram features followed by Bayes classifier for ground glass and micronodule detection. Also, Song et al. [18] have used sparse representation and dictionary learning and further compared k-nearest neighborhood (k-NN) and support vector machine (SVM) classifier. The classification accuracy of the k-NN classifier is more dependent on a distance metric. To overcome this disadvantage, they have proposed patch-adaptive sparse approximation method for classification of ILD patterns. Sparse representation is an enhanced k-NN model; k-NN selects the nearest neighbors by similarity ranking using distance metric, whereas in sparse representation nearest neighbors are computed by a weighted linear combination of reference dictionary. Nevertheless, the performance of sparse representation relies on the quality of reference dictionary. Parametric classifiers (such as ANN) help to learn the image texture and improve the classification accuracy. So, to improve computer-aided diagnosis, Lilla et al. [23] combined genetic algorithm followed by SVM classifier. Ishida et al. [19] used artificial neural network (ANN) for detection of ILD patterns. Also, their work has been extended by Ashizawa et al. [20] for differential diagnosis of ILD patterns. Recently, Maris et al. [21] used the convolutional neural network for the classification of seven ILD patterns. Moreover, Yang et al. [18, 22] proposed modified sparse representation and dictionary leaning-based ILD classification.

In this study, we have classified ILD patterns using patch-based approach and novel Multi-Scale Directional Mask Pattern (MSDMP) followed by feed-forward neural network. Further to prove the robustness of MSDMP, we have used our approach on the state-of-the-art database for medical image retrieval.

3 Proposed Method

3.1 Feature Extraction

Multi-Scale Directional Masks Pattern (MSDMP)

Multi-scale and multi-directional representation of conventional Gabor filter gives discriminative information about the image texture. Still, because Gabor filter is rotational variant; which limits its use in medical imaging. To overcome this problem, we chose rotational-invariant Gabor filters [15]. However, scale-invariant Gabor filter introduces aliasing effect [15]. So, we have introduced directional masks to combine multi-scale information of Gabor filters. On the other hand, local operator extracts information from local regions of an image; histogram of these patterns extracts edge distribution in an image. Collectively, in this study, we proposed a new multi-scale directional mask pattern (MSDMP) texture descriptor to merge multi-scale information of Gabor filters and most prominent advantages of local operators.

Figure 2 shows the use of proposed directional masks to obtain multi-scale directional 3D grid (MSD-3D), whereas Fig. 2a, b illustrates the obtained patch from annotated ROI, which is then convolved with rotational-invariant Gabor filter bank as in [15]; Gabor-filtered output is shown in Fig. 2c; further, Fig. 2e displays proposed directional mask (Mask) along four directions, multiplied with the local region extracted as shown in Fig. 2d. After multiplication with directional masks finally, obtained multi-scale directional 3D grid is shown in Fig. 2f–i. The operation of directional mask on a sample of local region extracted from each scale is shown in Fig. 3. Whereas Fig. 3a shows a sample of local region, Fig. 3b demonstrates multiplication of 0° directional mask to obtain binary pattern as shown in Fig. 3c using proposed MSDMP.

Let $G^{p,r}(x, y)$ denote Gabor filter bank, where p and r represent the number of scales and orientation correspondingly. The rotational-invariant Gabor filter (RIGF) [15] is represented in Eq. (1):

$$G^p = \sum_{r=0}^{R-1} G^{p,r}(x, y) \tag{1}$$

where $p = 1, 2, 3$ represents scales of Gabor filter.

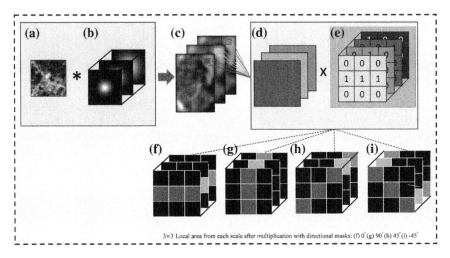

Fig. 2 **a** Original ROI, **b** rotational-invariant Gabor filter bank, **c** Gabor-filtered output, **d** 3 × 3 local region from each scale of Gabor-filtered output, **e** directional masks, **f–i** multi-scale directional 3D grid

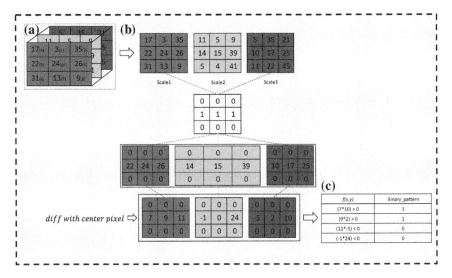

Fig. 3 Calculation of MSDM pattern for sample local area. **a** Sample 3 × 3 local area from each scale, **b** calculation of MSDM pattern using 0° directional mask, **c** binary pattern calculation

However, Eq. (2) represents convolution operation of image with Gabor filters.

$$I^p = I * G^p; p = 1, 2, 3 \tag{2}$$

where G^p, I^p represents RIGF and Gabor-filtered output, respectively.

$$Mask \in \begin{bmatrix} 0 & 0 & 0 \\ 1 & 1 & 1 \\ 0 & 0 & 0 \end{bmatrix} \begin{bmatrix} 0 & 1 & 0 \\ 0 & 1 & 0 \\ 0 & 1 & 0 \end{bmatrix} \begin{bmatrix} 0 & 0 & 1 \\ 0 & 1 & 0 \\ 1 & 0 & 0 \end{bmatrix} \begin{bmatrix} 1 & 0 & 0 \\ 0 & 1 & 0 \\ 0 & 0 & 1 \end{bmatrix}$$

Fig. 4 3 × 3 directional masks (dir) used for combining multi-scale information. From left to right: 0°, 90°, 45°, 135° directions, respectively

Next, we multiplied directional masks with local regions of I^p to combine multi-scale information. Consider I_l^p a local region of size (3 × 3) of Gabor-filtered output I^p as shown in Fig. 3a. We have designed directional masks as shown in Fig. 4. So, given a directional mask, multi-scale directional 3D grid (MSD-3D) is given by Eq. (3).

$$A_l^p(g_i) = I_l^p(g_i) \times mask_{dir}(g_i); i = 1, 2, \ldots, 8 \, and \, p = 1, 2, 3 \tag{3}$$

where g_i are the neighborhoods of center pixel g_c as denoted in Fig. 3a. A_l^p is a local multi-scale directional 3D grid (MSD-3D) as shown in Fig. 2f–i, while Fig. 4 represents directional mask ($mask_{dir}$) along $dir \in 0°, 90°, 45°, 135°$.

Further, *MSDMP* for A_l^p having center pixel g_c is calculated by Eqs. (5), (6), (7), and (8), where $g_c = A_l^2(2, 2)$.

$$diff_l^p = A_l^p - g_c; \ p = 1, 2, 3 \tag{4}$$

$$pattern_l^{dir}(1, j) = f^{dir} \left(diff_l^1(g_i), diff_l^3(g_k) \right) \tag{5}$$

where $i = \begin{cases} 0, 1, 5 & if \ dir = 0° \\ 0, 3, 7 & elseif \ dir = 90° \\ 0, 2, 6 & elseif \ dir = 45° \\ 0, 4, 8 & elseif \ dir = 135° \end{cases}$, $k = mod(i + 4, 8), j = 1, 2, 3$

and $pattern_l^{dir}$ is a binary code for local region l and the directional mask dir. Also, $diff_l^1$ and $diff_l^3$ are the first and third plain of $diff_l^p$ obtained using Eq. (4).

$$pattern_l^{dir}(1, 4) = f^{dir} \left(diff_l^2(g_i), diff_l^2(g_{i+4}) \right) \tag{6}$$

where $i = \left(1 + \frac{dir}{45°}\right)$; $dir \in 0°, 45°, 90°, 135°$ and $pattern_l^{dir}$ is obtained using Eqs. (5) and (6) is a four-bit binary code for local region l and directional mask dir.

$$f^{dir}(x, y) = \begin{cases} 1 & if \ x \times y > 0 \\ 0 & else \end{cases} \tag{7}$$

Finally, *MSDMP* code for given MSD-3D grid A_l^p with a directional mask dir is given by Eq. (8).

$$MSDMP_l^{dir} = \sum_{w=1}^{4} 2^{w-1} \times pattern(1, w) \tag{8}$$

Similarly, $MSDMP_l^{dir}$ is obtained for each local region of I^p which will create a $MSDMP^{dir}$ coded image for an input image. Further, a histogram of $MSDMP^{dir}$ is obtained using Eq. (9).

$$H_{MSDMP}^{dir}(m) = \sum_{i=1}^{row} \sum_{j=1}^{col} f1\big(MSDMP^{dir}(row, col), m\big); \quad m \in [0, 15] \tag{9}$$

$$f1(x, y) = \begin{cases} 1 & if \ \ x = y \\ 0 & else \end{cases} \tag{10}$$

$$f_{map} = \begin{bmatrix} H_{MSDMP}^{dir=0°} & H_{MSDMP}^{dir=90°} & H_{MSDMP}^{dir=45°} & H_{MSDMP}^{dir=135°} \end{bmatrix} \tag{11}$$

where f_{map} is a final feature vector obtained by concatenating histogram obtained using four different directional masks $(dir = 0°, 90°, 45°, 135°)$. So, feature vector length for an image using MSDMP is (1×64).

Gray level Run-Length Matrix (GLRLM)

Run-length encoding extracts useful information from image texture [24]. In a coarse texture, comparatively longer gray level runs would occur, rather in fine texture, many times short gray level run occurs. We have obtained gray level run-length matrix from which twelve features are extracted as proposed in [25]. So, in total, we have extracted (1×76) features, out of which 64 features are extracted using MSDMP and remaining 12 features are obtained using GLRLM.

3.2 Two-Layer Block Nets

In this study, we have used patch-based approach for ILD pattern classification (as 2D ROI is marked by experienced radiologists). The patch-based approach helps to extract the detailed information from HRCT scans. However, to achieve advantages of patch-based approach where the ROIs are not marked, we have introduced two-layer block net method. Figure 5 shows two-level block net.

$$N1_{p_n \times p_m}^{level\_0} = \left(\frac{n}{p_n}\right) \times \left(\frac{m}{p_m}\right) \tag{12}$$

where n, m are a number of rows and columns in an image, p_n, p_m are a number of rows and columns in an image block, and $N_{p_n \times p_m}^{level\_0}$ is the total number of $(p_n \times p_m)$ sized blocks obtained at a given level zero.

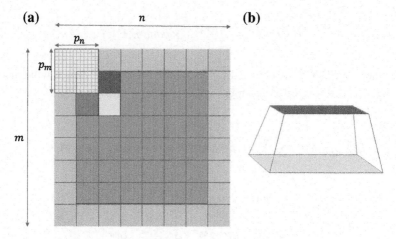

Fig. 5 Representation of two-level block net. **a** Represents a two-level block net and small block at level_0. **b** 3D view of two-level block net

$$N1^{level\_1}_{p_n \times p_m} = \left(\frac{n'}{p_n}\right) \times \left(\frac{m'}{p_m}\right) \tag{13}$$

where $n' = n - \frac{p_n}{2}$, $m' = m - \frac{p_m}{2}$ which is a size of the image at $level\_1$, and $N1^{level\_1}_{p_n \times p_m}$ is the total number of $(p_n \times p_m)$-sized blocks obtained at a given level one.

Further, *MSDMP* is used to extract features from each block obtained at $level\_0$ and $level\_1$. Feature vector length for an image using two-level block net is $(64 \times (N + N1))$

3.3 Artificial Neural Network (ANN)

ANN is a very superior and famous classifier used in pattern recognition as well as in medical diagnosis [14, 20, 21, 26]. In this work, we have used two layered feed-forward neural network trained with scaled conjugate gradient backpropagation algorithm for classification of ILD patterns. A number of input layer neurons are equal to a feature dimensionality, whereas hidden layer comprises of 120 neurons. Also, output layer has five neurons indicating five classes of ILD patterns.

4 Result and Discussion

To analyze the performance of proposed method, we have used recall, precision, F-score, and accuracy [18]. In this study, we have carried out two experiments on state-of-the-art database.

4.1 Experiment #1

In this experiment, we have used a publicly available dataset of ILD cases (DB1) [27]. This database has 2062 region of interest (ROI) with pattern label marked by three experienced radiologists. We have considered 109 HRCT sets for ILD classification. Each HRCT scan is of (512 × 512) size having resolution 16 bits/pixel. A brief summary of the dataset used is shown in Table 1. Detailed information about the scanning protocol is: Spacing between the consecutive slices is 10–15 mm, slice thickness is 1–2 mm, and scan time between two successive slices is 1–2 s. We have divided ROI into half overlapping patches having (31 × 31) pixel size.

ROI patches which are having 100% overlap with lung field and 75% overlap with marked ROI were selected for feature extraction. We have used this database for ILD classification. Correspondingly, Table 2 shows confusion matrix of ILD classification system where 70 and 10% ILD patches are used for training and validation, respectively, while 20% of ILD patches are used for testing the accuracy of the classification system. Further, proposed system is evaluated by varying percentage of training and testing sets. Figure 6a illustrates the performance of system framework on DB1, and Fig. 6b shows receiver operating characteristic (ROC) with 80% ILD patches for training and validation whereas 20% for testing. Also, Fig. 7a shows performance of k-NN classifier on DB1, and Fig. 7b shows performance comparison between k-NN and ANN classifier by using F-score as a performance measure. Figure 7b shows that proposed MSDMP in combination with ANN gives better accuracy than k-NN classifier. Table 3 shows that proposed ILD classification scheme outperforms other existing state-of-the-art methods in terms of F-score as well as accuracy.

Table 1 Database information

Tissue category	# Patches generated	# Training	# Testing
Emphysema	516	412	104
Fibrosis	1362	1090	272
Ground glass	684	547	137
Normal	1280	966	242
Micronodule	5262	4210	1052

Table 2 Confusion matrix of ILD classification by proposed method on DB1

DB1	TE	TF	TG	TN	TM
Emphysema (TE)	**459**	9	0	11	6
Fibrosis (TF)	2	**1284**	30	8	25
Ground glass (TG)	0	31	**581**	18	14
Normal (TN)	22	9	46	**1115**	108
Micronodule (TM)	33	29	27	128	**5103**

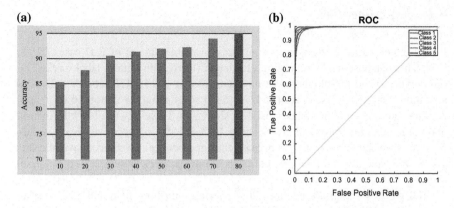

Fig. 6 Performance of the MSDMP on DB1. **a** ARR with different training set (%) and testing set (%); if X% of patches are used for training, then (90-X) % of patches are used for testing. **b** Receiver operating characteristic (ROC) for DB1 with training set (70%), validation set (10%), and testing set (20%)

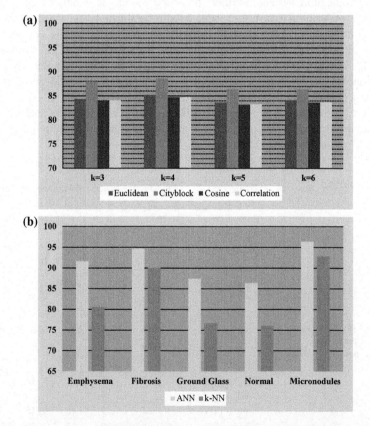

Fig. 7 Performance of the MSDMP on DB1. **a** Performance of k-NN classifier: ARR with different nearest neighbors (k) and using different distance measures. **b** Classwise comparison between k-NN and ANN classifiers with F-score as a performance measure

Table 3 Comparison of the proposed method and state-of-the-art methods

Method	Features	Classifier	F_{avg}	Accuracy
Song [18]	LBP + Intensity histogram + gradient Descriptor	PASA	0.8146	0.8264
Song [22]	Large margin local estimate classification		0.84	0.86
Marios [21]	CNN		0.8547	0.8561
Proposed method	MSDMP + GLRLM	ANN	0.9135	90.44

Fig. 8 Sample images from VIA-ELCAP CT database. From left to right one image from each category

Table 4 Results of (MSDMP + Two-layer block net) in terms of ARR on DB2

Database	Distance metric		
	L1	Euclidean	d1
VIA-ELCAP_CT	99.07	98.43	99.12

4.2 Experiment # 2

In this experiment, we have used VIA/I-ELCAP database [28] which is jointly created by vision and image analysis (VIA) and early lung cancer action program (ELCAP) research groups. This database provides HRCT scans of the lung having a resolution of (512 × 512). To check robustness of proposed MSDMP descriptor, we have used subset of this database in medical image retrieval system. This subset comprised of ten scans. Each scan contains 100 HRCT images. We have collected these cropped HRCT images from Murala et al. [12]. Figure 8 illustrates the sample-cropped HRCT scans from VIA/I-ELCAP database. Result of medical image retrieval system in terms of ARR using combination of proposed MSDMP feature descriptor and two-layer block nets is shown in Table 4.

5 Conclusion

This paper presents a classification scheme for ILD pattern using patch-based approach and ANN classifier. A new feature descriptor, Multi-Scale Directional Mask Pattern (MSDMP), is proposed for feature extraction. MSDMP combines multi-scale information from rotation-invariant Gabor transform and extracts information between the center pixel and directional multi-scale neighborhoods.

Further, the robustness of proposed feature descriptor is tested on VIA-ELCAP_CT with application to medical image retrieval system (MIRS). Also, results are compared with k-NN classifier. Performance evaluation shows that combination of proposed MSDMP descriptor, two-level block nets, and ANN classifier outperforms other existing methods.

References

1. Lung Disease & Respiratory Health Center. http://www.webmd.com/lung.
2. A. H. Mir, M. Hanmandlu, and S. N. Tandon(1995) Texture analysis of CT images. Eng. Med. Biol. Mag. IEEE, vol. 14, no. 6, pp. 781–786.
3. R. Uppaluri, T. Mitsa, M. Sonka, E. A. Hoffman, and G. McLennan(1997) Quantification of pulmonary emphysema from lung computed tomography images. American journal of respiratory and critical care medicine, vol. 156, no. 1, pp. 248–254.
4. Y. Xu, M. Sonka, G. McLennan, J. Guo, and E. A. Huffman (2006) MDCT-based 3-D texture classification of emphysema and early smoking related lung pathologies. IEEE transactions on medical imaging, vol. 25, no. 4, pp. 464–475.
5. I. C. Sluimer, P. F. van Waes, M. A. Viergever, and B. van Ginneken(2003) Computer-aided diagnosis in high resolution CT of the lungs. Medical physics, vol. 30, no. 12, pp. 3081–90.
6. Y. Uchiyama, S. Katsuragawa, H. Abe, J. Shiraishi, F. Li, Q. Li, C.-T. Zhang, K. Suzuki, and K. Doi (2003) Quantitative computerized analysis of diffuse lung disease in high-resolution computed tomography. Medical Physics, vol. 30, no. 9, pp. 2440–54.
7. A. Depeursinge, D. Sage, A. Hidki, A. Platon, P.-A. Poletti, M. Unser, and H. Müller (2007) Lung tissue classification using wavelet frames. In: 29th Annual International Conference of the IEEE. EMBS 2007, pp. 6259–62.
8. A. Depeursinge, D. Ville, P. A., A. Geissbuhler, P. Poletti, and H. Muller (2012) Near-Affine-Invariant Texture Learning for Lung Tissue Analysis Using IsotropicWavelet Frames. IEEE Transactions on Information Technology in Biomedicine, vol. 16, no. 4, pp. 665–675.
9. A. Depeursinge, P. Pad, A. S. Chin, A. N. Leung, D. L. Rubin, H. Muller, and M. Unser (2015) Optimized steerable wavelets for texture analysis of lung tissue in 3-D CT: Classification of usual interstitial pneumonia. In: 12th International Symposium on Biomedical Imaging (ISBI) 2015, pp. 403–6.
10. T. Ojala, M. Pietikäinen, and T. Mäenpää (2002) Multiresolution gray-scale and rotation invariant texture classification with local binary pattern. IEEE Transactions on pattern analysis and machine intelligence, vol. 24, no. 7, pp. 971–87.
11. S. Murala, R. P. Maheshwari, and R. Balasubramanian (2012) Local Tetra Patterns : A New Feature Descriptor for Content-Based Image Retrieval. IEEE Transactions on Image Processing, vol. 21, no. 5, pp. 2874–86.
12. S. Murala and Q. M. J. Wu (2015) Spherical symmetric 3D local ternary patterns for natural, texture and biomedical image indexing and retrieval. Neurocomputing, vol. 149, pp. 1502–14.
13. B. Manjunath and W. Ma (1996) Texture features for browsing and retrivieval of image data. IEEE Transactions on pattern analysis and machine intelligence, vol. 18, no. 8, pp. 837–42.
14. A Dudhane, G Shingadkar, P Sanghavi, B Jankharia and S Talbar (2017) Interstitial Lung Disease Classification Using Feed Forward Neural Networks. In: ICCASP, Advances in Intelligent Systems Research, vol. 137, pp. 515–521.
15. J. Han and K. K. Ma (2007) Rotation-invariant and scale-invariant Gabor features for texture image retrieval. Image and vision computing, vol. 25, no. 9, pp. 1474–81.

16. S. N. Talbar, R. S. Holambe, and T. R. Sontakke (1998) Supervised texture classification using wavelet transform. In 4th international conference on Signal Processing Proceedings ICSP '98, pp. 1177–80.

17. M. Nagao, K. Murase, Y. Yasuhara, and I. Junpei (1998) Quantitative Analysis of Pulmonary Emphysema : Three-DimensionalFractal Analysis of Single-Photon Emission ComputedTomography ImagesObtained with a Carbon ParticleRadioaerosol. American journal of roentgenology, vol. 171, no. 6, pp. 1657–63.

18. Y. Song, W. Cai, Y. Zhou, and D. D. Feng (2013) Feature-based image patch approximation for lung tissue classification. IEEE transactions on medical imaging, vol. 32, no. 4, pp. 797–808.

19. T. Ishida, S. Katsuragawa, K. Ashizawa, H. MacMahon, and K. Doi (1998) Application of artificial neural networks for quantitative analysis of image data in chest radiographs for detection of interstitial lung disease. Journal of digital imaging, vol. 11, no. 4, pp. 182–192.

20. K. Ashizawa, T. Ishida, H. MacMahon, C. J. Vyborny, S. Katsuragawa, and K. Doi (1999) Artificial neural networks in chest radiography: Application to the differential diagnosis of interstitial lung disease. Academic radiology, vol. 6, no. 1, pp. 2–9.

21. M. Anthimopoulos, S. Christodoulidis, L. Ebner, A. Christe, and S. Mougiakaou (2016) Lung Pattern Classification for Interstitial Lung Diseases Using a Deep Convolutional Neural Network. IEEE transactions on medical imaging, vol. 35, no. 5, pp. 1207–1216.

22. Y. Song, W. Cai, H. Huang, Y. Zhou, D. D. Feng, Y. Wang, M. J. Fulham, and M. Chen (2015) Large margin local estimate with applications to medical image classification. IEEE transactions on medical imaging, vol. 34, no. 6, pp. 1362–1377.

23. L. Böröczky, L. Zhao, and K. P. Lee (2006) Feature subset selection for improving the performance of false positive reduction in lung nodule CAD. IEEE Transactions on Information Technology in Biomedicine, vol. 10, no. 3, pp. 504–511.

24. M. M. Galloway (1975) Texture analysis using gray level run lengths. Computer graphics and image processing, vol. 4, no. 2, pp. 172–179.

25. X. Tang (2002) Texture information in run-length matrices. IEEE transactions on image processing, vol. 7, no. 11, pp. 1602–1609.

26. O. Friman, U. Tylén, H. Knutsson, M. Borga, and M. Lundberg (2002) Recognizing emphysema - a neural network approach. In: 16th International Conference on Pattern Recognition, vol. 1, pp. 512–515.

27. A. Depeursinge, A. Vargas, A. Platon, A. Geissbuhler, P. A. Poletti, and H. Müller (2012) Building a reference multimedia database for interstitial lung diseases. Computerized medical imaging and graphics, vol. 36, no. 3, pp. 227–238.

28. VIA/ELCAP CT Lung Image Dataset, available from [online]: https://veet.via.cornell.edu/lungdb.html.

Enhanced Characterness for Text Detection in the Wild

Aarushi Agrawal, Prerana Mukherjee, Siddharth Srivastava
and Brejesh Lall

Abstract Text spotting is an interesting research problem as text may appear at any random place and may occur in various forms. Moreover, ability to detect text opens the horizons for improving many advanced computer vision problems. In this paper, we propose a novel language agnostic text detection method utilizing edge-enhanced maximally stable extremal regions (MSERs) in natural scenes by defining strong characterness measures. We show that a simple combination of characterness cues helps in rejecting the non-text regions. These regions are further fine-tuned for rejecting the non-textual neighbor regions. Comprehensive evaluation of the proposed scheme shows that it provides comparative to better generalization performance to the traditional methods for this task.

Keywords Text detection · HOG · Enhanced MSER · Stroke width

1 Introduction

Text co-occurring in images and videos serves as a warehouse for valuable information for image description and thus assists in providing suitable annotations. Typical practical applications involve extracting street names and numbers, textual indications such as '*diversion ahead*,' etc., from road signs in natural scenes. Such

A. Agrawal (✉)
Department of Electrical Engineering, Indian Institute of Technology Kharagpur,
Kharagpur, India
e-mail: aarushiagrawal1995@gmail.com

P. Mukherjee · S. Srivastava · B. Lall
Department of Electrical Engineering, Indian Institute of Technology Delhi,
Hauz Khas, Delhi, India
e-mail: eez138300@ee.iitd.ac.in

S. Srivastava
e-mail: eez127506@ee.iitd.ac.in

B. Lall
e-mail: brejesh@ee.iitd.ac.in

© Springer Nature Singapore Pte Ltd. 2018   359
B. B. Chaudhuri et al. (eds.), *Proceedings of 2nd International Conference
on Computer Vision & Image Processing*, Advances in Intelligent Systems
and Computing 703, https://doi.org/10.1007/978-981-10-7895-8_28

information can be further stored in geo-tagged databases [1]. Autonomous vehicles are also heavily dependent on efficiency and accuracy of such methods to effectively follow traffic rules. Another area where text detection is applied is indexing and tagging images/videos where text in images helps in better understanding of the content [2]. Performing the above tasks is trivial for humans, but segregating it against a challenging background still remains as a complicated task for machines. Traditional methods for text detection employ the use of blob detection schemes like maximally stable extremal regions (MSERs) [3, 4], edge-based analysis, stroke width transform (SWT) [5, 6], strokelets [7] and features like histogram of oriented gradients (HOGs) [1, 8], Gabor-based features [6], text covariance descriptors [5, 9], and shape descriptors (e.g., Fourier descriptors [10, 11], Zernike moments [12]). The reason behind great popularity of using MSERs and SWT is their $O(n)$ time complexity for performing efficient segmentation which helps in detecting the text regions. MSERs are very effective in detecting the text components, but it is extremely sensitive to noise. So, most of the techniques concentrate on pruning the non-text regions using some heuristics or geometric properties. Despite the advent of deep learning-based techniques [13, 14] which have resulted in tremendous progress in machine-driven text detection, the traditional methods still hold relevance primarily owing to their simplicity and comparable generalization capability to different languages.

Authors in [15] utilize text specific saliency detection measure termed as *characterness*. The authors demonstrate that due to the presence of contrasting objects, saliency alone cannot be an effective indicator of textual region. They overcome this limitation by introducing saliency cues which accentuate the boundary information in addition to saliency [16]. Deriving motivation from this work, we propose a simple combination of various characterness cues for generating candidate bounding boxes for text regions. We use these characterness cues (HOG, stroke width variance, pyramid histogram of oriented gradients (PHOGs)) to refine the blobs generated by edge-enhanced MSERs (eMSERs) [15] for generating text candidates. This is followed by rejection of non-text regions by incorporating difference of entropy as a discriminating factor. The last step is the refinement step, where we combine the smaller blobs into one single text region by concatenating blobs with similar stroke width variance and characterness cue distribution. As per the above discussion, the key contributions of the paper are listed below:

1. We develop a language agnostic text identification framework using text candidates obtained from edge-based MSERs and combination of various characterness cues. This is followed by a entropy-assisted non-text region rejection strategy. Finally, the blobs are refined by combining regions with similar stroke width variance and distribution of characterness cues in respective regions.
2. We provide comprehensive evaluation on popular text datasets against recent text detection techniques and show that the proposed technique provides equivalent or better results.

Organization of the paper is as follows: The proposed methodology is discussed in Sect. 2. The experimental results and discussions are detailed in Sect. 3. Finally, the conclusion is provided in Sect. 4.

Fig. 1 Workflow of the proposed methodology

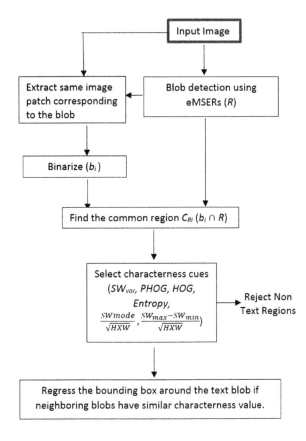

2 Proposed Methodology

The workflow of the proposed method is shown in Fig. 1. In the following subsections, we describe in detail the components of the proposed method.

2.1 Text Candidate Generation Using eMSERs

We begin by generating initial set of text candidates using edge-enhanced maximally stable extremal regions (eMSERs) approach [15]. MSER is a method for blob detection which extracts the covariant regions in the image. It is based on aggregation of regions which have similar intensity values at various thresholds which makes it a suitable candidate for detecting regions with text in images. It efficiently detects the characters in case of distinctive boundaries but fails in the presence of blur. In order to handle this, eMSERS are computed over the gradient amplitude-based image. It divides the image into two sets of regions: dark and bright; dark regions are those

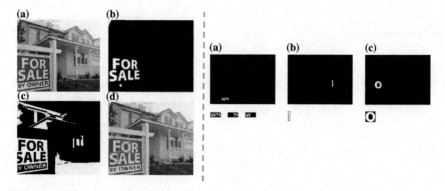

Fig. 2 Left column: **a** original image **b** bright regions **c** dark regions **d** after processing on these regions final set of blobs detected by eMSERs; right column: top row: **a–c** eMSER region; bottom row: **a** binarized region obtained from original image **b** binarized region neglected due to size constraints **c** binarized image—refined object (alphabet) obtained with less disturbance which gives us better results

which have lower intensity than their surroundings and vice versa. Initially, non-text regions are rejected based on geometric properties like aspect ratio, number of pixels, and skeleton length followed by connected component analysis for combining the text regions. Figure 2 shows instances of bright and dark regions formed during text candidate generation using eMSERs. As can be observed, in the bright regions the color of the text is lighter as compared to dark background (red) while in the dark regions the dark text was highlighted against the light colored background.

2.2 Elimination of Non-text Regions

The regions are further refined based on the property that text usually appears on a surrounding having a distinctive intensity. Utilizing this property, we refine textual regions while reject non-textual regions. To achieve this, we find corresponding image patches for the blobs identified by eMSERs. As the image patches contain spurious data along with the information in the form of text, we perform binarization over these image patches using Otsu's threshold [17] for that region and obtain a common region C_{Ri} between the binarized image patch b_i and the blob obtained by eMSER R (where $b_i \cap R > 90\%$) for image i. A blob is rejected, if it is not contained in the binarized image patch. Figure 2 shows some examples of this rejection strategy. We then define various characterness cues [15] for common regions C_{Ri}. Apart from stroke width and HOG used in [15], we check the values of pyramid histogram of oriented gradients (PHOGs) features and entropy for the blobs. During the experiments, we found that PHOG is a good measure of similarity over HOG. In case of alphabets, i.e., textual regions, we observe that the HOG and PHOG values for C_{Ri} are very less. We now briefly explain these cues.

1. Stroke width variance: A stroke is effectively a continuous band of same width in an image. Stroke width transform (SWT) [18] is defined as a local operator which gives the most likely stroke for every pixel in the image. In SWT, all the pixels are initialized with infinity as their stroke width. A Canny-based edgemap is then calculated followed by calculation of gradient direction for all the edge pixels. If the gradient direction (g_p) of an edge pixel p is opposite to the gradient direction (g_q) of next edge pixel q, then the distance between p and q is the stroke width, else the ray tracing p and q is discarded. The pixels having similar stroke widths are grouped using connected component analysis. The letter candidates are chosen after some postprocessing based on the stroke width variance and aspect ratio. The letter candidates are grouped to give text regions. The idea is to segregate text from other high-frequency content that might be present in the scene, e.g., trees branches etc. We perform a bottom-up aggregation by merging pixels with similar stroke widths into connected components which allows in detecting characters across wide range of scales. It is able to identify near-horizontal text candidates. Stroke width of a region (r) is defined as [15]

$$SW(r) = \frac{SW_{var}(l)}{Mean(l)^2} \tag{1}$$

where l defines the shortest path between every pixel p in the skeletal image of region (r) to the boundary of the region, SW_{var} is the stroke width variance, and *Mean* gives the stroke width mean. We utilize the stroke width variance only which should be less for text candidates. We also store the values of stroke width as $\frac{SW_{mode}}{\sqrt{HXW}}$ and $\frac{SW_{max}-SW_{min}}{\sqrt{HXW}}$ (stroke width deviation) where H and W denote height and width of the common region, respectively.

2. HOG and PHOG: PHOG consists of a histogram of orientation gradients over every subregion in the image for every resolution level. The HOG vectors computed over each pyramid in the grid cells are concatenated. As compared to HOG, PHOG is more efficient. HOG is invariant to geometric and photometric transformations. In addition to this, PHOG helps in providing a spatial layout for the local shape of the image. Therefore, we utilize their combination as a characterness cue.

3. Entropy: We calculate the entropy as Shannon's entropy for the common regions ($b_i \cap R$) given as

$$H = -\sum_{i=1}^{N-1} p_i log(p_i) \tag{2}$$

where N denotes the number of gray levels and p_i refers to the probability associated to the gray level i. In information theory, entropy is the measure of average information of a signal given its probability distribution. Higher entropy indicates higher disorder. In our scenario, text candidates show lower variation in color values; thus, typically there is a dominating color in histogram having one sharp peak. However, for non-character candidates, its color values span the histogram

Fig. 3 **a** Smaller regions in the blobs detected by eMSERs **b** final result after postprocessing

as result of color variation. This corresponds to the entropy of the text candidates yielding smaller values than that of the non-text candidates and hence acts as an important cue in distinguishing among them and rejecting non-text candidates as described in the next section.

2.3 Bounding Box Refinement

The remaining set of regions are refined by calculating a set of parameters as stroke width distribution, pretrained characterness cue distribution, and stroke width difference. We define a characterness cue distribution by computing the characterness cue values on ICDAR 2013 dataset. Additionally, we use this distribution to combine the neighboring candidate regions and aggregate them into one larger text region. We recompute the neighbors if they have similar distribution and reject otherwise. Finally, we combine all the neighboring regions into a single text candidate. Figure 3 shows the results of this postprocessing step.

3 Experimental Results and Discussions

3.1 Experimental Setup and Datasets

The experiments were performed on a 32 GB RAM machine with Xeon 1650 processor and 1GB NVIDIA graphics card. MATLAB 2015b was used as the programming platform. The datasets used for evaluation of the proposed methodology are publicly available text datasets: MSRATD500 [19] and KAIST [20]. MSRATD500 consists of 500 images (indoor and outdoor scenes). The standard size of image varies between 1296×864 and 1920×1280. It consists of scenes capturing sign-boards with text in Chinese, English, and mixed. The diversity and complex background in the images make the dataset challenging. The KAIST scene text dataset consists of 3000 images captured in different environmental settings (indoor and outdoor) with varying lighting conditions. The images are of size 640×480. It consists of scenes with English, Korean, and mixed texts. The majority of scenes are of shop and street numbers.

3.2 Evaluation Methodology

Metrics. The proposed technique is evaluated with precision, recall, and F-measure metrics on the chosen datasets. The input for computing these metrics is Intersection over Union (IoU) score, given as

$$IoU = \frac{|S_1 \cap S_2|}{|S_1 \cup S_2|} \tag{3}$$

where S_1 indicates the set of white pixels inside the blobs detected by our strategy before the elimination step (smaller individual blobs), S_2 indicates the set of white pixels inside the ground truth region, and $|\cdot|$ is the cardinality. The performance metrics in this paper are reported on blobs with majority of region being text, i.e., having IoU > 0.5.

Training and Testing. We perform training on ICDAR 2013 [21] dataset while the test set consists of MSRATD and KAIST datasets. This is unlike earlier methods where, in general, the training and testing samples are drawn from the same dataset. Moreover, such a setting makes the evaluation potentially challenging as well as allows us to evaluate the generalization ability of various techniques. The results on Characterness [15] and Blob Detection [22] methods with training and testing sets as described earlier are reported using the publicly available source code.

3.3 Results

Qualitative Results. Figure 4 shows qualitative results on a few example images from MSRATD and KAIST datasets. It can be observed that the images obtained after region refinement demonstrate better localization of textual regions while those on MSRATD dataset (Fig. 4(i)) show tighter localization as compared to other techniques. One of the aims of the proposed technique is to reduce false positives, which can be observed from the second row of Fig. 4(i) where the proposed method provides a tight bounding box on text regions while there are false positives with other techniques except Characterness. The signboard in the image does not consist of any

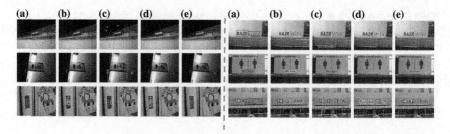

Fig. 4 Results on (i) MSRATD and (ii) KAIST dataset images: **a** ground truth, **b** Characterness [15], **c** Blob Detection [22], **d** proposed approach (before refinement step), and **e** proposed approach

Table 1 Performance measures on MSRATD dataset

Method/metric	Precision	Recall	F-measure
Proposed	0.85	0.33	0.46
Characterness [15]	0.53	0.25	0.31
Blob detection [22]	0.80	0.47	0.55
Epshtein et al. [18]	0.25	0.25	0.25
Chen et al. [23]	0.05	0.05	0.05
TD-ICDAR [19]	0.53	0.52	0.50
Gomez et al. [24]	0.58	0.54	0.56

text data still the contemporary methods detect it as a text candidate. This could be due the fact that the signboard consists of a rounded sketch which may correspond to alphabets such as 'O', 'Q'. Since the proposed technique strictly encodes the stroke width variance along with other characterness cues, we are able to avoid detection of such false candidates. Similar findings are observed for the KAIST dataset as well.

Quantitative Results. Tables 1 and 2 show empirical results on MSRA and KAIST scene datasets, respectively. From the empirical results, it can be seen that on MSRATD dataset, the proposed method achieves significantly higher precision and F-measure as compared to Characterness while having a 28% (precision) and 64% (F-measure) gain and a slightly lower (~6%) recall rate with blob detection. The proposed technique outperforms the compared methods on precision while performs close in terms of F-measure and recall. It is important to note here that the proposed technique does not involve any explicit training allowing the technique to be directly extensible to domains such as symbol identification, road sign identification. On KAIST dataset, the proposed method consistently outperforms characterness on all benchmarks with average improvement of 36%, 17%, and 29% in precision, recall, and F-measure, respectively. The proposed technique also achieves better precision as compared to blob detection. The results show that the proposed method is able to generalize better on a test set while being trained on an entirely distinctive character set. For completeness in comparison, we also provide performance of other techniques on KAIST dataset. However, it should be noted that the objective of these

Table 2 Performance measures on KAIST dataset

Method/metric	Precision	Recall	F-measure
KAIST-English			
Proposed	0.8485	0.3299	0.4562
Characterness	0.5299	0.2476	0.3136
Blob detection	0.8047	0.4716	0.5547
KAIST-Korean			
Proposed	0.9545	0.3556	0.4994
Characterness	0.7263	0.3209	0.4083
Blob detection [22]	0.9091	0.5141	0.6269
KAIST-Mixed			
Proposed	0.9702	0.3362	0.4838
Characterness	0.8345	0.3043	0.4053
Blob detection	0.9218	0.4826	0.5985
KAIST-All			
Proposed	0.9244	0.3407	0.4798
Characterness [15]	0.6969	0.2910	0.3757
Blob detection [22]	0.8785	0.4894	0.5933
Gomaz et al. [24]	0.66	0.78	0.71
Lee et al. [20]	0.69	0.60	0.64

techniques is generally to maximize text detection specifically for a script or to attain script independence with curated training examples with the mixture of scripts to be detected. This possibly makes the comparison with proposed technique tougher as the objective is to obtain better generalization ability.

4 Conclusion

This paper proposed an effective text detection scheme by utilizing stronger characterness measure. A postprocessing step is used to reject the non-textual blobs and combine smaller blobs obtained by eMSERs into one larger region. The effectiveness of the proposed scheme has been analyzed with precision, recall, and F-measure evaluation measures showing that the proposed scheme performs better than the traditional text detection schemes.

References

1. Minetto, R., Thome, N., Cord, M., Leite, N.J., Stolfi, J.: Snoopertext: A text detection system for automatic indexing of urban scenes. Computer Vision and Image Understanding 122, 92–104 (2014)
2. Ye, Q., Doermann, D.: Text detection and recognition in imagery: A survey. IEEE transactions on pattern analysis and machine intelligence 37(7), 1480–1500 (2015)
3. Chen, H., Tsai, S.S., Schroth, G., Chen, D.M., Grzeszczuk, R., Girod, B.: Robust text detection in natural images with edge-enhanced maximally stable extremal regions. In: 2011 18th IEEE International Conference on Image Processing. pp. 2609–2612. IEEE (2011)
4. Neumann, L., Matas, J.: Real-time scene text localization and recognition. In: Computer Vision and Pattern Recognition (CVPR), 2012 IEEE Conference on. pp. 3538–3545. IEEE (2012)
5. Huang, W., Lin, Z., Yang, J., Wang, J.: Text localization in natural images using stroke feature transform and text covariance descriptors. In: Proceedings of the IEEE International Conference on Computer Vision. pp. 1241–1248 (2013)
6. Yi, C., Tian, Y.: Localizing text in scene images by boundary clustering, stroke segmentation, and string fragment classification. IEEE Transactions on Image Processing 21(9), 4256–4268 (2012)
7. Yao, C., Bai, X., Shi, B., Liu, W.: Strokelets: A learned multi-scale representation for scene text recognition. In: Proceedings of the IEEE Conference on Computer Vision and Pattern Recognition. pp. 4042–4049 (2014)
8. Hanif, S.M., Prevost, L.: Text detection and localization in complex scene images using constrained adaboost algorithm. In: 2009 10th International Conference on Document Analysis and Recognition. pp. 1–5. IEEE (2009)
9. Sivic, J., Zisserman, A.: Video google: A text retrieval approach to object matching in videos. In: Computer Vision, 2003. Proceedings. Nineth IEEE International Conference on. pp. 1470–1477. IEEE (2003)
10. De, S., Stanley, R.J., Cheng, B., Antani, S., Long, R., Thoma, G.: Automated text detection and recognition in annotated biomedical publication images. International Journal of Healthcare Information Systems and Informatics (IJHISI) 9(2), 34–63 (2014)
11. Fabrizio, J., Marcotegui, B., Cord, M.: Text detection in street level images. Pattern Analysis and Applications 16(4), 519–533 (2013)
12. Kan, C., Srinath, M.D.: Invariant character recognition with zernike and orthogonal fourier-mellin moments. Pattern recognition 35(1), 143–154 (2002)
13. He, T., Huang, W., Qiao, Y., Yao, J.: Text-attentional convolutional neural network for scene text detection. IEEE Transactions on Image Processing 25(6), 2529–2541 (2016)
14. Jaderberg, M., Simonyan, K., Vedaldi, A., Zisserman, A.: Reading text in the wild with convolutional neural networks. International Journal of Computer Vision 116(1), 1–20 (2016)
15. Li, Y., Jia, W., Shen, C., van den Hengel, A.: Characterness: An indicator of text in the wild. IEEE Transactions on Image Processing 23(4), 1666–1677 (2014)
16. Mukherjee, P., Lall, B., Shah, A.: Saliency map based improved segmentation. In: Image Processing (ICIP), 2015 IEEE International Conference on. pp. 1290–1294. IEEE (2015)
17. Otsu, N.: A threshold selection method from gray-level histograms. Automatica 11(285-296), 23–27 (1975)
18. Epshtein, B., Ofek, E., Wexler, Y.: Detecting text in natural scenes with stroke width transform. In: Computer Vision and Pattern Recognition (CVPR), 2010 IEEE Conference on. pp. 2963–2970. IEEE (2010)
19. Yao, C., Bai, X., Liu, W., Ma, Y., Tu, Z.: Detecting texts of arbitrary orientations in natural images. In: Computer Vision and Pattern Recognition (CVPR), 2012 IEEE Conference on. pp. 1083–1090. IEEE (2012)
20. Lee, S., Cho, M.S., Jung, K., Kim, J.H.: Scene text extraction with edge constraint and text collinearity. In: Pattern Recognition (ICPR), 2010 20th International Conference on. pp. 3983–3986. IEEE (2010)

21. Karatzas, D., Shafait, F., Uchida, S., Iwamura, M., i Bigorda, L.G., Mestre, S.R., Mas, J., Mota, D.F., Almazan, J.A., de las Heras, L.P.: Icdar 2013 robust reading competition. In: Document Analysis and Recognition (ICDAR), 2013 12th International Conference on. pp. 1484–1493. IEEE (2013)
22. Jahangiri, M., Petrou, M.: An attention model for extracting components that merit identification. In: 2009 16th IEEE International Conference on Image Processing (ICIP). pp. 965–968. IEEE (2009)
23. Chen, X., Yuille, A.L.: Detecting and reading text in natural scenes. In: Computer Vision and Pattern Recognition, 2004. CVPR 2004. Proceedings of the 2004 IEEE Computer Society Conference on. vol. 2, pp. II–II. IEEE (2004)
24. Gomez, L., Karatzas, D.: Multi-script text extraction from natural scenes. In: Document Analysis and Recognition (ICDAR), 2013 12th International Conference on. pp. 467–471. IEEE (2013)

Denoising of Volumetric MR Image Using Low-Rank Approximation on Tensor SVD Framework

Hawazin S. Khaleel, Sameera V. Mohd Sagheer, M. Baburaj
and Sudhish N. George

Abstract In this paper, we focus on denoising of additively corrupted volumetric magnetic resonance (MR) images for improved clinical diagnosis and further processing. We have considered three dimensional MR images as third-order *tensors*. MR image denoising is solved as a low-rank tensor approximation problem, where the non-local similarity and correlation existing in volumetric MR images are exploited. The corrupted images are divided into 3D patches and similar patches form a group matrix. The group matrices exhibit low-rank property and is decomposed with *tensor* singular value decomposition (t-SVD) technique, and reweighted iterative thresholding is performed on core coefficients for removing the noise. The proposed method is compared with the state-of-the-art methods and has shown improved performance.

Keywords MR image · Denoising · Tensor singular value decomposition
Low-rank approximation

1 Introduction

Magnetic resonance imaging (MRI) is a widely used non-invasive technique based on nuclear magnetic resonance (NMR) phenomenon, in which the resonance property of atoms is utilized to produce high-resolution images of interior parts

H. S. Khaleel (✉) · S. V. Mohd Sagheer · M. Baburaj · S. N. George
Department of Electronics and Communication Engineering,
National Institute of Technology Calicut, Kerala 673601, India
e-mail: hawazin.khaleel@gmail.com

S. V. Mohd Sagheer
e-mail: sameeravm@gmail.com

M. Baburaj
e-mail: baburajmadathil@gmail.com

S. N. George
e-mail: sudhish@nitc.ac.in

© Springer Nature Singapore Pte Ltd. 2018
B. B. Chaudhuri et al. (eds.), *Proceedings of 2nd International Conference on Computer Vision & Image Processing*, Advances in Intelligent Systems and Computing 703, https://doi.org/10.1007/978-981-10-7895-8_29

of human body for clinical diagnosis of different diseases. In image processing perspective, MR imaging system involves forward and inverse Fourier transform [1].

However, MR images are perturbed by serious random noise due to limitations in the scanning times, hardware of MR imaging system, etc., which degrades the quality of images. This adversely affect the further processing of data which includes the proper diagnosis of diseases. The term noise in the context of MRI may refer to any physiological distortions, thermal noise from the subject, or electronic noise while obtaining the signal [2]. The MR images are acquired as complex data in the Fourier space/k-space consisting of real and imaginary components. Each of the quadrants is assumed to be interfered with additive Gaussian noise of equal variance [3]. Thus, the magnitude of image M is given by

$$M = \sqrt{(M_R + n_R)^2 + (M_I + n_I)^2} \tag{1}$$

where M_R and M_I are the real and imaginary components of complex MR image, respectively, and n_R and n_I are the additive noise corrupting both the quadrants.

The state-of-the-art methods for MR image denoising can be widely categorized as filtering approaches, transform domain techniques, statistical methods, algorithms based on sparsity, and self similarity, low-rank approximation, etc. [4]. In [5], Henkelman et al. introduced averaging, spatial, and temporal filtering for MR image denoising. As spatial filters caused blurring, anisotropic diffusion filter and its variants were proposed in [6, 7]. Later, Hasanzadeh et al. [8] solved denoising as linear minimum mean square error (LMMSE) estimation. Since this method failed to exploit data redundancy in the 3D MR data, non-local LMMSE estimation was proposed in [9]. A set of algorithms based on non-local means filter (NLM) [10] were proposed in [11–14], which produced the state-of-the-art results. Later, in [15], Coupe et al. approached MRI denoising based on principal component analysis (PCA). Another effective non-local denoising method is the well-known BM3D [16]. In this technique, similar patches are grouped into 3D data arrays, followed by shrinking/filtering in the 3D transform domain. In [17, 18], BM3D was extended to volumetric data, popular as BM4D technique. However, both BM4D and BM3D methods make use of orthogonal transforms and hence cannot adapt to varying image contents [19].

The low-rank structure of volumetric image/video is explored in many signal processing applications. In [20], the MRI denoising problem was addressed using sparse and low-rank matrix decomposition method. This method exploited rank deficiency of multichannel coil images and sparsity of artifacts. Nguyen et al. [21] addressed MR spectroscopic image (MRSI) denoising with low-rank approximation techniques that exploit low structure of MRSI data due to linear predictability and partial separability.

In [22], 2D natural image denoising was modeled as weighted nuclear norm minimization (WNNM) problem. Here the image model is $P = M + N$, M is the clean image of size $n_1 \times n_2$, and N is the additive noise, and the minimization problem can be modeled as

$$\hat{M} = \min_{M} \frac{1}{\sigma^2} \|P - M\|_F^2 + \lambda \|M\|_{w,*} \tag{2}$$

where $\|M\|_{w,*}$ is the weighted nuclear norm and is given as $\|M\|_{w,*} = \sum_i |w_i \sigma_i(M)|_1$ where $w = [w_1, w_2 \ldots w_n]$ is the weight assigned to ith singular value $\sigma_i(M)$ and $n = \min(n_1, n_2)$. Recently, many tensor-based low-rank approximation were proposed in the context of multi-frame image denoising. Since volumetric MR image can be treated as a multi-frame image [23], tensor-based approaches can be used for denoising the image. Dong et al. [24] proposed low-rank tensor approximation (LRTA) framework with Laplacian scale mixture (LSM) modeling which gave better results for multi-spectral images. Higher-Order singular value decomposition (HOSVD)/TUCKER decomposition and PARAFAC decomposition based tensor denoising techniques have appeared in [19, 25–27]. However, TUCKER and PARAFAC decompositions have several disadvantages [28]. In the case of PARAFAC decomposition calculation of approximation of a tensor for a given fixed rank is numerically unstable. The TUCKER decomposition is a general form to guarantee the existence of an orthogonal decomposition. These traditional methods rely on tensor flattening and hence lack flexibility.

An alternative representation known as *tensor*-singular value decomposition (*t*-SVD) has been proposed in [29] for building approximations to a given tensor. The algorithm for computing *t*-SVD is based on fast Fourier transform and hence is more efficient compared to the computation of full HOSVD [29]. In this paper, we propose denoising technique for 3D MRI in *t*-SVD framework, exploiting the non-local self similarity in an MR image. We considered the process of denoising of additively corrupted MR images for improving the accuracy of clinical diagnosis and further processing.

The organization of this paper is as follows. Section 2 overviews the basic definitions and notations used in this paper and the third-order *tensor* singular value decomposition utilized in the proposed algorithm. In Sect. 3, detailed explanation of proposed denoising method for MR images is given. Section 4 explains experimental analysis of the algorithm. In Sect. 5, we conclude the paper along with discussion of future works.

2 Basic Theory

2.1 Tensor Definitions and Notations

The term *tensor* denotes an n-dimensional array of elements. For example, a third-order tensor can be thought as a "cube" of data as shown in Fig. 1. *Slices* of a tensor refer to the two-dimensional structure by holding two indices of a third-order tensor. If $\mathcal{P}$ is a third-order tensor, then $\mathcal{P}(n, :, :)$ corresponds to nth horizontal slice, $\mathcal{P}(:, n, :)$ corresponds to nth lateral slice, and $\mathcal{P}(:, :, n)$ corresponds to nth frontal

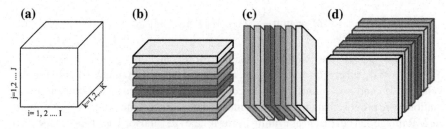

Fig. 1 Slices of third-order tensor [26]: **a** A third-order tensor **b** Horizontal **c** Lateral **d** Frontal slices

slice, which is shown in Fig. 1. $\mathcal{P}(:, m, n)$ is the (m, n)th mode-1 fiber, $\mathcal{P}(m, :, n)$ gives (m, n)th mode-2 fiber, and $\mathcal{P}(m, n, :)$ denotes the (m, n)th mode-3 *fibers*.

A new framework of *t*-SVD was proposed for tensor completion and denoising of multilinear data in some of the recent works [28]. The *t*-SVD representation of tensors has shown promising performance with respect to the tensor approximation problem.

2.1.1 Third-Order *tensor* Singular Value Decomposition (*t*-SVD)

To define *t*-SVD, we need to first understand the notion of *t*-product.

Definition 1 *t*-product: If $\mathcal{M} = \mathcal{N} * \mathcal{P}$ where $\mathcal{N} \in \mathbb{R}^{k_1 \times k_2 \times k_3}$, $\mathcal{P} \in \mathbb{R}^{k_2 \times k_4 \times k_3}$ and $*$ denotes *t*-product, then $\mathcal{M}$ is a tensor of size $k_1 \times k_4 \times k_3$, where the (l, m)th fiber of the tensor $\mathcal{M}$ is given by $\sum_{p=1}^{n_2} \mathcal{N}(l, p, :) * \mathcal{P}(p, m, :)$ for $l = 1, 2, ..., k_1$ and $m = 1, 2, ..., k_4$.

It is to be noted that, in *t*-product, matrix multiplication between elements is replaced by circular convolution of fibers.

Definition 2 *t*-SVD: For $\mathcal{C} \in \mathbb{R}^{(n_1 \times n_1 \times n_3)}$, the t-SVD of $\mathcal{C}$ is given by

$$\mathcal{C} = \mathcal{U} * \mathcal{S} * \mathcal{V}^T \tag{3}$$

where, $\mathcal{U}$ and $\mathcal{V}$ are orthogonal tensors of size $n_1 \times n_1 \times n_3$ and $n_2 \times n_2 \times n_3$ respectively. Here $\mathcal{S}$ is a rectangular tensor with each frontal slices diagonal (i.e. f-diagonal [28]) of size $n_1 \times n_2 \times n_3$, and $*$ denotes the *t*-product.

This decomposition is obtained by finding matrix SVDs of frontal slices in Fourier domain as explained in **Algorithm 1** [28]. An illustration of *t*-SVD decomposition for the 3D case is shown in Fig. 2.

Algorithm 1 Generalized t-SVD computation of Nth order tensor

---

Input: $\mathcal{A} \in \mathbb{R}^{n_1 \times n_2 \times n_3 \ldots \times n_N}$

    $\eta = n_3 n_4 \ldots n_N$

1: **for** $k = 3$ to N **do**

2:    $\mathcal{A}_f \leftarrow$ **fft**$(\mathcal{A}, [\ ], k)$;

3: **end for**

4: **for** $m = 1$ to η **do**

5:    $[\mathbf{U}, \mathbf{S}, \mathbf{V}] = $ **SVD**$(\mathcal{A}_f(:, :, m))$

6:    $\mathcal{U}_f(:, :, m) = \mathbf{U}$; $\mathcal{S}_f(:, :, m) = \mathbf{S}$; $\mathcal{V}_f(:, :, m) = \mathbf{V}$;

7: **end for**

8: **for** $m = 3$ to η **do**

9:    $\mathcal{U} \leftarrow$ **ifft**$(\mathcal{U}_f, [\ ], m)$; $\mathcal{S} \leftarrow$ **ifft**$(\mathcal{S}_f, [\ ], m)$; $\mathcal{V} \leftarrow$ **ifft**$(\mathcal{V}_f, [\ ], m)$;

10: **end for**

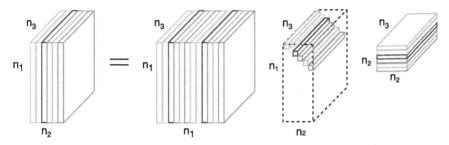

Fig. 2 The t-SVD of $n_1 \times n_2 \times n_3$ tensor [28]

Definition 3 Tensor Nuclear Norm The tensor nuclear norm of tensor $\mathcal{A}^{n_1 \times n_2 \times n_3}$ is defined as follows: $\| A \|_{TNN} = \sum_{k=1}^{n_3} \sum_{i=1}^{min(n_1, n_2)} |\mathcal{S}_f(i, i, k)|$, where $\mathcal{S}_f$ follows the same definition as in **Algorithm 1**.

3 Proposed Method: t-SVD Denoising for MR images

We model the denoising problem as recovering original MR image from its corrupted observations utilizing the low-rank structure of grouped image patches. For deriving the low-rank approximation, the noise degradation model is given by [19],

$$y = x + n \tag{4}$$

where y is the corrupted MR image, x is the original MR image and n is the additive noise. The major steps of proposed method is illustrated in Fig. 3. t-SVD denoising is performed on grouped image patches where the core coefficients of t-SVD decomposition undergo iterative thresholding for the removal of noise. The noisy image y is divided into 3D patches $\{y_j\}_{j=1}^M$ of size $p \times p \times L$. For a given reference patch y_j, K similar patches are found using block matching technique employing Euclidean distance as the similarity metric. The similar cubic patches are stacked into a 3D array, by vectorizing each slice of 3D patch and grouping the similar patches as

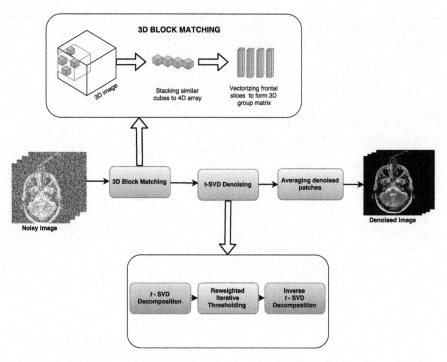

Fig. 3 Illustration of proposed t-SVD denoising

columns of 3D array. Hence group matrix $\{P_j\}_{j=1}^{K}$ is formed. The steps are outlined in **Algorithm 2**. As the images in 3D MR data are correlated, the tensor formed by similar cubic patches will exhibit low-rank property. For a given grouped tensor P_j, the estimate of denoised group $\hat{X}_j$ can be obtained by low-rank approximation (LRA) model [19]. Thus,

$$\hat{X}_j = \min_{X_j} \| P_j - X_j \|_F^2$$

$$\text{s.t. multi-rank}(X_j) \leq \{r_{j,1}, r_{j,2}, r_{j,3}, ..., r_{j,L}\} \tag{5}$$

Algorithm 2 Formation of similar patches

Input: Reference patch y_j, $y_j \in \mathbb{R}^{p \times p \times L}$

Output: 3D group of similar patches P_j, $P_j \in \mathbb{R}^{p^2 \times K \times L}$

1: **for** $j = 1, 2 \ldots, n$ **do**

2: $\quad \mathbb{U}_j = \{y_c \mid S(y_j, y_c) = \| y_j - y_c \|_2^2 < \tau\}$ for $c = 1, 2 \ldots, n$ where $|\mathbb{U}_j| = K$, K is the number of similar patches.

3: $\quad P_j = \{\{y_{k,l}(:)\}_{l=1}^{L}\}_{k=1}^{K}$, where $y_{k,l}(:) \rightarrow$ vectorization of each frontal slice of 3D patch $y_k \in \mathbb{U}_j$.

4: **end for**

The t-SVD decomposition of $X_j = \mathcal{U}_j * \mathcal{S}_j * \mathcal{V}_j^T$ where $*$ denotes the t-product, $\mathcal{U}_j$ and $\mathcal{V}_j$ are orthogonal tensors, and $\mathcal{S}_j$ is a rectangular f-diagonal tensor (i.e., each frontal slice is diagonal) [28]. Thus, the model in Eq. (5) can be expressed as

$$(\hat{\mathcal{U}}_j, \hat{\mathcal{S}}_j, \hat{\mathcal{V}}_j) = \min_{X_j}\{\| \ P_j - \mathcal{U}_j * \mathcal{S}_j * \mathcal{V}_j^T \ \|_F^2 + \tau \ \| \ X_j \ \|_{TNN}\} \qquad (6)$$

where, $\| \ X_j \ \|_{TNN}$ is the tensor nuclear norm (TNN). This model can be solved by iterative thresholding of t-SVD coefficients [19]. Firstly, t-SVD decomposition of each group matrix P_j is computed. The singular values of each group matrix exhibit a decaying pattern. Hence, iterative thresholding is applied on $\mathcal{S}_j$ values to denoise 3D MR images. To improve the flexibility of nuclear minimization problem in Eq. (6), weighted tensor nuclear norm (WTNN) is considered. Thus, problem Eq. (6) becomes

$$(\hat{\mathcal{U}}_j, \hat{\mathcal{S}}_j, \hat{\mathcal{V}}_j) = \min_{X_j}\{\| \ P_j - \mathcal{U}_j * \mathcal{S}_j * \mathcal{V}_j \ \|_F^2 + \tau \ \| \ X_j \ \|_{W,TNN}\} \qquad (7)$$

where $\| \ X_j \ \|_{W,TNN}$ is the weighted tensor nuclear norm. The estimate of denoised group $\hat{P}_j$ is obtained by solving Eq. (7). We decompose P_j in t-SVD domain as $\mathcal{U}_j * \mathcal{S}_j * \mathcal{V}_j^T$ where $\mathcal{U}_j \in \mathbb{R}^{p^2 \times p^2 \times L}$, $\mathcal{V}_j \in \mathbb{R}^{K \times K \times L}$ and $\mathcal{S}_j \in \mathbb{R}^{p^2 \times K \times L}$. The closed form solution to Eq. (7) is the weighted iterative thresholding of coefficient matrix $\mathcal{S}_j$ in Fourier domain, i.e., $\mathcal{S}_j^{\mathcal{F}}$, which will shrink less the larger values and shrink more the smaller values. $\mathcal{S}_j^{\mathcal{F}}$ is obtained from an intermediate step in the computation of t-SVD, i.e.,

$$[U_i, S_i, V_i] = \text{SVD}(\mathcal{P}_j^{\mathcal{F}}(:,:,i)), \text{ for } i = 1, 2, \ldots, L \text{ where } \mathcal{P}_j^{\mathcal{F}} \leftarrow 3D\text{-}fft(P_j) \quad (8)$$

$$\mathcal{U}_j^{\mathcal{F}}(:,:,i) = U_i ; \ \mathcal{S}_j^{\mathcal{F}}(:,:,i) = S_i ; \ \mathcal{V}_j^{\mathcal{F}}(:,:,i) = V_i, \text{ for } i = 1, 2, \ldots, L \quad (9)$$

The thresholding operation of core tensor $\mathcal{S}_j^{\mathcal{F}}$ is given by,

$$\hat{\mathcal{S}}_j^{\mathcal{F}} = \left[diag(\mathcal{S}_{i,i,k}^{\mathcal{F}} - w_i^k)_+\right]_{k=1}^L \qquad (10)$$

where $w_i^k = \left[\dfrac{C}{|\mathcal{S}_{i,i,k}^{\mathcal{F}}| + \epsilon}\right]_{k=1}^L$, $\epsilon > 0$ and C is empirically set as $2\zeta\sqrt{n}\sigma_w^2$, where σ_w is the updated variance as σ is getting reduced at each iteration and ζ is the tuning factor. The value of σ_w at tth iteration is given by,

$$\sigma_w = \gamma\sqrt{\sigma^2 - \| \ y - \hat{x}^{t-1} \ \|_2^2} \qquad (11)$$

Algorithm 3 t-SVD-based denoising by weighted TNN minimization for additive noise removal

Input: Noisy MR image y
Output: Denoised MR image $\hat{x}$
1: $\hat{x}^0 = y$
2: **for** $t = 1$ to $iter$ **do**
3:    $\hat{x}^{t-1} = \hat{x}^{t-1} + \delta(y - \hat{x}^{t-1})$
4:    $\sigma_w = \gamma\sqrt{\sigma^2 - \parallel y - \hat{x}^{t-1} \parallel_2^2}$, σ is the variance of noise.
5:    $P_j = \textbf{Algorithm 2}(y_j),\ y_j \in \mathbb{R}^{p \times p \times L}$,
        $j = 1, 2, \ldots, n$
6:    **for** $j = 1, 2 \ldots, n$ **do**
7:       $\mathcal{P}_j^{\mathcal{F}} \leftarrow fft(P_j, [\ ], 3)$;
8:       **for** $j = 1$ to L **do**
9:          $[U_j, S_j, V_j] = \text{SVD}(\mathcal{P}_j^{\mathcal{F}}(:,:,j))$
10:         $\mathcal{U}_j^{\mathcal{F}}(:,:,j) = U_j;\ \mathcal{S}_j^{\mathcal{F}}(:,:,j) = S_j;\ \mathcal{V}_j^{\mathcal{F}}(:,:,j) = V_j$;
11:      **end for**

12:      $\hat{\mathcal{S}}_j^{\mathcal{F}} = \left[diag(\mathcal{S}_{i,i,k}^{\mathcal{F}} - w_i^k)_+ \right]_{k=1}^{L}$ where $w_i^k = \left[\dfrac{C}{|\mathcal{S}_{i,i,k}^{\mathcal{F}}| + \epsilon} \right]_{k=1}^{L}$,

$$C = 2\zeta\sqrt{n}\sigma_w^2, \epsilon > 0, \zeta > 0, \text{ for } i = 1, 2, \ldots, min(p^2, K)$$

13:      $\hat{\mathcal{U}}_j \leftarrow ifft(\hat{\mathcal{U}}_j^{\mathcal{F}}, [\ ], 3);\ \mathcal{S} \leftarrow ifft(\hat{\mathcal{S}}_j^{\mathcal{F}}, [\ ], 3);\ \hat{\mathcal{V}}_j \leftarrow ifft(\hat{\mathcal{V}}_j^{\mathcal{F}}, [\ ], 3)$;
14:      Estimate of jth cubic patch $\hat{X}_j = \hat{\mathcal{U}}_j * \hat{\mathcal{S}}_j * \hat{\mathcal{V}}_j^T$.
15:   **end for**
16:   The denoised image $\hat{x}^t = \dfrac{\sum_M w_M \{\hat{X}_j\}_M}{\sum_M w_M}$.
17: **end for**

where σ^2 is the variance of noise, γ is the scaling factor, and $\hat{x}^{t-1}$ is the denoised image from previous iteration. The denoised patches are then weighted averaged to procure the denoised image. A regularization step is included in each iteration i.e.,

$$\hat{x}^t = \hat{x}^t + \delta(y - \hat{x}^t) \tag{12}$$

where δ is the regularization parameter. Thus, detailed information from the input MRI is added with denoised data to avoid losing much information in each iteration. The detailed procedure of the proposed method is described in **Algorithm 3**.

4   Analysis and Discussions

The efficiency of the proposed t-SVD-based denoising algorithm is assessed by conducting simulations on synthetic 3D MRI data from BrainWeb database [30] and Aukland database [31]. BrainWeb data includes noise-free T1 weighted (T1w) and

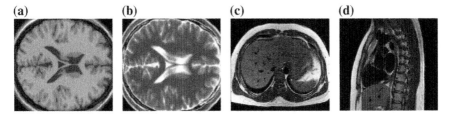

Fig. 4 Test images from BrainWeb and Aukland database; **a** BrainWeb T1w , **b** BrainWeb T2w, **c** Aukland set 1, **d** Aukland set 2

T2-weighted (T2w) data, and Aukland data consists of MR images of cardiac region, shown in Fig. 4. The data were corrupted with simulated additive white Gaussian at different values of noise variance. The variance σ is ranging from 1 to 15% of maximum intensity, i.e., $\sigma = 2.5, 5, 7, 10, 12, 18 \ldots 38$. The performance of denoising method was analyzed quantitatively using the quality metrics—peak signal to noise ratio (PSNR), edge preserving index (EPI), and structural similarity index (SSIM). The parameters of the algorithm were chosen empirically for different noise settings and fixed according to the best results obtained. The patch size was chosen differently for various noise levels and was set as 7×7 for $\sigma \le 20$ and 9×9 for $\sigma > 20$ with a reasonable trade-off between accuracy and speed. The number of similar patches K in each group was chosen in the range [60, 120]. K cannot be too small as few similar patches will be grouped, or dissimilar patches will be grouped if K is too large. The regularization parameter δ controls the quantum of residual image summed to

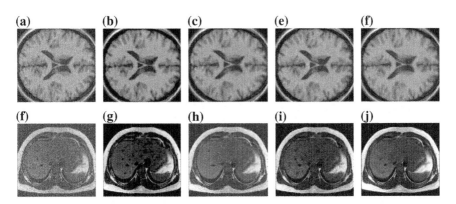

Fig. 5 Denoised data at $\sigma = 28$ corresponding to noise level 11% of maximum intensity; top row: BrainWeb dataset, bottom row: Aukland dataset; **a, f** Noisy image; **b, g** PRINLM [12]; **c, h** HOSVD [19]; **d, i** ANLM [14]; **e, j** Proposed

Table 1 Comparison of PSNR values for BrainWeb and Aukland datasets for σ = 5, 7.5, 10 and 12 corresponding to noise levels 2, 3, 4, and 5% of maximum intensity

Data set	Noise level	PRINLM [12]	HOSVD [19]	ANLM [14]	Proposed	Data set	Noise level	PRINLM [12]	HOSVD [19]	ANLM [14]	Proposed
T1w 1 mm	5	41.92	41.20	39.87	**42.05**	Aukland 1	5	36.26	36.56	34.87	**37.65**
	7.5	39.45	39.27	37.13	**39.85**		7.5	33.90	34.48	33.35	**35.39**
	10	37.59	37.45	35.08	**38.34**		10	32.11	32.54	31.98	**33.86**
	12	36.04	36.49	33.42	**37.20**		12	30.59	31.56	30.77	**32.71**
T2w 5 mm	5	37.61	37.88	36.94	**37.82**	Aukland 2	5	37.40	37.51	35.81	**38.61**
	7.5	35.19	35.59	34.58	**35.19**		7.5	34.67	35.21	33.97	**36.20**
	10	33.55	33.56	32.76	**33.39**		10	32.70	33.84	32.42	**34.62**
	12	32.19	32.42	31.29	**32.02**		12	31.10	32.26	31.10	**33.46**

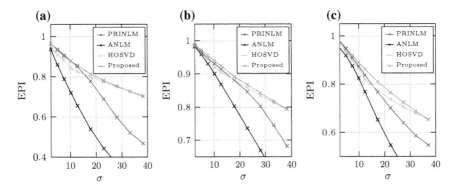

Fig. 6 EPI values for increasing noise variance σ: **a** T1w **b** T2w **c** Aukland set

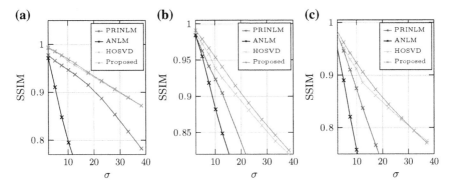

Fig. 7 SSIM values for increasing noise variance σ: **a** T1w **b** T2w **c** Aukland set

the output. The scaling factor γ will control the estimation of noise variance in each iteration. The parameters δ and γ were chosen in the interval $(0, 1)$ and $[0, 1]$, respectively. For all cases, we have chosen tuning factor $\zeta = \sqrt{2K}$. The proposed algorithm is compared with some of the well-known denoising techniques in the literature such as PRINLM [12], ANLM [14] and HOSVD [19]. The results are compared with respect to the performance metrics PSNR, SSIM and EPI for Brainweb and Aukland data. The denoised data of various methods at a noise level of 11% of maximum intensity are shown in Fig. 5. Table 1 shows PSNR obtained for T1w, T2w, PDw Brainweb data and Aukland data at different noise levels. The average PSNR is tabulated by testing four different images over 15 test cases. It can be observed that the proposed method gives highest average PSNR. To verify our method does not degrade MR data in a visual perspective, SSIM and EPI of denoised images were calculated for different noise levels. Figure 6 shows the performance of algorithm with respect to EPI. It can be inferred that, as noise variance increases, the EPI deterioration is slow as required for medical images. SSIM values at different noise settings are shown in Fig. 7. As observed, the proposed method is advantageous with respect to objective metrics and preserves the features as expected.

5 Conclusion and Future Works

In this paper, we have developed a novel MR image denoising technique using low-rank approximation. Here, similar cubic patches were grouped together to form a third-order tensor which is factorized using t-SVD and solved with the low-rank approximation model by weighted thresholding of core coefficients. The results were compared with various state-of-the-art approaches mentioned in the literature. The simulations were performed on test images from BrainWeb and Aukland database under different noise settings. The implementation of algorithms was performed for additively corrupted MR images, and different quality metrics such as PSNR, SSIM, and EPI were compared. As observed, the proposed method has shown quantitative and qualitative improvement in the results. Also, proposed method can be extended for removing rician noise which is common in MR images. The automatic determination of optimal parameters and robustness of algorithm to varying noise levels is warranted in a future study.

References

1. Zhu, H.: Medical image processing overview. University of Calgary (2003)
2. Aja-Fernández, S., Tristán-Vega, A.: A review on statistical noise models for Magnetic Resonance Imaging. LPI, ETSI Telecomunicacion, Universidad de Valladolid, Spain, Tech. Rep (2013)
3. Gudbjartsson, H., Patz, S.: The Rician distribution of noisy MRI data. Magnetic resonance in medicine 34(6), 910–914 (1995)
4. Mohan, J., Krishnaveni, V., Guo, Y.: A survey on the magnetic resonance image denoising methods. Biomedical Signal Processing and Control 9, 56–69 (2014)
5. McVeigh, E., Henkelman, R., Bronskill, M.: Noise and filtration in magnetic resonance imaging. Medical Physics 12(5), 586–591 (1985)
6. Perona, P., Malik, J.: Scale-space and edge detection using anisotropic diffusion. IEEE Transactions on pattern analysis and machine intelligence 12(7), 629–639 (1990)
7. Krissian, K., Aja-Fernández, S.: Noise-driven anisotropic diffusion filtering of MRI. IEEE transactions on image processing 18(10), 2265–2274 (2009)
8. Golshan, H.M., Hasanzadeh, R.P.: An optimized LMMSE based method for 3D MRI denoising. IEEE/ACM Transactions on Computational Biology and Bioinformatics 12(4), 861–870 (2015)
9. Sudeep, P., Palanisamy, P., Kesavadas, C., Rajan, J.: Nonlocal linear minimum mean square error methods for denoising MRI. Biomedical Signal Processing and Control 20, 125–134 (2015)
10. Buades, A., Coll, B., Morel, J.M.: A non-local algorithm for image denoising. In: Computer Vision and Pattern Recognition, 2005. CVPR 2005. IEEE Computer Society Conference on. vol. 2, pp. 60–65. IEEE (2005)
11. Manjón, J.V., Coupé, P., Martí-Bonmatí, L., Collins, D.L., Robles, M.: Adaptive non-local means denoising of MR images with spatially varying noise levels. Journal of Magnetic Resonance Imaging 31(1), 192–203 (2010)
12. Manjón, J.V., Coupé, P., Buades, A., Collins, D.L., Robles, M.: New methods for MRI denoising based on sparseness and self-similarity. Medical image analysis 16(1), 18–27 (2012)
13. Manjón, J.V., Carbonell-Caballero, J., Lull, J.J., García-Martí, G., Martí-Bonmatí, L., Robles, M.: MRI denoising using non-local means. Medical image analysis 12(4), 514–523 (2008)

14. Coupé, P., Manjón, J.V., Robles, M., Collins, D.L.: Adaptive multiresolution non-local means filter for three-dimensional magnetic resonance image denoising. IET Image Processing 6(5), 558–568 (2012)
15. Manjón, J.V., Coupé, P., Buades, A.: MRI noise estimation and denoising using non-local pca. Medical image analysis 22(1), 35–47 (2015)
16. Dabov, K., Foi, A., Katkovnik, V., Egiazarian, K.: Image denoising by sparse 3-D transform-domain collaborative filtering. IEEE Transactions on image processing 16(8), 2080–2095 (2007)
17. Maggioni, M., Katkovnik, V., Egiazarian, K., Foi, A.: Nonlocal transform-domain filter for volumetric data denoising and reconstruction. IEEE transactions on image processing 22(1), 119–133 (2013)
18. Foi, A.: Noise estimation and removal in MR imaging: the variance-stabilization approach. In: Biomedical Imaging: From Nano to Macro, 2011 IEEE International Symposium on. pp. 1809–1814. IEEE (2011)
19. Fu, Y., Dong, W.: 3D magnetic resonance image denoising using low-rank tensor approximation. Neurocomputing 195, 30–39 (2016)
20. Xu, L., Wang, C., Chen, W., Liu, X.: Denoising multi-channel images in parallel MRI by low rank matrix decomposition. IEEE transactions on applied superconductivity 24(5), 1–5 (2014)
21. Nguyen, H.M., Peng, X., Do, M.N., Liang, Z.P.: Denoising MR spectroscopic imaging data with low-rank approximations. IEEE Transactions on Biomedical Engineering 60(1), 78–89 (2013)
22. Gu, S., Zhang, L., Zuo, W., Feng, X.: Weighted nuclear norm minimization with application to image denoising. In: Proceedings of the IEEE Conference on Computer Vision and Pattern Recognition. pp. 2862–2869 (2014)
23. Vannier, M.W., Pilgram, T.K., Speidel, C.M., Neumann, L.R., Rickman, D.L., Schertz, L.D.: Validation of magnetic resonance imaging (MRI) multispectral tissue classification. Computerized Medical Imaging and Graphics 15(4), 217–223 (1991)
24. Dong, W., Li, G., Shi, G., Li, X., Ma, Y.: Low-rank tensor approximation with laplacian scale mixture modeling for multiframe image denoising. In: Proceedings of the IEEE International Conference on Computer Vision. pp. 442–449 (2015)
25. Rajwade, A., Rangarajan, A., Banerjee, A.: Image denoising using the higher order singular value decomposition. IEEE Transactions on Pattern Analysis and Machine Intelligence 35(4), 849–862 (2013)
26. Kolda, T.G., Bader, B.W.: Tensor decompositions and applications. SIAM review 51(3), 455–500 (2009)
27. Zhang, X., Xu, Z., Jia, N., Yang, W., Feng, Q., Chen, W., Feng, Y.: Denoising of 3D magnetic resonance images by using higher-order singular value decomposition. Medical image analysis 19(1), 75–86 (2015)
28. Zhang, Z., Ely, G., Aeron, S., Hao, N., Kilmer, M.: Novel methods for multilinear data completion and de-noising based on tensor-svd. In: Proceedings of the IEEE Conference on Computer Vision and Pattern Recognition. pp. 3842–3849 (2014)
29. Kilmer, M.E., Braman, K., Hao, N., Hoover, R.C.: Third-order tensors as operators on matrices: A theoretical and computational framework with applications in imaging. SIAM Journal on Matrix Analysis and Applications 34(1), 148–172 (2013)
30. BrainWeb: Simulated Brain Database. http://www.bic.mni.mcgill.ca/brainweb/
31. The Auckland Cardiac MRI Atlas. https://atlas.scmr.org/download.html

V-LESS: A Video from Linear Event Summaries

Krishan Kumar, Deepti D. Shrimankar and Navjot Singh

Abstract In this paper, we propose a novel V-LESS technique for generating the event summaries from monocular videos. We employed Linear Discriminant Analysis (LDA) as a machine learning approach. First, we analyze the features of the frames, after breaking the video into the frames. Then these frames are used as input to the model which classifies the frames into active frames and inactive frames using LDA. The clusters are formed with the remaining active frames. Finally, the events are obtained using the key-frames with the assumption that a key-frame is either the centroid or the nearest frame to the centroid of an event. The users can easily opt the number of key-frames without incurring the additional computational overhead. Experimental results on two benchmark datasets show that our model outperforms the state-of-the-art models on *Precision* and *F-measure*. It also successfully abates the video content while holding the interesting information as events. The computational complexity indicates that the V-LESS model meets the requirements for the real-time applications.

Keywords Event · Key-frames · LDA · Video summarization

1 Introduction

In this multimedia era, a rapid growth in the amount of digital video data around the clock is recorded by numerous cameras in the last few years. Thus, a large volume of the video content is rapidly produced by different applications such as video semantic

K. Kumar (✉) · D. D. Shrimankar
Department of Computer Science & Engineering, VNIT, Nagpur, India
e-mail: kkberwal@students.vnit.ac.in

D. D. Shrimankar
e-mail: dshrimankar@cse.vnit.ac.in

K. Kumar · N. Singh
Department of Computer Science & Engineering, NIT, Srinagar, Uttarakhand, India
e-mail: navjot.singh.09@gmail.com

© Springer Nature Singapore Pte Ltd. 2018
B. B. Chaudhuri et al. (eds.), *Proceedings of 2nd International Conference on Computer Vision & Image Processing*, Advances in Intelligent Systems and Computing 703, https://doi.org/10.1007/978-981-10-7895-8_30

annotation and saliency [1], video retrieval and browsing [2]. Therefore, a technique is promptly required, which can summarize the videos in order to help us for storing and accessing the content in minimum time as well as minimum storage space. Video summarization (*VS*) is one of the primary elements for the success of the above applications which can reduce the size of the video adequately and comprehensively which can be divided into two categories: key-frame-based VS and skimming-based VS [3].

Key-frames-based VS mainly keeps the overall content of a video through holding the important frames which is also referred as *Storyboard-based VS*. It was scrutinized by employing various clustering algorithms [4, 5] where different clusters are formed on the basis of the similarity between frames. This type of VS can be further categorized into two groups. (a) *Local perspective storyboard* where the temporal segment of video becomes free from the redundancy by selecting the frames as *key-frames* which are dissimilar from their neighboring frames [6]. However, the performance of such technique may be declined due to consideration of the local perspective only [7]. (b) *Global perspective storyboard* involves the extracted *key-frames* that cover the whole video in order to boost up the performance [8–10]. Moreover, various unsupervised learning approaches [11] have been employed for global representation-based VS. These techniques mainly summarize the video content by structural analysis with multiple features.

On the other hand, *skimming-based VS* often preserve the summaries of a video with interesting events along with their semantics in the form of highlights. This video highlights can be referred as *Event Summarization (ES)*. In this era, *ES* applications are widely used in sports videos, surveillance videos, etc. However, the main issue with *ES* process is to determine the correct integration model. The model should include the identification of the detected event boundaries so that the succinctness of the skim must be verifiable while holding the sufficient events. Although, *ES* has benefited over *storyboard VS* [12]. ES is the ability to include audio and motion elements that potentially enhance both the expressiveness and information in the abstract. On the other hand, *Key-frames* are also useful in abating computational time for various video analysis and retrieval applications. It is very important to observe that although *video skimming and storyboard based VS* is often generated differently, these also can be renovated from one to the other. Moreover, a good *VS* technique is urgently required, which helps us to get utmost video content about the target video sequence in a limited time.

We decided to employ an algorithm (LDA) to provide better classification in comparison to principal component analysis (PCA) [13]. PCA is better in feature classification, but for data classification, LDA does better. Moreover, LDA has the more powerful capability in representing nonlinear patterns than PCA. This is expected to be helpful for exploring the nonlinear structure in video data. In this paper, a novel V-LESS scheme-based on LDA is proposed and realized. By using LDA, the coarse and fine structures in the video are efficiently characterized by the components extracted from the feature space. The application for key-frame extraction and skimming formation is also scrutinized to demonstrate the advantages of the representational forms. Our contributions are in threefold:

- We convert the *VS* problem into the LDA-based data classification problem and also propose a novel V-LESS technique which is robust to extract the potential key-frames and also to remove the unimportant frames from the final summaries of the events.
- We design an integrated framework for both *key-frame* extraction and video skimming in a unified framework which offers the flexibility for practical applications in real time (average computing cost is 141 ms/frame).
- The V-LESS model offers us to select the numbers of *key-frames* without incurring an additional computational cost, in contrast to state of the art, which require presetting the number of *key-frame*.

2 Problem Formulation

Video **V** is divided into N frames by size of $W \times H$, where W and H represent the width and height of a frame, respectively. On an average, the videos are used from two benchmark datasets which comprise about 3000 *RGB* frames. The variation in the visual content of the successive frames is not much, but still, computation in processing these frames requires processing all the three planes of a frame. In order to reduce the computation time, all *N-colored* frames of a video are converted into gray scale frames. Each frame is then resized into the one-dimensional vector as an input vector for the LDA as data is classified into two categories: positive (active) frames and negative (low inactive) frames.

Video frame classification with LDA: The existing supervised approaches [7–9] have been scrutinized to generate the summaries of videos, with an aim of the outputs based on each input vector, where these models learn to produce the required such outputs. We observed that the PCA technique [13] can be employed as the unsupervised approach which is mainly used to obtain the directions (also referred as principal components) that maximize the variance in a dataset. On the other hand, LDA is supervised where it computes the directions (linear discriminants). LDA also represents the axes that maximize the separation between classes. Even, both LDA and PCA are linear transformation methodologies that are typically employed for dimensionality reduction. However, it was observed that LDA is the much better approach for multi-class classification task than PCA where class labels are well known. Moreover, LDA easily deals the case where the frequency within-class is unequal and their performances have been investigated on the randomly generated test dataset.

We got inspired by the above-supervised training techniques and decided to use a supervised approach (LDA). In order to formulate the existing VS problem as frame classification technique, initially, two target classes for a video is built for positive and negative frames independently. Then, LDA is used to search for a linear combination of variables (frames) that best separates the two classes (targets). In addition to this, it maximizes the fraction of between-class variance and the within-class variance for a particular dataset with the guarantee of maximal separability [14]. It uses the beneath Fisher score function to capture the notion of separability:

$$F_{score}(\alpha) = \frac{\alpha^T \mu_1 - \alpha^T \mu_2}{\alpha^T Cv \, \alpha} = \frac{\bar{\gamma}_1 - \bar{\gamma}_2}{Variance \ of \ \gamma \ within \ groups} \tag{1}$$

where μ_1 and μ_2 are the mean vectors of the frames, γ is $\alpha_1 f_1 + \alpha_2 f_2 + \cdots + \alpha_N f_N$. $f_1, f_2, ..., f_{n-1}$ and f_N are the N video frames. Moreover, the each frame is represented in the vector form with d features ($= H \times W$) where H and W are the height and width of frames, respectively. Given the Fisher score function requires to estimate the linear coefficients $\alpha = Cv^{-1}(\mu_1 - \mu_2)$ that maximize the score, where $Cv = \frac{1}{n_1 + n_2}(n_1 Cv_1 + n_2 Cv_2)$ is the pooled covariance matrix, and Cv_1 and Cv_2 are the covariance matrices. Here, we observed that the estimation of the Mahalanobis distance (Md) between two classes may be used to assess the effectiveness of the discrimination between the native and positive frames. Md value greater than 3 indicates that the two averages differ by more than 3 standard deviations. It means that the overlap, i.e., probability of misclassification becomes quite small. Md is defined as:

$$Md = \alpha^T (\mu_1 - \mu_2) \tag{2}$$

Finally, a new frame is classified by projecting it onto the maximally separating direction and classifying it as Cv_1 if:

$$\log \frac{p(cv_1)}{p(cv_2)} < \alpha^T(x - (\frac{\mu_1 + \mu_2}{2})) \ where \ p \ is \ class \ probability \tag{3}$$

Predictors: A simple linear correlation between the Fisher scores and predictors was employed for text which predictors contribute significantly to the discriminant function. Cv, correlation varies from -1 to 1, where -1 and 1 represent the highest contribution, but in opposite directions, and 0 indicates no contribution at all. However, it is very difficult to understand the analogy between the VS problem and an LDA approach directly.

We divide our work into two stages: phase I (Classification) and phase II (Summarization) as shown in Fig. 1. At phase I, a video is divided into two classes positive frames and negative frames (see Algorithm 1) to filter out unimportant frames. The resulting frames of phase I are then processed at phase II (Event summarization), where the Event summarization process is employed to extract the final

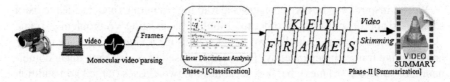

Fig. 1 Major component of the V-LESS model

Algorithm 1 Event summarization using LDA (ELDA)

1: **procedure** $ELDA(\{f_1, f_2, , f_N\})$ N VIDEO FRAMES WITH 1-DIMENSION)
2:      Calculating N-dimensional mean vectors and assuming the number of classes (c)=2.
3:      Calculating scatter matrix within-class $SM_W = \sum\limits_{i=1}^{c} S_i$ Phase I

   where $S_i = \sum\limits_{f \in D_i}^{N} (f - \mu_i)(f - \mu_i)^T$, $\mu_i = \frac{1}{n_i} \sum\limits_{f \in D_i}^{n} f_k$

4:      Calculating scatter matrix between-class $SM_B = \sum\limits_{i=1}^{c} N_i(\mu_i - \mu)(\mu_i - \mu)^T$

   where μ-overall mean, μ_i-sample mean and N_i sizes of the respective classes.
5:      Solving the matrix $\mathbf{A} = SM_W^{-1} SM_B$ as the generalized eigenvalue (λ) problem
         and calculating the v=eigenvector from $\mathbf{A}v = \lambda v$.
6:      Selecting linear discriminants for the new frames:
         • Sorting eigenvectors by decreasing λ value and pick k top eigenvectors i.e.,
         eigenvectors v with the lowest λ value, bear the least information (negative/
         inactive frames) and dropped using equation 1.
         • Filter the positive frames using equation 3.
7:      Form the maximum clusters of the resultant positive frames using equation 2.
8:      Announce the frames close to centroids of the clusters as *key-frames*.   Phase II
9:      **return** E <– final *key-frames* obtained using as discussed in section 2.1.
10: **end procedure**

---

key-frames (see Sect. 2.1). The benefit of the two-phase LDA-based model is very few (positive) frames need to process at phase II which helps us to obtain faster Event summarization.

2.1   Event Summarization: Video Skimming Using Key-Frames

In order to achieve a good ES as video skimming through *key-frames*. We observe that event boundaries play a vital role in the summarization process. It hardly occurs that an event is with a single frame. Thus, a parameter $[b_{min}, b_{max}]$ boundary frame number is estimated for detecting the boundary of an event, where b_{min} and b_{max} indicate the minimum and maximum boundary frame number, respectively. Moreover, the parameter $[b_{min}, b_{max}]$ value may or may not be at equidistant from the centroid of the cluster. The parameter $[b_{min}, b_{max}]$ value is obtained based on the similarity (*Euclidean distance*) between all frames of a cluster and *key-frame* of the cluster. Therefore, a parameter is required to fix the boundaries (b_{min}, b_{max}) of an event. Here, this parameter is also known as Event Boundary Threshold (*EBT*) which is experimentally calculated based on the size (H: Height × W: Width) of a frame, i.e., $EBT\,(\%) = \frac{H \times W \times 90}{100}$. If a frame either on the left-hand side or on the right-hand side of the key-frame number in the video should have 90% match with the key-frame, then the frame will be counted in the event. The value of b_{min} and b_{max} are fixed by the lowest frame number and with the largest frame number which has the 90%

match with key-frame, respectively. Finally, all potential events are detected based on the extracted key-frames. Then, these events are arranged in temporal order to achieve the video skimming. Therefore, the V-LESS model does not only allow the users to access the key-frame-based VS but also allow to access ES without incurring additional cost.

3 Experiments and Discussions

In this section, we discuss the several experiments and assessments to verify the efficacy and competence of our V-LESS model-based *VS* technique. We implemented LDA technique on two benchmark datasets. We selected first dataset of 50 videos from the Open Video Project (*OVP*)[1] which is the one employed in [6, 15, 16]. The second dataset[2] of 50 videos which contains several categories like *"sports, advertisements, TV shows, and home videos"*. The duration of videos varies from 1 to 10 min as reported in paper *VSUMM* [16].

Qualitative Analysis: Human-detected ground truth *key-frames* (ground truth) for both datasets from five users are accessed from VSUMM [16] official Web site. The output summaries of Delaunay Clustering *DT* [6], *STIMO* (*VISTO* approach extension) [15], *OVP* [17], and *VSUMM* [16] and are accessed from the same Web site. The average of the five ground truth set results in *key-frames* (summary) is counted while measuring the quantitative metrics. We compared our proposed LDA-based *VS* approach with *DT* [6], *STIMO* [15], *OVP* [17], *VSUMM* [16], Sparse Dictionary (*SD*)-based approach [18], Keypoint-Based Key-frame Selection (*KBKS*) [19], and Minimum Sparse Reconstruction (*MSR*) [20]. In the *SD* algorithm, the frames are elected as per the importance of curve corresponding to the top 10 local maximums. In the *KBKS* algorithm, the *key-frames* are elected automatically till 85% coverage. *MSR*-based *VS* algorithms selected a *key-frame* when experimental parameter TPOR crosses over 85% in order to perform a fair comparison. Our proposed V-LESS technique extracted the *key-frames* after applying LDA on the video and then removed the negative frames using Eq. 1. Then *key-frames* are declared as the central frame of a cluster after the formation of the different clusters using Eq. 2. A sample *key-frame* for fifth video of first dataset is shown in Fig. 2 with following observations:

- The existing *DT, VSUMM1, VSUMM2, OVP*, and *OnMSR* techniques failed to extract the last *key-frame* of the ground truth where some existing approach extracts some duplicate *key-frames*. However, our proposed *V-LESS* approach extracted better *key-frames* than the existing models without duplicate *key-frames*.

A sample resultant *key-frame* for forty-ninth video of the second dataset is depicted in Fig. 3. Here, we observed that

---

[1]http://www.open-video.org.
[2]https://sites.google.com/site/vsummsite/download.

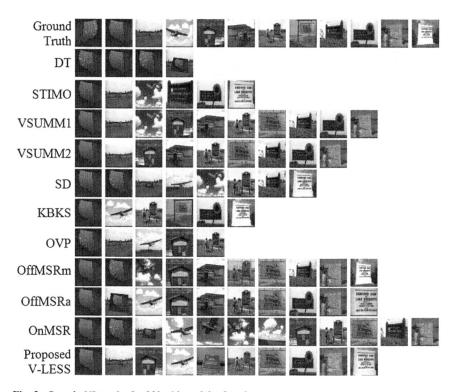

Fig. 2 Sample *VS* results for fifth video of the first dataset

- The existing *VSUMM1, VSUMM2, SD*, and *KBKS* techniques failed to extract more than 60% *key-frame* of the ground truth. However, our proposed *V-LESS* technique extracted most of the *key-frames* as selected in the ground truth without duplicate key-frames.

Quantitative Analysis: The process of extracting *key-frame* to attain the best *VS* had been intensively investigated, but there is no standard or best approach to assess their performance. Even, *key-frame* extraction by different techniques is equated with all the user summaries (*ground truth summaries*) by employing three assessment metrics. $Precision = \frac{TP}{TP+FP}$, $Recall = \frac{TP}{TP+FN}$, and F-measure $(F_\beta) = \frac{(1+\beta^2) \times Precision \times Recall}{\beta^2 \cdot Precision + Recall}$ metrics are usually estimated to gain the quantitative calculation with marked ground truth in frame unit of the following definitions:

- *True Positive (TP)*: an algorithm selects a frame and the frame is a part of an important event.
- *False Positive (FP)*: an algorithm selects a frame, but the frame is not a part of an important event.
- *True Negative (TN)*: an algorithm does not select a frame and the frame is not a part of an important event.

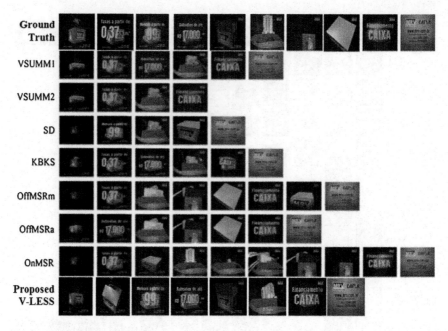

Fig. 3 Sample *VS* results for forty-ninth video of the second dataset

- *False Negative (FN)*: an algorithm does not select a frame, but the frame is a part of an important event.

Moreover, high *Precision* represents a larger fraction of the unimportant frames, which are removed by the algorithm. High *Recall* represents a larger fraction of the important frames, which are returned by the algorithm. In all, *F-measure* can be interpreted as the weighted average of *Precision* and *Recall*. For a user who considers equal priorities to both *Precision* and *Recall*, $\beta = 1$, i.e., F_1 becomes the Harmonic mean of *Precision* and *Recall*. It is also referred as the balanced *F-score*. Consequently, the maximum value of *F-measure* means more accurate technique. We compared our resultant *key-frames* with the ground truth summary and $\delta = \frac{H \times W}{d}$ value is used to compare the similarity (Euclidean distance) between extracted *key-frame* and ground truth summary form. Here, H and W are height and width of a frame, respectively, and $d = 10$ is fixed experimentally.

If similarity is greater than or equal to δ only, then we count such *key-frame* as a matched frame in the quantitative analysis for calculation of *Precision* and *Recall*. The quantitative performance of the V-LESS model is compared with state-of-the-art models for both datasets using $\beta = 1$ in Tables 1 and 2 where the best results are shown in bold. From Tables 1 and 2, it is observed that,

- Our proposed LDA-based summarization technique gains relatively a better performance as compared to other models.

Table 1 Performance of the model with state-of-the-art models for 1st dataset

Algorithm	Precision (%)	Recall (%)	F-measure (%)
DT [6]	47.0	50.0	48.5
STIMO [15]	39.0	65.0	48.8
OVP [17]	43.0	64.0	51.4
VSUMM1 [16]	42.0	77.0	54.4
VSUMM2 [16]	48.0	63.0	54.5
SD [18]	40.0	61.0	48.3
KBKS [19]	31.0	**89.0**	46.0
OffMSRm [20]	58.0	58.0	58.0
OffMSRa [20]	60.0	57.0	58.5
OnMSR [20]	50.0	66.0	56.9
Proposed V-LESS	**68.9**	61.6	**65.0**

Table 2 Performance of the model with state-of-the-art models for 2nd dataset

Algorithm	Precision (%)	Recall (%)	F-measure (%)
VSUMM1 [16]	38.0	**72.0**	49.7
VSUMM2 [16]	44.0	54.0	48.5
SD [18]	37.0	53.0	43.6
KBKS [19]	37.0	60.0	45.5
OffMSRm [20]	52.0	45.0	48.2
OffMSRa [20]	54.0	47.0	50.2
OnMSR [20]	47.0	54.0	50.2
Proposed V-LESS	**54.6**	51.8	**53.2**

- *Precision* of our proposed *V-LESS* technique on both datasets is maximum among all other models.
- *Recall* of our proposed *V-LESS* technique on both datasets is better than the most of the models except *STIMO, VSUMM, KBKS, OVP*, and *OnMSR*.
- *F-measure* of our V-LESS model is also maximum among all the other models which mean our proposed summarization model attains better performance than existing models.

Computational Complexity: For the experiments, the frame size of the videos from the Open Video Project *OVP* [17] is *352 × 240*. With a standard 3.0 GHz dual-core desktop computer, for a video shot of 300 frames (i.e., 10 s). In order to abate the computational cost, we utilized the LDA-based VS technique in two phases. Roughly, 35 s is required with taking account the time on feature extraction at phase I and less than 7.5 s at phase II. Hence, the time cost is not more than 42.5 s, which is much less than the *KBKS* [19]-based technique taking about 150 s (about 3.5 times faster). Therefore, the proposed *V-LESS* technique (i.e., 0.14 s per frame) can be used

Table 3 Computational time comparison

Algorithm	Sampling rate (frame/s)	Total time (s)
DT [6]	30	92.5
VSUMM [16]	30	71.8
SD [18]	24–30	90
KBKS [19]	30	150
MSR [20]	5	72.3
Proposed V-LESS	30	**42.4**

for the real-time applications like videos for entertainment, educational purposes, surveillance and security systems. The computational time of our V-LESS model for a video shot of 300 frames is compared with the existing models in Table 3.

4 Conclusion

In this paper, we have suggested a machine learning, supervised LDA based summarization V-LESS model for producing the concisely and intelligently video abstraction. This model mainly facilitated for the users to access the huge volumes of video content in an effective and efficient manner which is based on extraction of *keyframes*. So that number of frames in the summary should be minimized as possible. The experimental results on two benchmark datasets with various types of videos demonstrate that the V-LESS technique outperforms the state-of-the-art models with the best *Precision* and *F-measure*. Moreover, computational cost indicates that our approach can be used in real-time applications.

References

1. Singh N., et al., "A Novel Position Prior Using Fusion of Rule of Thirds and Image Center for Salient Object Detection," *MTAP* (2016), pp. 1–18.
2. Vermaak J., et al., "Rapid summarization and browsing of video sequences," *BMVC*, (2002), pp. 1–10.
3. Krishan K., et al., "Event BAGGING: A novel event summarization approach in multi-view surveillance videos," IEEE IESC'17.
4. Truong B.T., Venkatesh S., "Video abstraction: a systematic review and classification," *ACM Trans. Multimed. Comp. Comm. App.*, 3, 1, (2007), 37 pages.
5. Chowdhury A. S., et al., "Video Storyboard Design using Delaunay Graphs," *Int. Conf. on Pattern Recog.*, (2012), pp. 3108–3111.
6. Mundur P., Rao Y., Yesha Y., "Keyframe-based video summarization using Delaunay clustering," *Int. J. Digit. Libr.*, 6, 2, (2006), pp. 219–232.
7. Kumar K., et al., "Equal Partition based Clustering approach for Event Summarization in Videos," *SITIS*, 2016, pp. 119–126.

8. Chang H.S., et al., "Efficient video indexing scheme for content-based retrieval," *IEEE TCSVT*, 9, 8, (1999), pp. 1269–1279.
9. Gong Y., Liu X., "Video summarization using singular value decomposition," *IEEE CVPR*, 2, (2000), pp. 174–180.
10. K. Kumar, et al., "Eratosthenes sieve based key-frame extraction technique for event summarization in videos," *MTAP* (2017), pp. 1–22.
11. Zhuang Y., et al., "Adaptive key frame extraction using unsupervised clustering," *IEEE ICIP*, 1, (1998), pp. 866–870.
12. K. Kumar, et al., "SOMES: An efficient SOM technique for Event Summarization in multiview surveillance videos," Springer ICACNI'17.
13. Sahouria, E., et al., "Content analysis of video using principal components." *IEEE TCSVT*, 9, 8, (1999), pp. 1290–1298.
14. Altman, E. I., et al. "Corporate distress diagnosis: Comparisons using linear discriminant analysis and neural networks (the Italian experience).", *Journal of banking & finance*, 18, 3, (1994), pp. 505–529.
15. Furini M., et al., "Stimo: still and moving video storyboard for the web scenario," *Multimed. Tools Appl.* 46, 1, (2010), pp. 47–69.
16. Avila S., et al., "Vsumm: a mechanism designed to produce static video summaries and a novel evaluation method," *Pattern Recognit. Lett.*, 32, 1, (2011), pp. 56–68.
17. OVP, "Video Open Project Storyboard," https://open-video.org/results.php?size=extralarge, Retrieved December, 2016.
18. Cong Y., et al., "Towards scalable summarization of consumer videos via sparse dictionary selection," *IEEE TMM*, 14, 1, (2012), pp. 66–75.
19. Guan G., et al., "Keypoint based keyframe selection," *IEEE Trans. Circuits Syst. Video Tech.*, 23, 4, (2013), pp. 729–734.
20. Mei S., et al., "Video summarization via minimum sparse reconstruction," *Pattern Recog.*, 48, 2, (2015), pp. 522–533.

Action Recognition from Optical Flow Visualizations

Arpan Gupta and M. Sakthi Balan

Abstract Optical flow is an important computer vision technique used for motion estimation, object tracking and activity recognition. In this paper, we study the effectiveness of the optical flow feature in recognizing simple actions by using only their RGB visualizations as input to a deep neural network. Feeding only the optical flow visualizations, instead of the raw multimedia content, ensures that only a single motion feature is used as a classification criterion. Here, we deal with human action recognition as a multi-class classification problem. In order to categorize an action, we train an AlexNet-like Convolutional Neural Network (CNN) on Farneback optical flow visualization features of the action videos. We have chosen the KTH data set, which contains six types of action videos, namely walking, running, boxing, jogging, hand-clapping and hand-waving. The accuracy obtained on the test set is **84.72%**, and it is naturally less than the state of the art since only a single motion feature is used for classification, but it is high enough to show the effectiveness of optical flow visualization as a good distinguishing criterion for action recognition. The AlexNet-like CNN was trained in Caffe on two NVIDIA Quadro K4200 GPU cards, while the Farneback optical flow features were calculated using OpenCV library.

Keywords Optical flow · Convolutional Neural Networks · KTH data set
Action recognition

A. Gupta (✉) · M. S. Balan
Department of Computer Science and Engineering, The LNM Institute
of Information Technology, Jaipur, India
e-mail: arpan@lnmiit.ac.in
URL: http://lnmiit.ac.in/

M. S. Balan
e-mail: sakthi.balan@lnmiit.ac.in
URL: http://sakthibalan.in/

1   Introduction

Activity recognition in videos is an important area of research. Some of its applications are as follows: automated recognition in a video surveillance system, which ensures diligent operation as against an actual human (security guard) monitoring the videos, which can be tiresome. Automated anomaly detection in a crowded scene and traffic monitoring are also important applications of the same. It also has applications in elderly care and patient monitoring, though contact sensor-based gadgets and vital measuring instruments are preferred over the video feeds for activity recognition purposes. Moreover, the huge amount of video data available with the online or offline service companies needs to be efficiently and automatically indexed, searched and annotated. These multimedia content-based analysis tasks require scalable algorithms that can be applied to a wide range of videos.

Researchers have been developing new and improved feature representation techniques for multimedia data so as to ensure accurate and reliable recognition of entities and actions from images and videos. Some of these so-called handcrafted features include Scale Invariant Feature Transform (SIFT), Histogram of Gradients (HOG), Motion History Images (MHI), GIST, etc. These features represent some spatial or temporal information of the multimedia content in a lower dimensional space that can be directly fed to a computationally feasible machine learning model for object/action recognition. A comprehensive survey of motion analysis, making use of such types of features, is presented in [3].

Lately, deep learning networks, such as Convolution Neural Network (CNN), and Recurrent Neural Network (RNN) have been proved to be better at generalized recognition and prediction tasks. They automatically extract relevant features from the raw high-dimensional data by using repeated convolution filters. The ImageNet Large Scale Visual Recognition Challenge (ILSVRC) [21] has attracted the attention of vision researchers from all over the world. ILSVRC provides the largest annotated image data set for a variety of analysis tasks like recognition, classification, object detection and object localization. The winners of this competition have successfully used modified versions of CNNs and improved upon the accuracy year after year.

The CNNs have, mostly, been applied to image processing tasks, where an image is independent of other images and they have only spatial dimensions. CNNs help in automatic extraction of features that are relevant for the classification task. There have not been many attempts of applying CNNs on videos, which adds a temporal dimension to the input data. The temporal features need to be considered for accurate human action recognition from a sequence of video frames.

Optical flow is one such important motion feature which can be used to recognize human actions. Humans are able to identify motion with ease. They can also predict the direction of future motion. A running person will tend to move forward, frame after frame. For example, in Fig. 2f, the person will, most likely, move to the right, in the subsequent frame. Thus, its corresponding optical flow will denote the person moving towards right. With the optical flow features, one can, mostly, identify simple actions like walking, boxing, hand-waving, etc. The examples given in Fig. 2 denote a single action frame and its corresponding optical flow visualization.

A question arises that if only the optical flow motion feature is taken into consideration, will this feature alone be enough to identify simple actions. If so, then how to develop the classification pipeline for this multi-class classification problem. In our work, we try to answer this question by developing a simple multi-class classification pipeline, which uses only the optical flow visualization feature for the task of action recognition. To do so, we train a CNN only on the optical flow visualizations of consecutive frame pairs of the action videos and use the trained model to categorize a test video by a simple voting mechanism. The same approach, due to its simplicity, can be applied for other spatio-temporal feature visualizations to compare their effectiveness on a classification task. The CNN automatically extracts the relevant features from the visualizations, like it does for the large image data sets. The only difference is, instead of extracting some general set of latent features, it extracts a set of latent features representing some motion encoded in the form of optical flow.

Section 2 covers the literature review. Section 3 describes our proposed approach for recognition of actions using only the optical flow visualization features. The KTH human action data set and the CNN architecture descriptions are given in Sects. 4.1 and 4.2, respectively. The results and the related discussion are presented, thereafter, in Sect. 5, which is followed by conclusion and references. Lastly, the Appendix contains some sample optical flow visualizations obtained from the training set.

2 Literature Review

Activity recognition from videos has been discussed in detail in [2, 24]. Optical flow, proposed by Horn and Schunck [12], has applications in video compression, segmentation, analysis and stabilization. If flow vectors are computed for a small subset of the spatial coordinates, then optical flow is said to be sparse, whereas if flow vectors are computed for all the locations, then it is known as dense optical flow. A number of methods have been proposed to calculate sparse and dense optical flow fields [6, 7, 9, 18]. Evaluations of various optical flow algorithms are provided on the KITTI[1] data set [10] and on the Middlebury[2] data set [4].

The authors of [23] have trained two separate CNNs for recognition of actions from videos. One CNN is used to learn spatial information in frames, while the other CNN learns the motion information across frames. In order to learn the motion information, the second CNN is trained on the optical flow displacement fields. Together, they are able to give state-of-the-art accuracy on UCF-101 and HMDB-51 data sets.

In [13], a 3D CNN for human action recognition has been developed by extending the 2D CNN. This helps in extracting features from the spatial as well as the temporal domain, thereby capturing motion information, which is then fed to a

---

[1]http://www.cvlibs.net/datasets/kitti/eval_stereo_flow.php?benchmark=stereo.

[2]http://vision.middlebury.edu/flow/eval/results/results-e1.php.

classifier for recognizing human actions. The authors applied their 3D CNN model on the TRECVID and KTH data sets.

Fischer et al. in [8] have developed FlowNet, a CNN which is able to predict future optical flow by taking only two image frames as input. Their model was learnt on a synthetic data set which they created by superimposing the images of 3D chairs on background images taken from Flickr. Further, they modified their CNN to include a "correlation" layer. Their model was tested on KITTI, MPI-Sintel, Middlebury and their own Flying Chairs data sets.

The problem of large displacements in optical flow has been tackled in [26]. The authors of [26] have developed DeepFlow, a technique involving variational approach and matching algorithm. The matching algorithm proposed by them is, specifically, for optical flow features, where they can efficiently recognize fast actions, denoting large displacements. They have trained a six-layer CNN for the same.

Mahbub et al. [19] have also performed action classification on the KTH and Weizmann data sets using Lucas–Kanade optical flow [18] and random sample consensus (RANSAC) methods. They obtain a low dimensional feature vector representation of motion with localization and apply a Euclidean-based model and SVM for classification.

A project in the Stanford by Pol Rosello [20] trained AlexNet-like CNNs for prediction of future optical flow vectors after applying spline interpolation on the network predictions.

The above methods mainly used some spatio-temporal feature along with traditional machine learning model for action recognition, or they have modified deep neural networks to make them learn from temporal frame sequences. A simple approach that has not been tried is making a deep neural network learn to recognize actions by training it on the image visualizations of motion features. This removes the burden of coming up with a modified version of CNN which takes 3D video input, with increased set of parameters. The standard CNN architectures can, thus, be effectively used.

In our work, we recognize human actions based on only the optical flow visualization[3] features. These visualizations encode the magnitude and direction information by mapping the 2D flow field vectors to the RGB colour space.

3   Methodology

We hypothesize that optical flow field visualizations are sufficient to categorize simple actions. Here, simple actions can be defined as having a (nearly) static background with uniform motion of single/similar foreground object(s). The optical flow field visualizations are mapping of 2D flow vectors (u, v), which are calculated for spatial locations, to HSV colour space, where "H" channel corresponds to direction, "V" channel corresponds to magnitude, and taking maximum saturation ("S") level.

---

[3]http://docs.opencv.org/3.2.0/d7/d8b/tutorial_py_lucas_kanade.html.

The mapping is done by a simple min–max normalization. The HSV pixel is, subsequently, mapped to RGB image space. This optical flow visualization is dark at spatial coordinates where there is no motion, and is bright where motion is involved. A drawback is that, in some frame pairs, noise is present due to camera jitter and, therefore, the visualization appears bright even for the portion of the frame denoting the background (refer Fig. 3). Also, as optical flow is not considering the entire history of motion and takes into account only two frames of the same video δt time apart, a single visualization would not provide enough information for the entire action video. For representing previously occurred motion in a video, with a single visualization, Motion History Image(MHI) [1] can be used.

A 2D flow field (u, v) denotes velocity of a pixel in a frame at time t to a frame at time $t + \delta t$, taking into consideration some assumptions, such as brightness constancy. Here, δt has to be small enough to ensure that pixel motions are tractable between frames, else the optical flow generated will be noisy. Therefore, a smaller value of δt has lower noise content. We consider $\delta t = 1$, i.e. only consecutive frame pairs, for creation of visualization features. More number of visualizations can be created by estimating an upper bound for δt. But care should be taken as increasing δt might make "jogging" look like "running". That is, similar types of actions might get mixed up.

These raw visualization features can be directly fed to a CNN. Training a CNN requires a large number of examples due to the huge amount of parameters involved. Creation of visualization from videos in the above manner can generate a large number of training samples, due to the many sequential frames available in video data.

A CNN trained in this manner will be able to make predictions on single visualization images. As such, we will get $N - 1$ action label predictions for a sequence of N frames of a video, by taking optical flow of only the consecutive frame pairs ($\delta t = 1$). Deciding on a single action label can be done by taking the action which gets the highest number of votes.

4 Experimentation

Our training and evaluation procedures are summarized in the following steps:

1. Calculation of Farneback dense optical flow [7] values from two consecutive frames of all training set videos. This is done only for the frame sequences where the action is occurring. The action markers, provided with the KTH data set, help in identification of action sequences in a video.
2. Conversion of optical flow matrix to HSV and then to BGR visualization.
3. Conversion of visualizations to LMDB with random key values. It ensures that the mini-batches of training samples, consisting of a contiguous set of samples, are not biased.
4. Training an AlexNet-like CNN (details in Sect. 4.2).
5. Evaluation of trained CNN on the validation set for determining the threshold of summed-up pixel values of background subtraction mask (*BGThreshold*) [27].

This helps in identifying frames of no action and, hence, these frames are not used for prediction.

6. Using the trained CNN and *BGThreshold* value of 110000 to make predictions for all visualizations (with $\delta t = 1$) of test set videos, except those frames where summed-up background subtraction mask is less than *BGThreshold*.
7. Getting the resultant label of each test set action video by taking the action with the highest number of votes.

4.1 Data set Description

KTH action recognition data set [22] is a well-known benchmark data set for the task of recognizing simple human actions. The data set has 6 actions namely boxing, hand-clapping, hand-waving, jogging, running and walking, which are performed by 25 different persons in varying conditions. Actions are performed in indoor and outdoor setting with static homogeneous background, having scale/lighting variations, and different clothes. Each video is of 25 fps frame rate, duration of around 15–30 s, and contains one action performed by a single person. The frames are down-sampled to 120×160 (H $\times$ W). The outdoor actions generally have more noise (sometimes due to zooming) than the indoor actions. Running, jogging and walking have individuals coming in from one side of the frame and leaving from the other side, as such, there are time sequences with no action, when the person is out of the frame. A text annotation of the frame markings, where the actions are occurring, has been provided along with the data set.

We create the optical flow field visualization data set from the raw action videos as explained in Sect. 3 by taking temporally consecutive occurring frame pairs. Thus, for an N frame sequence where an action occurs, we get $(N - 1)$ flow visualization images. The training set is formed taking into consideration the action sequence frame markings. This would reduce the noise in the training set, but will also reduce the number of examples on which to train the network. The total number of visualizations was 74016 in the training set (for 191 action videos, 1 action video is missing in the original data set). The validation and test sets have 192 and 216 videos, respectively.

The validation and test sets do not take into consideration the action sequence frame markings provided in annotation file. Therefore, the number of frame pairs for test set is 102312 (for 216 action videos). These include sequences of video where no action is occurring. We employ background subtraction to detect frames of no action. The mask values will be low for frames where only the background is visible, and will be high if a foreground object appears in the frame. We choose a threshold of 110000 for summed-up mask values by checking for a range of values on the validation set. Fig. 1b shows percentage accuracy on the validation set for different values of *BGThreshold*.

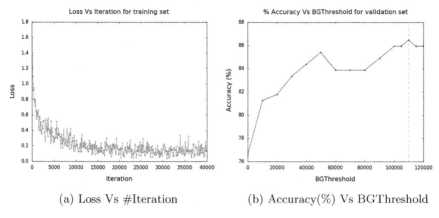

(a) Loss Vs #Iteration   (b) Accuracy(%) Vs BGThreshold

Fig. 1 AlexNet-like CNN. Train batch size = 64, epochs = 35

Some sample optical flow visualizations from the training set are shown in the Appendix. Figures 2 and 3 illustrate identifiable and noisy visualizations, respectively. In Fig. 2c, the hand motions are in separate directions, thus having different colours. A person walking, running or jogging in any direction cannot be easily distinguished by the visualization alone. Figure 3d, e has some bright colours even in the absence of the person. Usually, outdoor actions have more noise as compared to the indoor actions.

4.2 CNN Architecture Description

We used Caffe [14] for training an AlexNet-like [15] CNN and OpenCV [5] for finding dense Farneback optical flow features [7]. The BGR optical flow feature visualizations were converted into LMDB format before being fed to the CNN. The LMDB format is a standard format for improving the training performance of a CNN in Caffe.

Scaling of the pixel values was done at the input layer of the CNN, by dividing by 255, and mirroring was used as a data augmentation technique. The weights were initialized by Xavier initialization [11], with constant bias. The details of the CNN layers are provided in Table 1. Each convolution layer (conv) and fully connected layer (FC) was followed by a rectified linear unit layer (ReLU). The three FC layers had 128, 128 and 6 neurons, respectively. FC6 and FC7 used a dropout ratio of 0.5 at the time of training. The six neurons correspond to the six action labels. Final layer computed the softmax loss.

Table 1 AlexNet-like CNN architecture. Batch size = 64; input scaled by 1/255. Total parameters = 1771456

Layer	#Filters	Width	Stride	Pad	Output	#Parameters
Input	–	–	–	–	$3 \times 120 \times 160$	0
conv1	64	7	1	0	$64 \times 114 \times 154$	$64 \times 7 \times 7 \times 3$
MaxPool	–	2	2	0	$64 \times 57 \times 77$	0
conv2	128	3	2	0	$128 \times 28 \times 38$	$128 \times 3 \times 3 \times 64$
MaxPool	–	2	2	0	$128 \times 14 \times 19$	0
conv3	192	3	1	0	$192 \times 12 \times 17$	$192 \times 3 \times 3 \times 128$
conv4	128	3	1	0	$128 \times 10 \times 15$	$128 \times 3 \times 3 \times 192$
conv5	128	3	1	0	$128 \times 8 \times 13$	$128 \times 3 \times 3 \times 128$
MaxPool	–	3	1	0	$128 \times 6 \times 11$	0
FC6	–	–	–	–	128	$(128 \times 6 \times 11) \times 128$
dropout	–	–	–	–	Ratio = 0.5	0
FC7	–	–	–	–	128	128×128
dropout	–	–	–	–	Ratio = 0.5	0
FC8	–	–	–	–	6	128×6

We used Stochastic Gradient Descent with 40000 iterations for training the network, with base learning rate(α) = 0.01, momentum(μ) = 0.9, and learning rate is dropped by a factor of 10 ($\gamma = 0.1$) after every 10000 iterations.

The network was trained on two NVIDIA Quadro K4200 GPU cards.

5   Results and Discussion

After training the CNN for 40000 iterations and setting *BGThreshold* = 110000, the final accuracy on the test set was calculated in the same way as done for the validation set. This came out to be 84.72%.

Figure 1a illustrates the progress of loss during training of the CNN.

The data used is only the optical flow visualization data, which is quite often not identifiable, even by humans. As is clear from Fig. 2, the optical flows of running and jogging are very similar. Boxing and running involve background noise, probably, due to the outdoor scenario and camera jitter. As such, there will be noise in the BGR colour space mapping. Owing to these constraints, the CNN tends to misclassify such cases. "Running" is often misclassified and appears to be "jogging", while other actions have reasonable accuracy values.

The classification scores for individual classes on the test set are shown in Table 2. Boxing and hand-waving are easy to identify and, thus, have high accuracy scores. Jogging has high accuracy but most of the running samples are misclassified as jogging. Thus, it may be concluded that actions having similar motions, either slow or

Table 2 Evaluation on test set and comparison. Total 216 videos, out of which 183 are classified correctly (i.e. 84.72%). Rows are predictions, and columns are ground truth labels

	Ground Truth					
	Boxing	Hand-clapping	Hand-waving	Jogging	Running	Walking
Boxing	35	2	0	0	0	1
Hand-clapping	0	29	2	0	0	0
Hand-waving	0	5	34	0	0	0
Jogging	0	0	0	35	18	1
Running	1	0	0	1	18	2
Walking	0	0	0	0	0	32
Accuracy (%)	97.22	80.56	94.44	97.22	50.0	88.89
Mahbub et al. [19] (%)	83.784	86.496	86.486	82.432	83.784	83.784

fast, will need some extra information for classification, other than the optical flow information.

Our results are not among the best-reported results on this data set. The main reason being that we used only a single short-term motion feature for classification. Laptev et al. [16] have reported 91.8% accuracy. Similarly, Liu and Shah [17] have used 2D Gabor filter-based detection and reported an accuracy of 94.2%, but they both follow a leave-one-out-cross-validation approach and, hence, cannot be compared to our results. A broad study of the different local spatio-temporal features for action recognition is provided in [25].

6   Conclusion and Future Scope

In this work, we have trained an AlexNet-like CNN on dense optical flow visualization features of videos for the purpose of action recognition. The problem is treated as a multi-class classification problem, where the classes correspond to the types of human actions. We use the famous KTH human action data set which has six types of action videos. The training is performed using the optical flow visualizations of 191 action videos of training set, using frames where action is occurring. We obtained an accuracy of 84.72% on the test set, which is lower than the state of the art on this data set, due to the consideration of a single motion feature.

Our focus has been the simplicity of the method rather than improving the accuracy, as only one motion feature with consecutive frame pair optical flow is used. Variations of the CNN can be tried to improve the accuracy up to some extent, which is representative of the information available with the raw features. A CNN tends to automate the feature extraction process but those features would not be relevant if the data itself is not representative of the classes. Here, we explore the effectiveness of using only the optical flow visualizations to distinguish different actions.

In future, we will be trying to explore the effect of other such motion features like MHI and build a comparative study of the different features with respect to the activity recognition task. A simple method to improve the accuracy may also be tried, by using the multi-stream CNN architectures that learn from different high-dimensional subspaces and try to optimize the final score by combining the losses obtained from the different streams.

Appendix: Sample Optical Flow Visualizations

See Figs. 2 and 3.

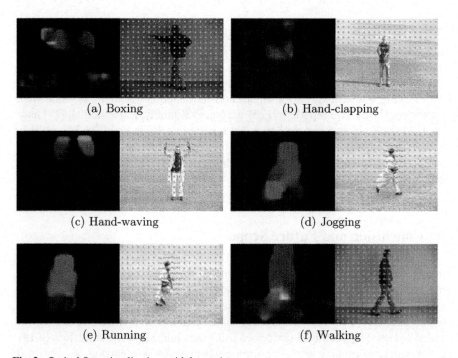

(a) Boxing	(b) Hand-clapping
(c) Hand-waving	(d) Jogging
(e) Running	(f) Walking

Fig. 2 Optical flow visualizations with less noise

(a) Boxing   (b) Hand-clapping

(c) Hand-waving   (d) Jogging

(e) Running   (f) Walking

Fig. 3 Optical flow visualizations with high noise and ambiguity

References

1. Aaron F. Bobick. Action Recognition Using Temporal Templates. *Journal of Chemical Information and Modeling*, 53(9):1689–1699, 2013.
2. J K Aggarwal and M S Ryoo. Human activity analysis. *ACM Comput. Surv.*, 43(3):1–43, 2011.
3. J.K. Aggarwal and Q. Cai. Human Motion Analysis: A Review. *Computer Vision and Image Understanding*, 73(3):428–440, 1999.
4. Simon Baker, Daniel Scharstein, J. P. Lewis, Stefan Roth, Michael J. Black, and Richard Szeliski. A database and evaluation methodology for optical flow. *International Journal of Computer Vision*, 92(1):1–31, 2011.
5. G. Bradski. The OpenCV Library. *Dr. Dobb's Journal of Software Tools*, 2000.
6. Thomas Brox, Nils Papenberg, and Joachim Weickert. High Accuracy Optical Flow Estimation Based on a Theory for Warping. *Computer Vision - ECCV 2004*, 4(May):25–36, 2004.
7. Gunnar Farnebäck. Two-frame Motion Estimation Based on Polynomial Expansion. In *Proceedings of the 13th Scandinavian Conference on Image Analysis*, SCIA'03, pages 363–370, Berlin, Heidelberg, 2003. Springer-Verlag.
8. Philipp Fischer, Alexey Dosovitskiy, Eddy Ilg, Philip Häusser, Caner Hazirbas, Vladimir Golkov, Patrick van der Smagt, Daniel Cremers, and Thomas Brox. Flownet: Learning optical flow with convolutional networks. *CoRR*, arXiv:1504.06852, 2015.
9. David Fleet and Yair Weiss. Optical Flow Estimation. *Mathematical models for Computer Vision: The Handbook*, pages 239–257, 2005.
10. Andreas Geiger, Philip Lenz, and Raquel Urtasun. Are we ready for autonomous driving? the KITTI vision benchmark suite. *Proceedings of the IEEE Computer Society Conference on Computer Vision and Pattern Recognition*, pages 3354–3361, 2012.

11. Xavier Glorot and Yoshua Bengio. Understanding the difficulty of training deep feedforward neural networks. *Proceedings of the 13th International Conference on Artificial Intelligence and Statistics (AISTATS)*, 9:249–256, 2010.
12. Berthold Horn and B Schunck. Determining optical flow. *Artificial Intelligence*, 17(1–2):185–203, 1981.
13. Shuiwang Ji, Ming Yang, and Kai Yu. 3D Convolutional Neural Networks for Human Action Recognition. *Pami*, 35(1):221–31, 2013.
14. Yangqing Jia, Evan Shelhamer, Jeff Donahue, Sergey Karayev, Jonathan Long, Ross Girshick, Sergio Guadarrama, and Trevor Darrell. Caffe: Convolutional Architecture for Fast Feature Embedding. *arXiv preprint* arXiv:1408.5093, 2014.
15. Alex Krizhevsky, Ilya Sulskever, and Geoffrey E Hinton. ImageNet Classification with Deep Convolutional Neural Networks. *Advances in Neural Information and Processing Systems (NIPS)*, pages 1–9, 2012.
16. Ivan Laptev, Marcin Marszałek, Cordelia Schmid, and Benjamin Rozenfeld. Learning realistic human actions from movies. *26th IEEE Conference on Computer Vision and Pattern Recognition, CVPR*, 2008.
17. J Liu and M Shah. Learning human action via information maximization. *Conference on Computer Vision and Pattern Recognition*, pages 2971–2978, 2008.
18. BD Lucas and T Kanade. An Iterative Image Registration Technique with an Application to Stereo Vision. *Ijcai*, 130:121–129, 1981.
19. Upal Mahbub, Hafiz Imtiaz, and Md Atiqur Rahman Ahad. An optical flow based approach for action recognition. *14th International Conference on Computer and Information Technology, ICCIT 2011*, (Iccit):646–651, 2011.
20. Pol Rosello. Predicting Future Optical Flow from Static Video Frames. 2016.
21. Olga Russakovsky, Jia Deng, Hao Su, Jonathan Krause, Sanjeev Satheesh, Sean Ma, Zhiheng Huang, Andrej Karpathy, Aditya Khosla, Michael Bernstein, Alexander C. Berg, and Li Fei-Fei. ImageNet Large Scale Visual Recognition Challenge. *International Journal of Computer Vision*, 115(3):211–252, 2015.
22. Christian Schuldt, Ivan Laptev, and Barbara Caputo. Recognizing human actions: A local svm approach. In *Proceedings of the Pattern Recognition, 17th International Conference on (ICPR'04) Volume 3 - Volume 03*, ICPR '04, pages 32–36, Washington, DC, USA, 2004. IEEE Computer Society.
23. Karen Simonyan and Andrew Zisserman. Two-Stream Convolutional Networks for Action Recognition in Videos. *arXiv preprint* arXiv:1406.2199, pages 1–11, 2014.
24. Michalis Vrigkas, Christophoros Nikou, and Ioannis a. Kakadiaris. A Review of Human Activity Recognition Methods. *Frontiers in Robotics and AI*, 2(November):1–28, nov 2015.
25. Heng Wang, Muhammad Muneeb Ullah, Alexander Klaser, Ivan Laptev, and Cordelia Schmid. Evaluation of local spatio-temporal features for action recognition. *BMVC 2009 - British Machine Vision Conference*, pages 124.1–124.11, 2009.
26. Philippe Weinzaepfel, Jerome Revaud, Zaid Harchaoui, and Cordelia Schmid. DeepFlow: Large displacement optical flow with deep matching. *Proceedings of the IEEE International Conference on Computer Vision*, (Section 2):1385–1392, 2013.
27. Zoran Zivkovic. Improved adaptive Gaussian mixture model for background subtraction. *Proceedings of the 17th International Conference on Pattern Recognition*, 2(2):28–31 Vol. 2, 2004.

Human Activity Recognition by Fusion of RGB, Depth, and Skeletal Data

Pushpajit Khaire, Javed Imran and Praveen Kumar

Abstract A significant increase in research of human activity recognition can be seen in recent years due to availability of low-cost RGB-D sensors and advancement of deep learning algorithms. In this paper, we augmented our previous work on human activity recognition (Imran et al., IEEE international conference on advances in computing, communications, and informatics (ICACCI), 2016) [1] by incorporating skeletal data for fusion. Three main approaches are used to fuse skeletal data with RGB, depth data, and the results are compared with each other. A challenging UTD-MHAD activity recognition dataset with intraclass variations, comprising of twenty-seven activities, is used for testing and experimentation. Proposed fusion results in accuracy of 95.38% (nearly 4% improvement over previous work), and it also justifies the fact that recognition improves with an increase in number of evidences in support.

Keywords Convolutional neural networks · Deep learning
Depth motion map · RGB-D sensors · Skeleton · UTD-MHAD
Motion history image and fusion

1 Introduction

Human activity recognition and related research aimed to automatically detect and analyze human activities from videos. It has applications in robotics, surveillance, security, industry automation, and health care among many others. There has been a

P. Khaire (✉) · P. Kumar
Department of Computer Science and Engineering, Visvesvaraya National Institute
of Technology, Nagpur, India
e-mail: pushpjitkhaire@gmail.com

J. Imran
Department of Computer Science and Engineering, Indian Institute of Technology,
Roorkee, India
e-mail: javed.dcs2016@iitr.ac.in

© Springer Nature Singapore Pte Ltd. 2018
B. B. Chaudhuri et al. (eds.), *Proceedings of 2nd International Conference on Computer Vision & Image Processing*, Advances in Intelligent Systems and Computing 703, https://doi.org/10.1007/978-981-10-7895-8_32

considerable increase in research interest in this area, mainly because of two reasons; first, availability of low-cost depth sensors and second, due to return of neural networks in deep form as convolutional neural networks. Success of other deep learning techniques also enhanced research in activity recognition. Availability of cost inexpensive sensors, which captures a sequence of RGB images, depth images, or other modalities such as skeleton, provides a fast and accurate multivariate data for analyzing activities. As explained in [1], there are several advantages of depth cameras as compared to traditional RGB cameras. For example, the output of depth cameras is insensitive to changes in lighting conditions. In addition, the 3D structure and shape information provided by the depth maps makes it easier to deal with problems like segmentation and detection. Depth images are able to provide three-dimensional structure and motion information toward distinguishing different actions [2]. Furthermore, the availability of well-known and diverse RGB-D datasets like MSR Action 3D [3], Berkeley MHAD [4], UTD-MHAD [2], and CAD 60 [5] among others, supports in extensive research for human activity recognition.

There are many works on depth data reported in the literature. For instance, a three-channel deep convolutional neural network with weighted hierarchical depth motion maps, for human action recognition, was proposed in [6]. The 2D spatial structures of weighted hierarchical depth motion maps are converted into pseudo-color images for additional enhancement and assistance in recognition. Three ConvNets are initialized with the models obtained from ImageNet and fine-tuned independently on the color-coded weighted hierarchical depth motion maps constructed in three orthogonal planes. The proposed method achieved better results on most of the individual datasets and corresponding methods at that juncture; similar technique with added modality is presented in [1].

In [7], an effective local spatiotemporal descriptor for action recognition from depth video sequences is discussed. The entire algorithm is carried out in three stages. In the first stage, a depth sequence is divided into temporally overlapping depth segments which are used to generate three depth motion maps (DMMs), capturing the shape and motion cues. A methodology to recognize human action as time series of representative 3D poses was proposed in [8], and the projected method takes 3D skeletal joint locations inferred from depth maps as input and a compact representation of postures named HOJ3D that characterizes human postures as histograms of 3D joint locations within a modified spherical coordinate system. Posture words were built by clustering and training of discrete HMMs to classify sequential postures into action types. A framework for human activity recognition using 3D posture data is discussed in [9]. In order to obtain a suitable representation of the human body, 11 relevant joints were detected and encoded as a relevant set of joints into postures. Thus, since each posture represents a recurrent pattern of joints positions, an activity can be described as a sequence of known postures.

Our previous work in [1] bears similarity with [6] in the sense that pretrained ImageNet model is applied to train depth motion maps (DMMs) in four projected views; Front, Side, Top, and Bottom. We add Motion History Images (MHIs) as a fourth modality in earlier work; MHIs were generated from RGB videos where the intensity of each pixel is a function of the regency of motion in a sequence. Four

CNNs were trained separately corresponding to Front view, Side view, Top view, and MHI. Individual results of four streams were fused separately to produce the final classification score.

A simple and effective recognition technique using skeleton joints is presented in [10]. Activity represented by skeleton joint sequences is converted to posture features by clustering; these features are then trained on multiclass support vector machine for classification. Computation and association of key poses are carried out using a clustering algorithm, without the need of a learning algorithm. However, it is effective only when number of classes is less and does not fair well for complex datasets. It is also ineffective on large classes with intraclass variations among activities constituted by similar gestures as in UTD-MHAD [2].

This work is an improvement of our earlier work done in [1]. Here we add skeletal data as new modality and explore different classification and fusion method with earlier modalities so as to improve the overall recognition rate. We attempted three different methodologies for this purpose whose details are given in the following section.

2 Proposed Classification and Fusion Approaches

As the main focus of the work is to utilize skeletal data from RGB-D sensors for possible fusion with depth data and traditional RGB data, three different methods were explored for combining skeletal modality. First is a simple technique of modified weighted fusion using skeletal variance along orthogonal planes. The variance of movement of skeleton joints along different coordinates is utilized to assign appropriate weights for the fusion of scores from Front, Side, Top, and MHI CNN streams instead of simple averaging of scores. In the second approach, clustering is applied to skeletal data and multiclass SVM is used for classification. This trained SVM model is applied as a fifth stream for final classification. Finally, in the third approach, artificial skeleton images are generated from skeletal data, which are then trained on pretrained VGG-16 CNN for recognition purpose. This newly trained CNN is used as a fifth stream for final classification. Details on each approach are given in following subsections.

2.1 Modified Weighted Fusion Using Skeletal Variance in Orthogonal Planes

Simple averaging of scores from different modalities tends to misclassify, when there are conflicting scores, i.e., classification from two or more streams is different. An attempt is made to handle such conflicts, by using the information from the skeleton stream. First a thresholding technique is employed to detect cases of conflicting scores. The threshold value is selected by calculating the difference between the first two maximum peak values in the fused scores of all the test samples, followed by taking the mean of the distribution of difference values. By experimentation, we set the threshold value equal to one-third of the mean value. All those cases where the difference in the score of the two peak values was less than the threshold value were considered to be a case of conflict, which was handled by a different scheme of weighted fusion. The idea is discussed through Fig. 1. In simple averaging scheme, a random test sample is classified by taking the maximum of the final/fused classification score. In Fig. 1, the final classification score belongs to class 1, being the maximum, test sample will be classified as class 1. But as observed in figure, there are secondary peaks/maximum with marginal difference from the largest peak value which have most probably arisen because of conflicting scores from the different streams (Front, Side, Top, and MHI). Thus, these cases should be considered as conflicting cases where simple averaging tends to mis-classify. These cases are further handled via weighted fusion after assessing the reliability of each stream score using the variance of skeleton stream data.

For conflicting scores above a certain threshold, i.e., for misclassified samples, a general intuition is applied by means of variance in skeletal data. For a test sample, statistical variance of all joints in skeletal data along the orthogonal planes X, Y, and Z is computed, denoted by σ_x^2, σ_y^2, and σ_z^2 respectively. For fusion streams Front, Side, Top, and MHI presented in our previous work [1], variance of skeletal data acts as evidence to support as a weighted factor for these streams.

Let m_1, m_2, m_3, and m_4 be the weighted factors for fusion streams of Front, Side, Top, and MHI, respectively. Using a general intuition of visibility, we can relate the variance and four stream views as inversely proportional to their corresponding variances in Z, Y, X, and Z planes, respectively. Changing the proportionality sign to equality sign by adding the constant of proportionality, equation for weighted factors can be written as:

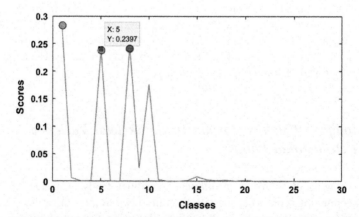

Fig. 1 Final scores obtained for a test sample after average fusion. X-axis denotes total 27 classes; final fusion scores are denoted on the Y-axis

$$m_1 = k/\sigma_z^2$$
$$m_2 = k/\sigma_y^2$$
$$m_3 = k/\sigma_x^2 \qquad (1)$$
$$m_4 = k/\sigma_z^2$$

For normalization of weighted factors, we consider the value of k as:

$$k = \frac{\sigma_z^2 \times \sigma_y^2 \times \sigma_x^2}{2 \times \sigma_x^2 \sigma_y^2 + \sigma_z^2 \sigma_y^2 + \sigma_z^2 \sigma_x^2} \qquad (2)$$

These modified weights are further combined with individual scores of *FScores*, *SScores*, *TScores,* and *MHIScores* of Front, Side, Top, and MHI, respectively. Each scores obtained from individual modality is a vector of size equal to the number of classes in consideration. Final scores obtained by weighted fusion using average rule and product rule are given in Eq. (3).

$$FinalScores = \frac{m_1 \times FScores + m_2 \times SScores + m_3 \times TScores + m_4 \times MHIScores}{4}$$

$$(3)$$

For normalization of weights, constant of proportionality k has more significance in average rule rather than in product rule. For product rule, fusion of scores can be obtained using Eq. (4)

$$FinalScores = FScores^{m_1} \times SScores^{m_2} \times TScores^{m_3} \times MHIScores^{m_4} \qquad (4)$$

FinalScores obtained by weighted fusion is a vector containing posterior probabilities (scores) for each class. A test sample is classified to a class, having maximum score within *FinalScores*.

2.2 Fusion of Scores Obtained by Trained SVM on Skeletal Data

An effort is made to train multiclass support vector machine for classification, using clustering of skeleton joint sequences. For creation of feature vector constituting an activity from skeletal data, normalization of joints followed by posture selection is performed. Normalization is done by calculating Euclidean distance between spine (torso) joint and neck joint. Posture selection representing an activity is carried out by applying k-means clustering on normalized joints sequences. Clustering results in formation of centroids; these centroids in sorted order are given as features for

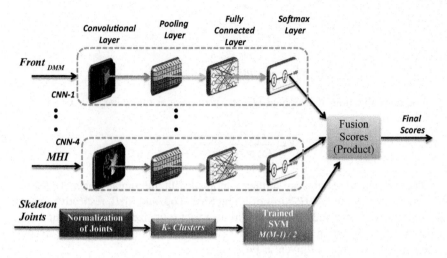

Fig. 2 Fusion of skeleton stream using SVM with 4-CNN streams of Front_{DMM}, Side_{DMM}, Top_{DMM}, and MHI

training multiclass SVM. Multiclass SVM is implemented with the "one-versus-one" approach; this approach is constructed by building of several binary classifiers and then by training them one against one. For classification of N number of classes, $N (N - 1)/2$ binary classifiers are needed; the final decision of the classification is taken by applying voting strategy among all the binary classifiers [10].

For classification, training and testing strategy presented in [1, 2] is followed. Training is performed on odd subject samples, and testing is performed on even subject samples of the dataset; trained SVM is utilized as a fifth stream for classification and combined with other trained modalities of depth and RGB data. The final classification score is acquired through average fusion as described in Fig. 2.

2.3   Skeleton Images with Convolutional Neural Networks

The earlier two methods require less number of computations. The third approach uses convolutional neural networks with skeleton images which are generated from skeleton joint sequences. These synthetically generated images are then trained on pretrained CNN for classification. Finally, classification score is obtained from softmax classifier which is later fused with other modalities. Details on creation of skeleton images followed by training on CNN are given as below.

Skeleton Sequences to Images: An activity in skeletal data constitutes of n frames (sequences) and k number of joints. Number of frames for an activity differs from

(a)

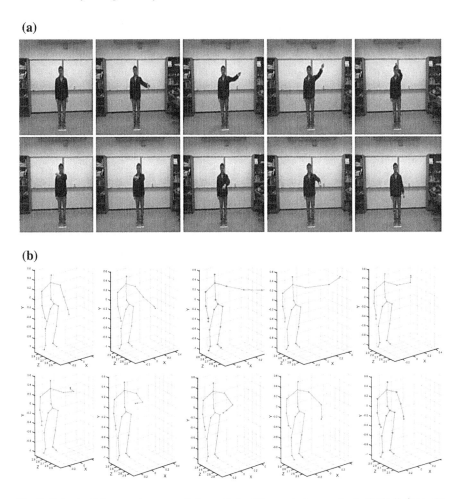

(b)

Fig. 3 Subset of frames (sequences) for activity "draw circle counterclockwise" from the UTD-MHAD dataset. **a** RGB frames. **b** Skeleton sequences

activity to activity; it also differs for same activity performed by different subjects. Number of joints throughout the activity mostly remains unchanged. Subset of RBG frames and skeleton frames (sequences) for activity "draw circle counterclockwise" from the UTD-MHAD dataset is shown in Fig. 3a and Fig. 3b, respectively. In each frame, a joint has three-dimensional coordinate values, i.e., x, y, and z values. To account for the variations in the height and positions of different human subjects for same or different activities requires normalization of joints. A straightforward solution is to compensate the position of the skeleton by centering the coordinate space on one skeleton joint. Based on this, a method for normalization is introduced here.

Let $J_{(p,\,q,\,c)}$ denotes the joint value of pth joint in qth frame and cth coordinate in an activity. For an activity, each joint of skeleton $J_{(p,\,q,\,c)}$ in their respective coordinate space is normalized to the distance between spine (torso) and hip joint of first frame. Here, $J_{(3,\,1,\,c)}$ and $J_{(4,\,1,\,c)}$ indicate spine joint and hip joint of first frame. Normalized joints are denoted by $J_{N(p,\,q,\,c)}$, values of c can be x, y, or z according to normalization done in coordinate space. Normalization of joints is done using Eq. (5)

$$J_{N(p,q,c)} = \frac{J_{(p,q,c)} - J_{(3,1,c)}}{\left\| J_{(4,1,c)} - J_{(3,1,c)} \right\|} \tag{5}$$

where,

$p = 1, 2, ..., k$ (*number of joints in an activity*)
$q = 1, 2, ..., n$ (*number of frames in an activity*)
$c = x$ *or* y *or* z (*coordinate values*)

Let Px and Py be the pixel locations in two-dimensional image to be filled with color associated with parts (joints). For an activity having q frames and p joints, pixel locations are obtained as:

$$
\begin{aligned}
Px(p,q) &= C1 + k1 \times J_{N(p,q,x)} + k2 \times J_{N(p,q,y)} + k3 \times J_{N(p,q,z)} \\
Py(p,q) &= C2 + k4 \times J_{N(p,q,x)} + k5 \times J_{N(p,q,y)} + k6 \times J_{N(p,q,z)}
\end{aligned}
\tag{6}
$$

where $C_1, C_2, k_1, k_2, k_3, k_4, k_5, k_6$ are constants, usually the values of C_1 and C_2 are the midpoints of size of the image in x and y directions, it also depends upon the joint selected for normalization. Other constant values may depend upon stretch required in x and y directions for pixel formation. $Px(p,q)$ and $Py(p,q)$ are matrices of size $(p \times q)$ which indicates pixel locations for subsequent joints in each frame, where row p corresponds to joints and column q corresponds to frames.

To transform skeleton sequences comprising an activity to a skeleton image, skeleton joints of the human body are divided into five main parts, i.e., two arms, two legs, and a trunk. Each part of the skeleton is represented by a different color in an image, like left arm, right arm, left leg, right leg, and trunk are represented by green, red, gray, yellow, and blue, respectively. Skeleton joints associated with body part have the same color, for instance; left shoulder, left elbow, left wrist, and left hand joints associated with left arm have green color. To create a motion template for storing the history of movement of body parts, the color intensity is varied by a constant factor proportional to the number of frames. Change in color intensity reflects the movement of part from initial frame to final frame along with change in location. Algorithm 1 describes the skeleton image formulation and colorization of an activity using pixel locations of skeleton joints obtained from Eq. (6)

Algorithm 1: Skeleton Image Creation and Colorization for an Activity

---

Input :

1. $p = 1, 2, ..., k.$	//total joints
2. $f = 1, 2, ..., n.$	//total frames
3. $I(x,y)$	//Template Image with uniform background
4. Px, Py	//Pixel locations of joints
5. $g = \{$"left arm", "right arm",....$\}$	//total parts
6. $gcolor = \{$"green", "red",....$\}$	//colors of body parts

$Skeleton\_Colorization\ (p, f, g, I, Px, Py, gcolor)$
  $i = 0, j = 0$ // initialization
  $repeat\ \ j = j+1$
          $repeat\ \ i = i+1$
              $if\ \ p\ (j) \in g\ (w),\ \exists\ w$ // joint belongs to part

                        $I\ (\ Px_{(i,j)}\ ,\ Py_{\ (i,j)}\ ) = gcolor - \Delta r$ //Change (decrease) in
                                                                             color intensity

                  end
                  $\Delta r = i \times d$ // d is small constant
              $until\ \ i > total\ frames\ in\ (f)$
      $until\ \ j > total\ joints\ in\ (p)$

Output :
  1. $I(\ x,\ y\ )$ //Skeleton image with uniform background color representing activity.

Figure 4 shows the sample images created for different activities using depth maps, motion history, and skeleton sequences utilized for training and testing. Figure 4(5) shows the skeleton image obtained for activity "draw circle counter-clockwise". Five body parts: left arm, right arm, left leg, right leg, and trunk are represented by green, red, gray, yellow, and blue, respectively. Changes in color of parts can be noted as frame changes in activity. Skeleton images thus created resembles with activity represented by other modalities such as depth motion maps and motion history image.

Training on pretrained VGG-16: For each activity with n frames, a single skeleton image is created from skeleton joints and number of frames. These synthetically generated skeleton images are given as input to the first layer of VGG-16 pretrained model for training. To match the size specifications of the input layer of pretrained model, skeleton images were resized to 224 × 224. Numbers of training samples were increased by changing different background color to skeleton images constituting same activity. Figure 5 shows fusion of 5-CNN streams. The fifth stream is trained separately using skeleton images as input and then later fused with other four CNN streams.

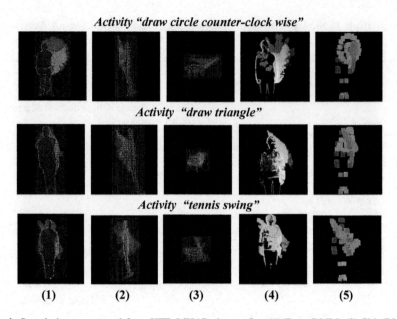

Fig. 4 Sample images created from UTD-MHAD dataset for: (1) Front-DMM, (2) Side-DMM, (3) Top-DMM, (4) MHI, and (5) skeleton image. Representing activities, "draw circle counterclockwise," "draw triangle" and "tennis swing"

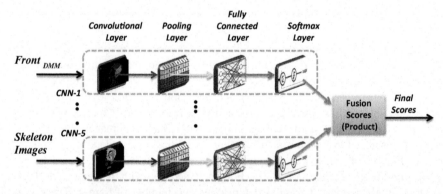

Fig. 5 Fusion of 5-CNN streams of Front$_{DMM}$, Side$_{DMM}$, Top$_{DMM}$, MHI, and skeleton images

3   Results and Discussion

Proposed methods were tested on UTD-MHAD [2], one of the challenging dataset for activity recognition. It comprises of 27 activities with average number of frames for each activity nearly 60 frames, the variations between activities like "draw circle clockwise", "draw triangle," "draw circle counterclockwise," and others were very minimal, thus making it difficult for a recognition technique to distinguish between them. Training is performed on odd subjects samples, and testing is performed on even subject samples as followed in [1, 2]. For all the approaches presented, training and testing is done on machine with GPU-Quadro m6000. Training samples were generated from subject 1 and subject 3, while testing was performed on 430 samples of even subjects of UTD-MHAD dataset.

The results obtained from all the three methods are compared along with results from other works in the literature. Comparison of different methods for activity recognition on UTD-MHAD dataset is given in Table 1. Comparing the three approaches presented here, multiclass SVM using clustering technique does not perform well for complex datasets with intraclass variations. Given method of clustering performs well on datasets like CAD-60, having large number of frames which distinctively form clustering points from joint sequences. With k-means clustering, best result was noted when k has value of 5. After fusion with 4-CNN streams, slight improvement in recognition was notifiable.

Method of modified weighted fusion, using variance of skeleton joints, was very simple with less number of computations, and no training of classifier is required. It acts as evidence to fusion strategy and provides assistance to other recognition methods arranged with depth and/or RGB data. Still, being computationally efficient, it improves the accuracy of the fusion strategy by nearly 2% over average fusion.

Out of the proposed three methods, generation of skeleton images from joint sequences is computationally expensive compared to other two. Training has been done on pretrained VGG-16 model with 2160 skeleton image samples. These samples are generated by replacing the uniform background color with different colors. On comparison, method of skeleton images with CNN gives better results compared to other two proposed methods. Highest accuracy achieved was 95.38%, with only 20 samples misclassified out of 430 tested. Class-specific accuracy of recognition for 430 test samples is given in Fig. 6.

Table 1 Comparison of activity recognition methods on UTD-MHAD

Sr. No	Approach	Accuracy (%)
1	C. Chen et al. [2]	79.1
2	Bulbul et al. [11]	88.4
3	Javed et al. [12] (Average Rule)	88.8
4	Javed et al. [12] (Product Rule)	91.2
5	Ours (Clustering and SVM) (Average Rule)	89.12
6	Ours (Variance-Fusion) (Average Rule)	90.27
7	Ours (Skeleton Images with CNN) (Average Rule)	92.82
8	Ours (Clustering and SVM) (Product Rule)	89.45
9	Ours (Variance-Fusion) (Product Rule)	91.89
10	Ours (Skeleton Images with CNN) (Product Rule)	95.38

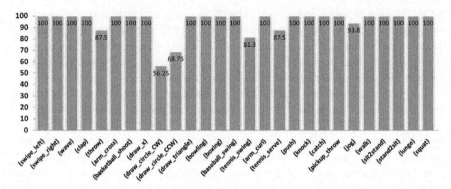

Fig. 6 Test results obtained after fusion of 5-Stream CNN using skeleton images, class-specific accuracy (%) of 27 activities of UTD-MHAD dataset

4   Conclusion and Future Work

Skeletal data of human body obtained from depth sensor have significant information, which can be used for tasks such as activity recognition, surveillance. Main focus of the presented paper was to utilize skeletal data for activity recognition along with other RGB and depth data. As shown in results, presented three approaches for fusion improves the overall accuracy of recognition. Out of the proposed methods, fusion of trained CNN on skeleton images as fifth CNN stream, achieved significant improvement in recognition with accuracy of 95.38%. More complex problems on recognition in constrained environments such as ATMs, surveillance, and other applications can be addressed in future.

Acknowledgements This research was supported by Science and Engineering Research Board (SERB) under project no. ECR/2016/000387, in cooperation with the Department of Science & Technology (DST), Government of India. The views and conclusions contained in this document are those of the authors and should not be interpreted as representing the official policies, either expressed or implied, of DST-SERB or the Government of India. The DST-SERB or Government of India is authorized to reproduce and distribute reprints for Government purposes notwithstanding any copyright notation thereon.

References

1. Imran, J., Kumar, P.: Human Action Recognition using RGB-D Sensor and Deep Convolutional Neural Networks. In: IEEE International Conference on Advances in Computing, Communications and Informatics (ICACCI), pp. 144–148. Jaipur, India (2016)
2. Chen, C., Jafari, R., Kehtarnavaz, N.: UTD-MHAD: A Multimodal Dataset for Human Action Recognition Utilizing a Depth Camera and a Wearable Inertial Sensor. In: IEEE International Conference on Image Processing (ICIP), pp. 168–172 (2015)

3. Li, W., Zhang, Z., and Liu, Z.: Action recognition based on a bag of 3D points. In: Proc. IEEE Conf. Comput. Vis. Pattern Recog. Workshops, San Francisco, CA, USA, pp. 9–14. Jun. (2010)

4. Ofli, F., Chaudhry, R., Kurillo, G., Vidal, R., and Bajcsy R.: Berkeley MHAD: A Comprehensive Multimodal Human Action Database. In: Proc. IEEE Workshop Appl. Comput. Vision, pp. 53–60. Jan. (2013)

5. Sung, J., Ponce, C., Selman, B., Saxena, A.: Unstructured Human Activity Detection from RGBD Images. In: IEEE International Conference on Robotics and Automation RiverCentre, Saint Paul, Minnesota, USA, pp. 842–849, (2012)

6. Wang, P., Li, W., Gao, Z., Zhang, J., Tang, C., and Ogunbona, P. O.: Action Recognition From Depth Maps Using Deep Convolutional Neural Networks, IEEE Transactions on Human-Machine Systems, Vol. 46, No. 4, pp. 498–509 August (2016)

7. Chen, C., Liu, M., Zhang, B., Han, J., Jiang, J., Liu, H.: 3D Action Recognition Using Multi-temporal Depth Motion Maps and Fisher Vector, In: Proceedings of the Twenty-Fifth International Joint Conference on Artificial Intelligence. International Joint Conferences on Artificial Intelligence, pp. 3331–3337 (2016)

8. Xia, L., Chen, C. C., and Aggarwal, J. K.: View Invariant Human Action Recognition Using Histograms of 3D Joints. In: Proceedings of IEEE Conference on Computer Vision and Pattern Recognition Workshops, pp. 20–27, Providence, RI, (2012)

9. Gaglio, S., Lo Re, G., and Morana, M.: Human Activity Recognition Process Using 3-D Posture Data. IEEE Transactions on Human-Machine Systems, Vol. 45, No. 5, pp. 586–597 (2015)

10. Cippitelli, E., Gasparrini, S., Gambi, E., and Spinsante, S.: A Human Activity Recognition System Using Skeleton Data from RGBD Sensors, Computational Intelligence and Neuroscience, Article ID 4351435, pp. 1–14, Volume 2016 (2016)

11. Farhad, M. B., Jiang, Y., and Ma, J.: DMMs- Based Multiple Features Fusion for Human Action Recognition. International Journal of Multimedia Data Engineering and Management (IJMDEM) Volume 6, Issue 4, pp. 23–39 (2015)

12. Aggarwal, J.K., Xia, L.: Human Activity Recognition from 3D Data: A Review. Pattern Recognition Letters, 48, pp. 70–80 (2014)

Author Index

© Springer Nature Singapore Pte Ltd. 2018
B. B. Chaudhuri et al. (eds.), *Proceedings of 2nd International Conference on Computer Vision & Image Processing*, Advances in Intelligent Systems and Computing 703, https://doi.org/10.1007/978-981-10-7895-8

Printed by Printforce, the Netherlands